The Geography of Transport Systems

Third edition

Jean-Paul Rodrigue
with Claude Comtois and Brian Slack

Routledge
Taylor & Francis Group

LONDON AND NEW YORK

First edition published 2006 by Routledge
Second edition published 2009 by Routledge

Third edition published 2013 by Routledge
2 Park Square, Milton Park, Abingdon, Oxon OX14 4RN

Simultaneously published in the USA and Canada
by Routledge
711 Third Avenue, New York, NY 10017

Routledge is an imprint of the Taylor & Francis Group, an informa business

© 2006, 2009, 2013 Jean-Paul Rodrigue, Claude Comtois and Brian Slack

British Library Cataloguing in Publication Data
A catalogue record for this book is available from the British Library

Library of Congress Cataloging in Publication Data
Rodrigue, Jean-Paul, 1967-
 The geography of transport systems / Jean-Paul Rodrigue, Claude Comtois,
 and Brian Slack. – [Third edition].
  pages cm
 Includes bibliographical references and index.
 1. Transportation geography. I. Comtois, Claude, 1954- II. Slack,
 Brian, 1939- III. Title.
 HE323.R63 2013
 388.01—dc23

2012048389

ISBN: 978–0–415–82253–4 (hbk)
ISBN: 978–0–415–82254–1 (pbk)
ISBN: 978–0–203–37118–3 (ebk)

Typeset in Times New Roman
by RefineCatch Limited, Bungay, Suffolk

To Gordana and Nikola

Contents

Photographs

Figures

 # Contributors

Dr John Bowen, Department of Geography, Central Washington University, 400 E. University Way, Ellensburg, WA, 98926, USA.

Dr Laetitia Dablanc, Institut Français des Sciences et Technologies, des Transports, de l'Aménagement et des Réseaux, SPLOTT Laboratory, 2 rue de la Butte Verte, Marne-la-Vallée, France.

Dr César Ducruet, CNRS (Centre National de la Recherche Scientifique), UMR 8504 Géographie-Cités, équipe P.A.R.I.S., 13 rue du Four, F-75006, Paris, France.

Dr Markus Hesse, University of Luxembourg, Faculté des Lettres, des Sciences Humaines, des Arts et des Sciences de l'Education, Route de Diekirch, L-7220 Walferdange, Luxemburg.

Dr Theo Notteboom, Institute of Transport and Maritime Management Antwerp, ITMMA House, Kipdorp 59, 2000, Antwerp, Belgium.

Dr Shih-Lung Shaw, Department of Geography, The University of Tennessee, Knoxville, Tennessee, 37996-0925, USA.

● **Preface**

The *Geography of Transport Systems* is now into its third edition. Substantial efforts have been made to build on the first and second editions' success by improving the content and its structure. Like the previous editions, we have elected for a more synthetic writing style, instead of a narrative, where the goal is to provide a structured framework to the reader. Great care has been made to avoid factual information so that the textbook can retain its relevance in spite of continuous and often unforeseeable changes in the transport industry. A large quantity of statistical information is available on the companion website, which is constantly updated.

A driving force of the global economy resides in the capacity of transport systems to ship large quantities of freight and to accommodate vast numbers of passengers. The world has become interconnected at several scales. This new geographical dimension transcends a more traditional perspective of transportation mainly focused on the city or the nation. At the beginning of the twenty-first century, the geography of transportation is thus fundamentally being redefined by global, regional and local issues.

Presenting these issues to students or the public remains a challenging task. This book has specifically been designed with this in mind. Its origins are rather unusual since it began in 1997 as an online initiative to provide material about transport geography and was simply titled "Transport Geography on the Web". The material was considerably revised and expanded over the years, often thanks to comments and queries received, as the site gained a wider audience. It has already endured for more than 15 years the test of being exposed to the scrutiny of a global audience including practitioners, policy makers, educators and, most importantly, students.

Like the previous two editions, the textbook is articulated along two core approaches to transport geography, one conceptual and the other methodological. The conceptual parts present what we think are some of the most relevant issues explaining contemporary transport geography. In addition to the more conventional topics related to transport modes, terminals, as well as urban transportation, the book also substantially focuses on emerging issues such as globalization, supply chain management, energy and the environment. Many of these issues have been superficially covered, if at all, in the past, but their importance cannot be underestimated in a transport geography that involves an increasingly integrated world.

The methodological parts address how transportation information is used to assist transport operators to allocate their resources (investments, vehicles) or to influence public policy. This includes a wide array of methods ranging from qualitative to quantitative. Since transport is a field of application, the use of methodologies is particularly relevant as they relate to real world issues. The merging of methodologies and information technologies has led to many new opportunities, notably with the emergence of transportation geographic information systems (GIS-T). It has become a very active field of investigation and application.

It is our hope that the reader will have a better understanding of the nature, function, importance and challenges of contemporary transportation systems. The online companion site will ensure that this book will not be a static endeavor and will be revised and updated as changes take place in this fascinating field which is transport geography. Unless otherwise credited the photographs are by Jean-Paul Rodrigue. Special thanks also to Elisabet Sinkie for the often unappreciated and underpaid work of putting a textbook together.

New York, November 2012

1 Transportation and geography

Movements of people, goods and information have always been fundamental components of human societies. Contemporary economic processes have been accompanied by a significant increase in mobility and higher levels of accessibility. Although this trend can be traced back to the industrial revolution, it significantly accelerated in the second half of the twentieth century as trade was liberalized, economic blocs emerged and the comparative advantages of global labor and resources were used more efficiently. However, these conditions are interdependent with the capacity to manage, support and expand movements of passengers and freight as well as their underlying information flows. Societies have become increasingly dependent on their transport systems to support a wide variety of activities ranging, among others, from commuting, supplying energy needs, to distributing parts between manufacturing facilities and distribution centers. Developing transport systems has been a continuous challenge to satisfy mobility needs, to support economic development and to participate in the global economy.

Concept 1 – What is transport geography?

The purpose of transportation

The unique purpose of transportation is to overcome space, which is shaped by a variety of human and physical constraints such as distance, time, administrative divisions and topography. Jointly, they confer a friction to any movement, commonly known as the friction of space. However, these constraints and the friction they create can only be partially circumscribed. The extent to which this is done has a cost that varies greatly according to factors such as the distance involved, the capacity of modes and infrastructures and the nature of what is being transported. There would be no transportation without geography and there would be no geography without transportation. The goal of transportation is thus to transform the geographical attributes of freight, people or information, from an origin to a destination, conferring on them an added value in the process. The convenience at which this can be done – transportability – varies considerably.

> **Transportability** refers to the ease of movement of passengers, freight or information. It is related to transport costs as well as to the attributes of what is being transported (fragility, perishable, price). Political factors can also influence transportability such as laws, regulations, borders and tariffs. When transportability is high, activities are less constrained by distance.

The specific purpose of transportation is to fulfill a demand for mobility, since transportation can only exist if it moves people, freight and information around. Otherwise it has no purpose. This is because transportation is dominantly the outcome of a **derived demand** (Figure 1.1).

In economic systems what takes place in one sector has impacts on another; demand for a good or service in one sector is derived from another. For instance, a consumer buying a good in a store will likely trigger the replacement of this product, which will generate demands for activities such as manufacturing, resource extraction and, of course, transport. What is different about transport is that it cannot exist alone and a movement cannot be stored. An unsold product can remain on the shelf of a store until a customer buys it (often with discount incentives), but an unsold seat on a flight or unused cargo capacity in the same flight remain unsold and cannot be brought back as additional capacity later. In this case an opportunity has been missed since the amount of transport being offered has exceeded the demand for it. The derived demand of transportation is often very difficult to reconcile with an equivalent supply and actually transport companies would prefer to have some additional capacity to accommodate unforeseen demand (often at much higher prices). There are two major types of derived transport demand:

> **Direct derived demand**. Refers to movements that are directly the outcome of economic activities, without which they would not take place. For instance, work-related activities commonly involve commuting between the place of residence and the workplace. There is a supply of work in one location (residence) and a demand of labor in another (workplace), transportation (commuting) being directly derived

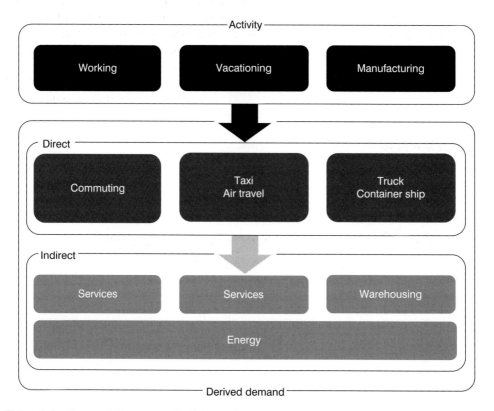

Figure 1.1 Transportation as a derived demand

from this relationship. For freight transportation, all the components of a supply chain require movements of raw materials, parts and finished products on modes such as trucks, rail or container ships. Thus, transportation is directly the outcome of the functions of production and consumption.

Indirect derived demand. Considers movements created by the requirements of other movements. For instance, fuel consumption from transportation activities must be supplied by an energy production system requiring movements from zones of extraction, to refineries and storage facilities and, finally, to places of consumption. Warehousing can also be labeled as an indirect derived demand since it is a "non movement" of a freight element. Warehousing exists because it is virtually impossible to move commodities directly from where they are produced to where they are consumed.

Distance, a core attribute of transportation, can be represented in a variety of ways, ranging from a simple Euclidean distance – a straight line between two locations – to what can be called logistical distance; a complete set of tasks required to be done so that distance can be overcome (Figure 1.2).

Any movement must thus consider its geographical setting which in turn is linked to spatial flows and their patterns. Three major representations can be used for distance and the friction it imposes on transportation:

Euclidean distance. A simple function of a straight line between two locations where distance is expressed in geographical units such as kilometers. Commonly used to provide an approximation of distance, but almost never has a practical use.

Transport distance. A more complex representation where a set of activities related to circulation, such as loading, unloading and transshipment, are considered. Additional elements such as costs and time are also part of the transport distance. On Figure 1.2, the transport distance between locations A and B includes pickup, travel by mode 1, transshipment, travel by mode 2 and finally, delivery. The same applies

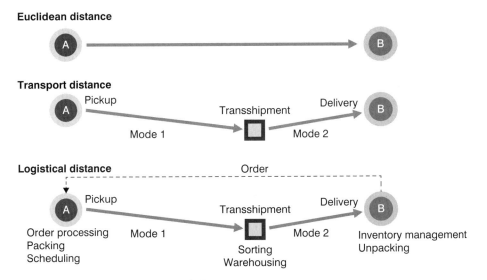

Figure 1.2   Different representations of distance

to the circulation of people, although the activities involved will be different. For instance, someone using air travel between two locations will require going to an airport, may transit through an intermediate hub airport and will finally need to reach his/her destination from the airport terminal. Transport distance is jointly expressed in geographical units, in cost and in time.

Logistical distance. A complex representation that encompasses all the tasks required so that a movement between two locations can take place. Logistical distance thus includes flows, but also a set of activities necessary for the management of these flows. For freight movements, among the most significant tasks are order processing, packing, sorting and inventory management. Geographical distance units are less relevant in its assessment, but the factors of costs and time are very significant. Time not only involves the delay related to management and circulation, but also how it is used to service the transport demand, namely the scheduling of pickups and deliveries. On Figure 1.2, the logistical distance between locations A and B, includes an order from B, which is processed, packed and scheduled to be picked up. At the intermediate transshipment location, sorting and warehousing are performed, and finally, at the destination the delivery will be unpacked and used. For the transportation of passengers, logistical distance also concerns a specific array of tasks. Taking again an air travel example, a ticket would first need to be purchased, commonly several weeks in advance. Other common time and cost tasks concern checking in, security checks, boarding and disembarking, and picking up luggage. Thus, a three-hour flight can in reality be a movement planned several weeks in advance and its full realization can take twice as much time if all the related logistical activities are considered.

Any movement must thus consider its geographical setting which in turn is linked to spatial flows and their patterns. The concept of flow has four major components:

- **Geographical**. Each flow has an origin and a destination and consequently a degree of separation. Flows with high degrees of separation tend to be more limited than flows with low degrees of separation.
- **Physical**. Each flow involves specific physical characteristics in terms of possible load units and the conditions in which they can be carried. Flows, depending on the transportation mode, can be atomized (smallest load unit) or massified (moving load units in batches).
- **Transactional**. The realization of each flow has to be negotiated with providers of transport services, such as booking a slot on a container ship or an air travel seat. Commonly, a flow is related to a monetary exchange between provider of transportation and the user.
- **Distribution**. Flows are organized in sequences where the more complex are involving different modes and terminals. Many transport flows are scheduled and routed to minimize costs or maximize efficiency, often through intermediary locations.

Urbanization, multinational corporations, the globalization of trade and the international division of labor are all forces shaping and taking advantage of transportation at different, but often related, scales. Consequently, the fundamental purpose of transport is geographic in nature, because it facilitates movements between different locations. Transport plays a role in the structure and organization of space and territories, which may vary according to the level of development. In the nineteenth century, the purpose of the emerging modern forms of transportation, mainly railways and maritime shipping, was to expand coverage with the creation, expansion and consolidation of national markets. In the twentieth century, the objective shifted to selecting itineraries, prioritizing transport modes, increasing the capacity of existing networks and responding to mobility needs and this at a scale that was increasingly global, with its own space of flows. In the twenty-first century, transportation must cope with a globally oriented economic system in a timely and cost-effective way, but also with several local problems such as congestion and capacity constraints.

The importance of transportation

Transport represents one of the most important human activities worldwide. It is an indispensable component of the economy and plays a major role in spatial relations between locations. Transport creates valuable links between regions and economic activities, between people and the rest of the world. Transport is a multidimensional activity whose importance is:

- **Historical**. Transport modes have played several different historical roles in the rise of civilizations (Egypt, Rome and China), in the development of societies (creation of social structures) and also in national defense (Roman Empire, American road network).
- **Social**. Transport modes facilitate access to healthcare, welfare, and cultural or artistic events, thus performing a social service. They shape social interactions by favoring or inhibiting the mobility of people. Transportation thus supports and may even shape social structures.
- **Political**. Governments play a critical role in transport as sources of investment and as regulators. The political role of transportation is undeniable as governments often subsidize the mobility of their populations (highways, public transit, etc.). While most transport demand relates to economic imperatives, many communication corridors have been constructed for political reasons such as national accessibility or job creation. Transport thus has an impact on nation building and national unity, but it is also a political tool.
- **Economic**. The evolution of transport has always been linked to economic development. It is an industry in its own right (car manufacturing, air transport companies, etc.). The transport sector is also an economic factor in the production of goods and services. It contributes to the value added of economic activities, facilitates economies of scale, influences land (real estate) value and the geographic specialization of regions. Transport is both a factor shaping economic activities, and is also shaped by them.
- **Environmental**. Despite the manifest advantages of transport, its environmental consequences are also significant. They include air and water quality, noise level and public health. All decisions relating to transport need to be evaluated taking into account the corresponding environmental costs. Transport is a dominant factor in contemporary environmental issues.

Transportation studies are therefore multidisciplinary, and can involve hard (e.g. engineering) or soft sciences (e.g. economics) depending on the dimension being investigated such as infrastructure provision, operational management or planning. Substantial empirical evidence indicates that the importance of transportation is growing. The following contemporary trends can be identified regarding this issue:

Growth of the demand. The years following the Second World War have seen a considerable growth of the transport demand related to individual (passengers) as well as freight mobility. This growth is jointly the result of larger quantities of passengers and freight being moved, but also the longer distances over which they are carried. Recent trends underline an ongoing process of mobility growth, which has led to the multiplication of the number of journeys involving a wide variety of modes that service transport demands.

Reduction of costs. Even if several transportation modes are very expensive to own and operate (ships and planes, for instance), costs per unit transported have dropped significantly over the last decades. This has made it possible to overcome larger distances and further exploit the comparative advantages of space. As a result, despite the lower costs, the share of transport activities in the economy has remained relatively constant in time.

Expansion of infrastructures. The above two trends have obviously extended the requirements for transport infrastructures both quantitatively and qualitatively. Roads, harbors, airports, telecommunication facilities and pipelines have expanded considerably to service new areas and add capacity to existing networks. Transportation infrastructures are thus a major component of land use, notably in developed countries.

Facing these contemporary trends, an important part of the spatial differentiation of the economy is related to where resources (raw materials, capital, people, etc.) are located and how well they can be distributed. Transport routes are established to distribute resources between places where they are abundant and places where they are scarce, but only if the costs are lower than the benefits. Consequently, transportation has an important role to play in the conditions that affect global, national and regional economic entities. It is a strategic infrastructure that is so embedded in the socioeconomic life of individuals, institutions and corporations that it is often invisible to the consumer, but always part of all economic and social functions. This is paradoxical, since the perceived invisibility of transportation is derived from its efficiency. If transport is disrupted or ceases to operate, the consequences can be dramatic, such as workers unable to reach their workplace or parts not being delivered to factories.

Transportation in geography

The world is obviously not a place where features such as resources, people and economic activities are randomly distributed. Geography seeks to understand the spatial order of things as well as their interactions. Transportation is an element of this spatial order as it is at the same time influenced by geography as well as having an influence on it. Transportation is not necessarily a science, but a field of application borrowing concepts and methods from a wide variety of disciplines. Transportation interests geographers for two main reasons. First, transport infrastructures, terminals, equipment and networks occupy an important place in space and constitute the basis of a complex spatial system. Second, since geography seeks to explain spatial relationships, transport networks are of specific interest because they are the main support of these interactions.

Transport geography is a sub-discipline of geography concerned about movements of freight, people and information. It seeks to understand their spatial organization by linking spatial constraints and attributes with the origin, the destination, the extent, the nature and the purpose of movements.

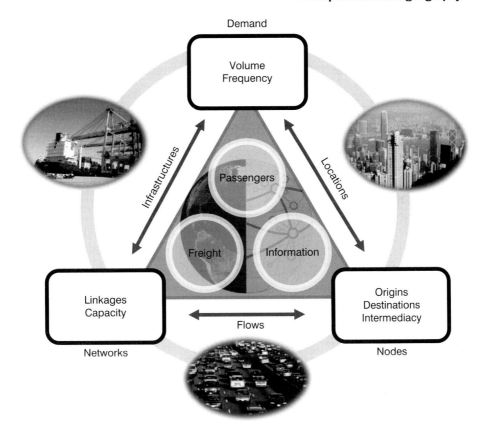

Figure 1.3 The transport system

Transport geography, as a discipline, emerged from economic geography in the second half of the twentieth century. Traditionally, transportation has been an important factor behind the economic representations of the geographic space, namely in terms of the location of economic activities and the monetary costs of distance. The growing mobility of passengers and freight justified the emergence of transport geography as a specialized field of investigation. In the 1960s, transport costs were recognized as key factors in location theories and transport geography began to rely increasingly on quantitative methods, particularly over network and spatial interactions analysis. However, from the 1970s globalization challenged the centrality of transportation in many geographical and regional development investigations. As a result, transportation became underrepresented in economic geography in the 1970s and 1980s, even if mobility of people and freight and low transport costs were considered as important factors behind the globalization of trade and production.

Since the 1990s, transport geography has received renewed attention, especially because the issues of mobility, production and distribution are interrelated in a complex geographical setting. It is now recognized that transportation is a system that considers the complex relationships between its core elements (Figure 1.3). Transport geography must be systematic as one element of the transport system is linked with numerous others; transport systems are complex. An approach to transportation thus involves several fields where some are at the core of transport geography while others are more peripheral. However, three central concepts to transport systems can be identified:

- **Transportation nodes**. Transportation primarily links locations, often characterized as nodes. They serve as access points to a distribution system or as transshipment/intermediary locations within a transport network. This function is mainly serviced by transport terminals where flows originate, end or are being transshipped from one node to the other. Transport geography must consider its places of convergence and transshipment.
- **Transportation networks**. Considers the spatial structure and organization of transport infrastructures and terminals. Transport geography must include in its investigation the structures (routes and infrastructures) supporting and shaping movements.
- **Transportation demand**. Considers the demand for transport services as well as the modes used to support movements. Once this demand is realized, it becomes an interaction which flows through a transport network. Transport geography must evaluate the factors affecting its derived demand function.

The analysis of these concepts relies on methodologies often developed by other disciplines such as economics, mathematics, planning and demography. Each provides a different dimension to transport geography. For instance, the spatial structure of transportation networks can be analyzed with graph theory, which was initially developed for mathematics. Further, many models developed for the analysis of movements were borrowed from physical sciences (such as the gravity model). Multidisciplinarity is consequently an important attribute of transport geography, as in geography in general.

The role of transport geography is to understand the spatial relations that are produced by transport systems. This gives rise to several fallacies about transportation in terms of the respective relations between access, accessibility, distance and time. A better understanding of spatial relations is essential to assist private and public actors involved in transportation to mitigate transport problems, such as capacity, transfer, reliability and integration of transport systems. There are three basic geographical considerations relevant to transport geography:

Location. As all activities are located somewhere, each location has its own characteristics conferring a potential supply and/or a demand for resources, products, services or labor. A location will determine the nature, the origin, the destination, the distance and even the possibility of a movement to be realized. For instance, a city provides employment in various sectors of activity in addition to consuming resources.

Complementarity. Locations must require exchanging goods, people or information. This implies that some locations have a surplus while others have a deficit. The only way an equilibrium can be reached is by movements between locations having surpluses and locations having demands. For instance, a complementarity is created between a store (surplus of goods) and its customers (demand of goods).

Scale. Movements generated by complementarity are occurring at different scales, pending the nature of the activity. Scale illustrates how transportation systems are established over local, regional and global geographies. For instance, home-to-work journeys generally have a local or regional scale, while the distribution network of a multinational corporation is most likely to cover several regions of the world.

Consequently, transport systems, by their nature, consume land and support the relationships between locations.

Concept 2 – Transportation and space

Authors: Jean-Paul Rodrigue and Claude Comtois

Physical constraints

Transport geography is concerned with movements that take place over space. The physical features of this space impose major constraints on transportation systems, in terms of what mode can be used, the extent of the service, its costs, capacity and reliability. Three basic spatial constraints of the terrestrial space can be identified:

Topography. Features such as mountains and valleys have strongly influenced the structure of networks, the cost and feasibility of transportation projects. The main land transport infrastructures are built usually where there are the least physical impediments, such as on plains, along valleys, through mountain passes, or when absolutely necessary through the digging of tunnels. Water transport is influenced by water depths and the location of obstacles such as reefs. Coastlines exert an influence on the location of port infrastructure. Aircraft require airfields of considerable size for takeoff and landing. Topography can impose a natural convergence of routes that will create a certain degree of centrality and may assist a location in becoming a trade center as a collector and distributor of goods. Topography can complicate, postpone or prevent the activities of the transport industry. Physical constraints fundamentally act as absolute and relative barriers to movements. An absolute barrier is a geographical feature that entirely prevents a movement while relative barriers impose additional costs and delays. Land transportation networks are notably influenced by topography, as highways and railways tend to be impeded by grades higher than 3 percent and 1 percent respectively. Under such circumstances, land transportation tends to be of higher density in areas of limited topography.

Hydrology. The properties, distribution and circulation of water play an important role in the transport industry. Maritime transport is influenced greatly by the availability of navigable channels through rivers, lakes and shallow seas. Several rivers such as the Mississippi, the St. Lawrence, the Rhine, the Mekong or the Yangtze are important navigable routes into the heart of continents and historically have been the focus of human activities that have taken advantage of the transport opportunities. Port sites are also highly influenced by the physical attributes of the site where natural features (bays, sand bars, and fjords) protect port installations. Since it is at these installations that traffic is transshipped, the location of ports is a dominant element in the structure of maritime networks. Where barriers exist, such as narrows, rapids or land breaks, water transport can only overcome these obstacles with heavy investments in canals or dredging. Conversely waterways serve as barriers to land transportation necessitating the construction of bridges, tunnels and detours, etc.

Climate. Its major components include temperature, wind and precipitation. Their impacts on transportation modes and infrastructure range from negligible to severe. Freight and passenger movement can seriously be curtailed by hazardous conditions such as snow, heavy rainfall, ice or fog. Air transportation is particularly vulnerable to weather disruptions, such as during winter when a snow storm can create cascading effects. Jet streams are also a major physical component that international air carriers must take into consideration. For an aircraft, the speed of wind can affect travel costs. Tailwind conditions can reduce flight time up to several hours for intercontinental flights.

Climate is also an influence over transportation networks by affecting construction and maintenance costs. Even volcanic eruptions can have an impact as was the case in 2010 when a volcanic eruption in Iceland released large amounts of ash in the atmosphere, which forced the closing of most airports in northwestern Europe as well as the cancellation of many transatlantic flights out of concern that the ash could damage jet engines.

From a geometrical standpoint, the sphericity of the earth determines the great circle distance; the least distance line between two points on a sphere. This feature explains the paths followed by major intercontinental maritime and air routes. For instance, Figure 1.4 shows the shortest path between New York and Moscow (about 7,540 km). This path corresponds to an air transportation corridor. Air travel over the North Atlantic between North America and Europe follows a similar path.

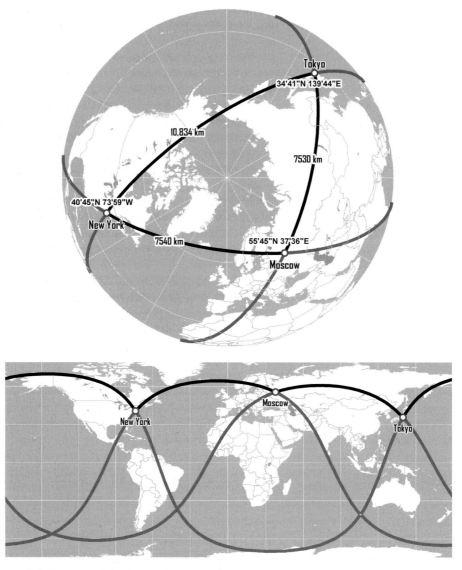

Figure 1.4 The great circle distance between New York, Moscow and Tokyo

Overcoming the physical environment

Rapid technological developments have enabled transportation to overcome the physical environment. Before the Middle Ages, road location was adapted to topography. Since then, efforts have been made to pave roads, bridge rivers and cut paths over mountain passes. Engineering techniques in terms of arch and vault used in Byzantine and Gothic church constructions in the twelfth century permitted bridge building across wider streams or deep river valleys. Road building has been at the core of technological efforts to overcome the environment. Roads have always been the support for local and even long distance travel. From the efforts to mechanize individual transport to the development of integrated highways, road building has transformed the environment. Land transportation was further facilitated with the development of technical solutions for preventing temporary interruptions in road transport provision through routeways protection. More recently, the development of road transport and the growth in just-in-time and door-to-door services have increased engineering demands for constructing multi-level and high speed highways.

Innovations in maritime transport can be found around the world. The earliest developments came in the transformation of waterways for transportation purposes through the development of canal locks. Further improvements in navigation came with the cutting of artificial waterways. Some of the earliest examples can be found in the Dutch canals, the Martesana canals of Lombardy, the canal de Briare in France or the Imperial canal of China. Further improvements in navigation technology and the nature of ships permitted the increase in speed, range and capacity of ocean transport. But the increasing size of ships has resulted in excluding canals such as Panama from servicing the largest, modern and efficient world's maritime carriers. Several canal authorities have thus embarked on expansion programs. Artificial islands are also created to permit port installations in deep waters. In China, it became clear that dredging the Yangtze River Delta was insufficient to ensure the competitiveness of the port of Shanghai. The development of a new port site in Hangzhou Bay and the modification of the Yangshan islands landscape have become indispensable.

Passages through the Arctic Ocean are being investigated with a view to creating new international connections (Figure 1.5). If this trend continues parts of the Arctic could be used more reliably for navigation, at least during summer months and for longer periods of time. The main trans-Arctic routes include:

The **Northern Sea Route** along the arctic coast of Russia. This is the maritime route that is likely to be free of ice first. In 2007 it was open during the summer months for the first time in recorded history, but it remains to be seen how stable this opening is. It would reduce a maritime journey between East Asia and Western Europe from 21,000 km using the Suez Canal to 12,800 km, cutting transit time by 10–15 days.
The **Northwest Passage** crossing Canada's Arctic Ocean could become usable on a regular basis by 2020, lessening maritime shipping distances substantially. The maritime journey between East Asia and Western Europe would take about 13,600 km using the Northwest Passage, while taking 24,000 km using the Panama Canal.
The **Arctic Bridge** linking the Russian port of Murmansk or the Norwegian port of Narvik to the Canadian port of Churchill could be used, mostly for the grain trade.
The **Transpolar Sea Route** would use the central part of the Arctic to link most directly the Strait of Bering and the Atlantic Ocean of Murmansk. This route is hypothetical for now as it involves ice-free conditions that are not yet observed.

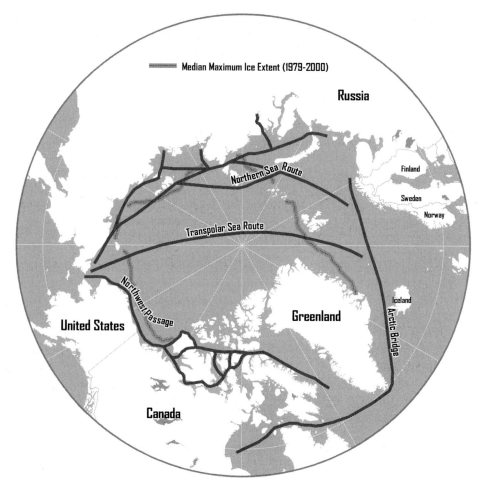

Figure 1.5 Polar shipping routes

Source: Adapted from *The New York Times*, October 10, 2005.

The consideration of Arctic routes for commercial navigation purposes remains a very speculative endeavor, mainly for three reasons:

First, it is uncertain to what extent the receding perennial ice cover is a confirmed trend or simply part of a long-term climatic cycle. Even if the Arctic routes became regularly open during the summer, in the medium term the Arctic would still remain closed to commercial navigation during the winter months. Since maritime shipping companies are looking for regular and consistent services, this seasonality has limited commercial appeal.

Second, there is very limited economic activity around the Arctic Circle, implying that shipping services crossing the Arctic have almost no opportunity to drop and pick up cargo as they pass through. Thus, unlike other long distance commercial shipping routes there is limited revenue generation potential for shipping lines along the Arctic route, which forbids the emergence of transshipment hubs. This value proposition could improve if resources (oil and mining) around the Arctic are extracted in greater quantities, which would favor bulk shipping.

Third the Arctic remains a frontier in terms of weather forecast, charting and building a navigation system, implying uncertainties and unreliability for navigation. This implies that substantial efforts have to be made to ensure that navigation can take place in a safe manner along well-defined navigation routes.

The role of technology has been determinant in the development of the air transport sector. From the experiments of the Montgolfier brothers to the advent of jet aircraft, aerial crossing of rugged terrain over considerable distance became possible. Technical innovation in the aeronautic industry has permitted planes to avoid adverse atmospheric conditions, improve speed, increase stage length and raise carrying capacity. With the rapid rise in air passenger and freight transport, emphasis has been given to the construction of airport terminals and runways. As airports occupy large areas, their environmental imprint is important. The construction of Chek Lap Kok airport in Hong Kong led to leveling mountainous land for the airport site. Kansai airport servicing Osaka has been built on an artificial island.

Transportation and the spatial structure

The concepts of site and situation are fundamental to geography and to transportation. While the site refers to the geographical characteristics of a specific location, its situation concerns its relationships in regard to other locations. Thus, all locations are relative to one another but situation is not a constant attribute as transportation developments change levels of accessibility, and thus the relations between locations. The development of a location reflects the cumulative relationships between transport infrastructure, economic activities and the built-environment. The following factors are particularly important in shaping the spatial structure:

- **Costs.** The spatial distribution of activities is related to factors of distance, namely its friction. Locational decisions are taken in an attempt to minimize costs, often related to transportation.
- **Accessibility.** All locations have a level of accessibility, but some are more accessible than others. Thus, because of transportation, some locations are perceived as more valuable than others.
- **Agglomeration.** There is a tendency for activities to agglomerate to take advantage of the value of specific locations. The more valuable a location, the more likely agglomeration will take place. The organization of activities is essentially hierarchical, resulting from the relationships between agglomeration and accessibility at the local, regional and global levels.

Many contemporary transportation networks are inherited from the past, notably transport infrastructures. Even if over the last two hundred years new technologies have revolutionized transportation in terms of speed, capacity and efficiency, the spatial structure of many networks has not much changed. This inertia in the spatial structure of some transportation networks can be explained by two major factors:

Physical attributes. Natural conditions can be modified and adapted to suit human uses, but they are a very difficult constraint to escape, notably for land transportation. It is thus not surprising to find that most networks follow the easiest (least cost) paths, which generally follow valleys and plains. Considerations that affected road construction a few hundred years ago are still in force today, although they are sometimes easier to circumscribe.

Historical considerations. New infrastructures generally reinforce historical patterns of exchange, notably at the regional level. For instance, the current highway network of France has mainly followed the patterns set by the national roads network built early in the twentieth century. This network was established over the Royal roads network, itself mainly following roads built by the Romans. At the urban level, the pattern of streets is often inherited from an older pattern, which itself may have been influenced by the pre-existing rural structure (lot pattern and rural roads).

While physical and historical considerations are at play, the introduction of new transport technology or the addition of new transport infrastructure may lead to a transformation of existing networks. Recent developments in transport systems such as container shipping, jumbo aircrafts and the extensive application of information technology to transport management have created a new transport environment and a new spatial structure. These transport technologies and innovations have intensified global interactions and modified the relative location of places. In this highly dynamic context, two processes are taking place at the same time:

Specialization. Linked geographical entities are able to specialize in the production of commodities for which they have an advantage, and trading for what they do not produce. As a result, efficient transportation systems are generally linked with higher levels of regional specialization. The globalization of production clearly underlines this process as specialization occurs as long as the incurred saving in production costs are higher than the incurred additional transport costs.

Segregation. Linked geographical entities may see the reinforcement of one at the expense of others, notably through economies of scale. This outcome often contradicts regional development policies aiming at providing uniform accessibility levels within a region.

The continuous evolution of transportation technology may not necessarily have expected effects on the spatial structure, as two forces are at play: concentration and dispersion. A common myth tends to relate transportation solely as a force of dispersion, favoring the spread of activities in space. This is not always the case. In numerous instances, transportation is a force of concentration and clustering, notably for business activities. Since transport infrastructures are generally expensive to build, they are established first to service the most important locations. Even if it was a strong factor of dispersion, the automobile has also favored the concentration of several activities at specific places and in large volumes. Shopping centers are a relevant example of this process where central locations emerge in a dispersed setting.

Space/time relationships

One of the most basic relationships of transportation involves how much space can be overcome within a given amount of time. The faster the mode, the larger the distance that can be overcome within the same amount of time. Transportation, notably improvements in transport systems, changes the relationship between time and space. When this relationship involves easier, faster and cheaper access between places, this result is defined as a space/time convergence because the amount of space that can be overcome for a similar amount of time increases significantly. It is, however, a spatially and socially uneven process since it will impact the accessibility of locations differently. For instance, infrastructure will not be laid up uniformly and segments of the population will experience a greater improvement in mobility because of their socioeconomic status (e.g. business people). In spite of these uneven processes, significant regional and continental gains were achieved during the eighteenth and nineteenth centuries with the establishment of national and continental railway systems as well as with the growth of maritime shipping, a process which continued into the twentieth century with air and road transport systems. The outcome has been significant differences in space/time relationships, mainly between developed and developing countries, reflecting differences in the efficiency of transport systems.

At the international level, globalization processes have been supported by improvements in transport technology. The result of more than 200 years of technological improvements has been a space/time convergence of global proportions in addition to the regional and continental processes previously mentioned. This enabled the extended exploitation of the advantages of the global market, notably in terms of resources and labor. Significant reductions in transport and communication costs occurred concomitantly. There is thus a relationship between space/time convergence and the integration of a region in global trade. Five major factors are of particular relevance in this process:

Speed. The most straightforward factor relates to the increasing speed of many transport modes, a condition that particularly prevailed in the first half of the twentieth century. More recently, speed has played a less significant role as many modes are not going much faster. For instance, an automobile has a similar operating speed today as it had 60 years ago and a commercial jet plane operates at a similar speed than one 30 years ago.

Economies of scale. Being able to transport larger amounts of freight and passengers at lower costs has improved considerably the capacity and efficiency of transport systems. For space/time convergence this implies that there is more capacity for a given quantity of passengers or freight being carried. Instead, the traffic can be handled with fewer trips implying that at the aggregate level it is moving faster.

Expansion of transport infrastructures. Transport infrastructures have expanded considerably to service areas that were not previously serviced or were insufficiently serviced. A paradox of this feature is that although the expansion of transport infrastructures may have enabled distribution systems to expand, it also increased the average distance over which passengers and freight are being carried.

Efficiency of transport terminals. Terminals, such as ports and airports, have shown a growing capacity to handle large quantities in a timely manner. Thus, even if the speed of many transport modes has not increased, more efficient transport terminals and a better management of flows have helped reduce transport time.

Information technologies (IT). Enabled several economic activities to bypass spatial constraints in a very significant manner as IT supports complex management structures. Electronic mail is an example where the transmission of information does not have a physical form (outside electrons or photons) once the supporting infrastructure is established. There is obviously a limit to this substitution, but several corporations are trying to use the advantages of telecommuting as much as they can because of the important savings involved.

Improvements in transport technology enabled a gradual space/time convergence of the global transport system (Figure 1.6). Before the industrial revolution, transport

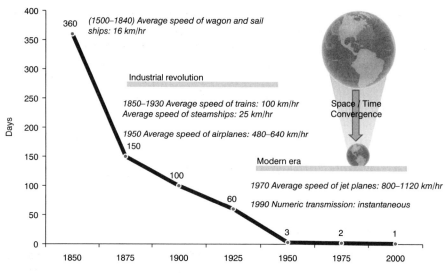

Figure 1.6 Global space/time convergence: days required to circumnavigate the globe

technology only permitted limited access to other regions of the world. Technological innovations in the domain of transportation were essentially used to increase the economic efficiency of advanced economies enabling them to have access to resources and markets. This began with mercantilism and gradually shaped the global economic space leaning on the transmission of information and fast and cheap transport systems. It is important to mention that the global space/time convergence is not spatially uniform, implying that some regions benefit more than others. For instance, space/time convergence in Western Europe and North America, and over the North Atlantic, has taken place at a faster rate than in other regions of the world, such as Latin America or Africa. It can be assumed that as economic and infrastructure development takes place around the world, space/time convergence is likely to become more uniform.

Circumnavigation is a good proxy for space/time convergence. Prior to the steamship, circumnavigating the globe would take about one sailing year, a journey greatly delayed by rounding the Cape of Good Hope and the Strait of Magellan. The late nineteenth and early twentieth centuries provided a series of innovations that would greatly improve circumnavigation, notably the construction of the Suez (1869) and Panama (1914) canals as well as steam propulsion. Circumnavigation was reduced to about 100 days (the "Jules Verne effect") at the beginning of the twentieth century and to 60 days by 1925 with fast liner services. The introduction of the jet plane in the second half of the twentieth century reduced circumnavigation to about 30 hours if two direct and connecting long range flights could be booked.

Yet, space/time convergence does not occur in a ubiquitous manner. In time, some locations gain more accessibility than others particularly if they experience the accumulation of transport infrastructures. After centuries of transport developments and their impacts on geography, global accessibility reflects a heterogeneous geography. Space/time convergence can also be inverted under specific circumstances, which means that a process of space/time divergence takes place. For instance, congestion is increasing in many metropolitan areas, implying additional delays for activities such as commuting. Traffic in congested urban areas is moving at the same speed that it did one hundred years ago in horse carriages. Air transportation, despite having dramatically contributed to the space/time convergence, is also experiencing growing delays. Flight times are getting longer between many destinations, mainly because of takeoff, landing and gate access delays. Airlines are simply posting longer scheduled flight times to factor in congestion. The termination of the Concorde supersonic jet service in 2003 can also be considered as a space/time divergence. More stringent security measures at airports have also imposed additional delays, which tend to penalize short distance flights. Additionally, direct transport services can be discontinued and replaced by a hub-and-spoke structure.

Concept 3 – Transportation and commercial geography

Authors: Jean-Paul Rodrigue and Claude Comtois

Trade and commercial geography

Historically, wealth was predominantly related to agricultural output implying that the largest economies were those with the largest populations. Trade patterns mostly followed demographics. The industrial revolution irremediably changed this relationship with mechanization and its multiplying effects on production and consumption. Yet, economic systems remain based on trade and transactions since specialization and efficiency require interdependency. People trade their labor for a wage, having to commute in the process, while corporations trade their output for capital. Trade is the transmission of a possession in return for a counterpart, generally money, which is often defined as a medium of exchange. This exchange involves a transaction and its associated flows of capital, information, commodities, parts or finished products. All this necessitates the understanding of commercial geography.

> **Commercial geography** investigates the spatial characteristics of trade and transactions in terms of their cause, nature, origin and destination. It leans on the analysis of contracts and transactions. From a simple commercial transaction involving an individual purchasing a product at a store, to the complex network of transactions maintained between a multinational corporation and its suppliers, the scale and scope of commercial geography varies significantly.

Trade, in terms of its origins and destinations, has a spatial logic. It reflects the economic, social and industrial structure of the concerned markets, but also implies other factors such as transport costs, distance, political ties, exchange rates and the reciprocal economic advantages proponents get from trade. For trade to occur, several conditions must be met:

> **Availability**. Commodities, from coal to computer chips, must be available for trade and there must be a demand for these commodities. In other terms, a surplus must exist at one location and a demand in another. A surplus can often be a simple matter of investment in production capabilities, such as building an assembly plant, or can be constrained by complex environmental factors like the availability of resources such as fossil fuels, minerals and agricultural products.
> **Transferability**. Transport infrastructures in allowing commodities to be moved from their origins to their destinations favor the transferability of goods. There are

three major impediments to transferability, namely policy barriers (tariffs, custom inspections, quotas), geographical barriers (time, distance) and transportation barriers (the simple capacity to move the outcome of a transaction). Distance often plays an important role in trade, as does the capacity of infrastructures to route and to tranship goods.

Transactional capacity. It must be legally possible to make a transaction. This implies the recognition of a currency for trading and legislations that define the environment in which commercial transactions are taking place, such as taxation and litigation. In the context of a global economy, the transactional environment is very complex but is important in facilitating trade at the regional, national and international levels. The fundamental elements of a commercial transaction involving the transportation of a good are the letter of credit and the bill of lading. The transport terms have been regulated since 1936 by international commercial terms that are regularly updated and revised.

Once these conditions are met, trade is possible and the outcome of a transaction results in a flow. Three particular issues relate to the concept of flow:

Value. Flows have a negotiated value and are settled in a common currency. The American dollar, which has become the major global currency, is used to settle and/or measure many international transactions. Further, nations must maintain reserves of foreign currencies to settle their transactions and the relationship between the inbound and outbound flows of capital is known as the balance of payments. Although, nations try to maintain a stable balance of payments, this is rarely the case.

Volume. Flows have a physical characteristic, mainly involving a mass. The weight of flows is a significant variable when trade involves raw materials such as petroleum or minerals. However, in the case of consumption goods, weight has little significance relatively to the value of the commodities being traded. With containerization, a new unit of volume has been introduced; the TEU (Twenty-foot Equivalent Unit), which can be used to assess trade flows.

Scale. Flows have a range that varies significantly based on the nature of a transaction. While retailing transactions tend to occur at a local scale, transactions related to the operations of a multinational corporation are global in scale.

Trends in commercial geography

Traditionally, commercial activities tended to develop where there was a physical break along transport chains as cargo needed to be transferred from one mode to the other and where a new actor took over its ownership, or its custody. The physical break imposed transactions, an important reason why most of the world's most important financial centers tend to be port cities or major load break centers. The contemporary commercial setting is marked by increasing free trade and profound technological, industrial and geopolitical changes. The liberalization of trade, as confirmed by the implementation of the World Trade Organization (WTO), has given a strong impetus and a positive trend in the growth rate of world trade and industrial production. This has led to strong competitive pressures and shifting competitive advantages. However, in a true free trade environment, regulatory agencies would not be required. But in spite of attempts at deregulation, transactions and trade are prone to disputes, litigations and perceived imbalances concerning who benefits the most. Although these issues mainly apply to international trade, there are also situations where trade is constrained between the provinces/states of a nation.

In spite of globalization, much trade is still dominantly regional. An overview of world trade flows indicate that trade within regions is more significant than trade between regions, but long distance trade is steadily growing. Figures indicate the increasing share of East Asia, especially China, in world trade both in terms of exports and imports. Flows of merchandise have also been accompanied by a substantial growth in foreign direct investments. There is thus a remarkable reallocation of production capacities through outsourcing and offshoring following changes in comparative advantages around the world. This trend goes in tandem with mergers and acquisitions of enterprises that are increasingly global in scope. Analysis of international trade thus reveals the need to adopt different strategies to adapt to this new trading environment. As production is being relocated, there is a continuous shift in emphasis in the structure of the export and import of world economies.

Major changes have occurred in the organization of production. There is a noticeable increase in the division of labor concerning design, planning and assembly in the manufacturing process of the global economy. Interlocking partnerships in the structure of manufacturing have increased the trade of parts and the supply of production equipment around the world. One-third of all trade takes place among parent companies and their foreign affiliates. A part of this dynamism resides in the adoption of standards, a process which began in the late nineteenth century to promote mass production. It permitted the rapid development of many sectors of activity, including railways, electricity, the automobile industry and telecommunications more recently (Internet, Electronic Data Interchange). The decline of manufacturing in its share of the global gross domestic product (GDP) is illustrative of the growing complexities that added value brings to the function of production. In the realm of globalization of economic activities, the International Organization for Standardization developed the ISO norms that serve as comparison between various enterprises around the world. These norms are applicable to the manufacturing and services industries and are a necessary tool for growth.

Another significant force of change in commercial geography is the growth of consumption, although this is not taking place uniformly. As a result, commercial geography is influenced by the market size, the consumption level of an economy (often measured in GDP; Figure 1.7), and also by the growth potential of different regions of the world. Economic growth taking place in East and Southeast Asia has been one of the most significant force-shaping changes in the contemporary commercial environment. The commodification of the economy has led to significant growth in retail and wholesale and the associated movements of freight.

The World Bank often uses GDP per capita to classify the level of economic development of nations. The wealthiest nations account for the largest markets in the world. The GDP is thus a reasonable approximation of the size of a market, but not necessarily of the standards of living (or quality of life). For instance, China has a much higher GDP than Korea, implying that China is a bigger market, but Korea is a more sophisticated economy with higher standards of living. The global generation of wealth remains highly concentrated. The four largest economies, the United States, Japan, China and Germany, alone accounted for more than 38 percent of the world's GDP in 2008. Thus, nine countries (G8 + China) generated more than half the global economic activity. Still, the dynamism is shifting with China overtaking Japan to become the world's second largest economy in 2010. Countries such as Brazil and India have also experienced a remarkable growth.

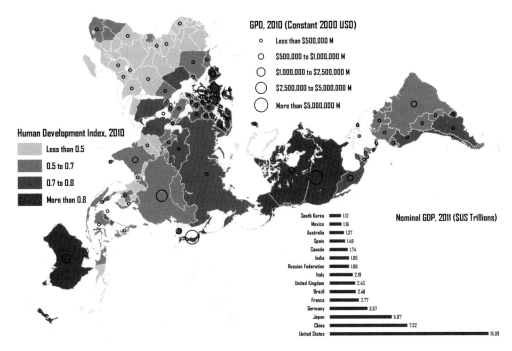

Figure 1.7 Global Gross Domestic Product and Human Development Index, 2010

Source: UNEP (2012) The UNEP Environmental Data Explorer, as compiled from World Development Indicators (WDI-The World Bank). United Nations Environment Programme.

The Human Development Index (HDI) is a composite measure ranging from 0 to 1 that includes life expectancy, education (literacy rate) and standards of living (GDP per capita). It is more representative of the commercial potential with countries with a HDI above 0.8 accounting for the world's main markets. This commercial potential and dynamism shapes global transactions and flows.

Commercialization of the transport industry

The liberalization of trade was accompanied by a growth of transportation since transactions involve movements of freight, capital, people and information. Developments in the transport sector are matched by global and regional interdependence and competition. Transportation, like commodities, goods and services, is traded, sometimes openly and subject to full market forces, but more often subject to a form of public control (regulation) or ownership. The core component of a transport-related transaction involves its costs that either have to be negotiated between the provider of the service and the user or are subject to some arbitrary decree (price setting such as public transit). Since transportation can be perceived as a service, its commercialization (how it is brought to the market) is an important dimension of its dynamics (Figure 1.8). Transport service providers tend to be private entities, particularly in the global freight sector. Local passenger transportation providers (transit) tend to be publicly owned.

Commercialization involving the extension of the operational scale of transport systems ensures that it reaches its optimal market potential for the passenger and freight markets. Although an optimal market size can never be attained due to regulations preventing monopolies and differences in consumer preferences (e.g. modal choice), the

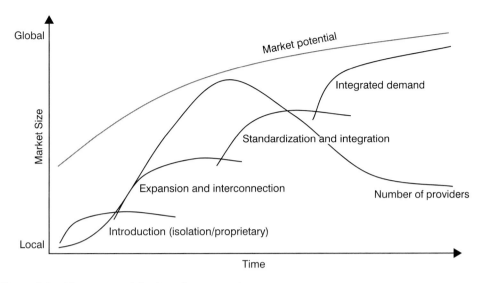

Figure 1.8   The commercialization of transportation

trend to ensure maximal market exposure is unmistakable. Within transport systems, four distinct cyclic phases of extension and functional integration can be identified:

Introduction. Initially, a transport system is introduced to service a specific opportunity in an isolated context. The technology is often "proprietary" and incompatible with other transport systems. Since they are not interconnected, this does not represent much of an issue.

Expansion and interconnection. As the marketability and the development potential of a transport system becomes apparent, a phase of expansion and interconnection occurs. The size of the market serviced by these transport systems consequently increases as they become adopted in new locations and as new providers are created to service those markets. At some point, independently developed transport systems connect. This connection is, however, often subject to a function of transshipment between two incompatible transport systems.

Standardization and integration. This phase often involves the emergence of a fully developed transport system servicing vast national markets. The major challenge to be addressed involves a standardization of modes and processes, further expanding the commercial potential. Modal flows are moving more efficiently over the entire network and are able to move from one mode to the other through intermodal integration. A process of mergers and acquisitions of transport providers often accompanies this phase for the purpose of rationalization and market expansion.

Integrated demand. The most advanced stage of extension of a distribution system involves a system fully able to answer mobility needs of passengers and freight under a variety of circumstances, either predicted or unpredicted demand. As this system tends to be global, it commonly operates close to market potential. In such a setting, a transport system expresses an integrated demand where transport supply is tuned with the demand in an interdependent system.

One important component of the commercialization of transportation concerns investments in infrastructure, modes and terminals, as well as marketing and financing.

Investments are performed either to expand the geographical extent and/or the capacity of a transport system or to maintain its operating conditions. The public and private sectors have contributed to the funding of transport investments depending on economic, social and strategic interests. For obvious reasons, the private sector seeks transport investments that promise economic returns while the public sector often invests for social and strategic reasons. In many cases private transport providers have difficulties acting independently in formulating and implementing their transport investments. Various levels of government are often lobbied by transport firms for financial and/or regulatory assistance in projects that are presented as of public interest and benefit. The consolidation of regional markets and the resulting increase in transborder traffic has led transport firms to seek global alliances and greater market liberalization in the transport and communication sector as a means to attract investments and to improve their productivity.

Deregulation and divestiture policy in the transport industry have led governments to withdraw from the management, operations and ownership of national carriers, ports and airports. This has given rise to a major reorganization of the international and national transport sectors with the emergence of transnational transport corporations that are governing the global flow of air, maritime and land trade and the management of airports, ports and railyards.

Concept 4 – The geography of transportation networks

Authors: Jean-Paul Rodrigue and César Ducruet

Transport networks

Transportation systems are commonly represented using networks as an analogy for their structure and flows. Transport networks belong to the wider category of spatial networks because their design and evolution are physically constrained, as opposed to non-spatial networks such as social interactions, corporate organization and biological systems.

> The term **network** refers to the framework of routes within a system of locations, identified as nodes. A **route** is a single link between two nodes that are part of a larger network that can refer to tangible routes such as roads and rails, or less tangible routes such as air and sea corridors.

The territorial structure of any region corresponds to a network of all its economic interactions. The implementation of networks, however, is rarely premeditated but the consequence of continuous improvements as opportunities arise, investments are made and as conditions change. The setting of networks is the outcome of various strategies, such as providing access and mobility to a region, reinforcing a specific trade corridor or technological developments making a specific mode and its network more advantageous over others. A transport network denotes either a permanent track (e.g. roads, rail and canals) or a scheduled service (e.g. airline, public transit, train). It can be extended to cover various types of links between points along which movements can take place.

In transport geography, it is common to identify several types of transport structures that are linked with transportation networks with key elements such as nodes, links, flows, hubs or corridors. Network structure ranges from centripetal to centrifugal in terms of the accessibility they provide to locations. A centripetal network favors a limited number of locations while a centrifugal network tends not to convey any specific locational advantages. The recent decades have seen the emergence of transport hubs, a strongly centripetal form, as a privileged network structure for many types of transport services, notably for air transportation. Although hub-and-spoke networks often result in improved network efficiency, they have drawbacks linked with their vulnerability to disruptions and delays at hubs, an outcome of the lack of direct connections. Evidence underlines that the emergence of hub-and-spoke networks is a transitional form of network development rationalizing limited volumes through a limited number of routes. When traffic becomes sufficient, direct point-to-point services tend to be established as they better reflect the preference of users.

Hubs, as a network structure, allow a greater flexibility within the transport system, through a concentration of flows. For instance, on Figure 1.9, a point-to-point network involves 16 independent connections, each to be serviced by vehicles and infrastructures. By using a hub-and-spoke structure, only 8 connections are required. The main advantages of hubs are:

- Economies of scale on **connections** by offering a high frequency of services. For instance, instead of one service per day between any two pairs in a point-to-point network, four services per day could be possible.
- Economies of scale at the **hubs**, enabling the potential development of an efficient distribution system since the hubs handle larger quantities of traffic.
- Economies of scope in the use of **shared transshipment facilities**. This can take several dimensions such as lower costs for users as well as higher quality infrastructures.

Many transportation services have adapted to include a hub-and-spoke structure. The most common examples involve air passenger and freight services which have developed such a structure at the global, national and regional levels, like those used by parcel carriers such as UPS, FedEx and DHL. However, potential disadvantages may also occur such as additional transshipment as less point-to-point services are offered, which for some connections may involve delays and potential congestion as the hub becomes the major point of transshipment.

Transport networks are better understood by the usage level (e.g. number of passengers, tons, vehicles, capacity) than by their sole topology based on a binary state (i.e. presence or absence of links). Inequalities between locations can often be measured by

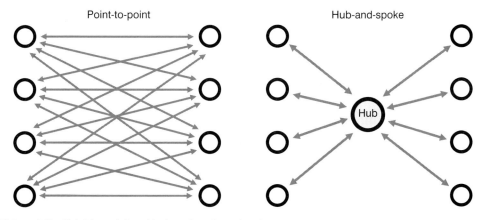

Figure 1.9 Point-to-point and hub-and-spoke networks

the quantity of links between nodes and the related revenues generated by traffic flows. Many locations within a network have higher accessibility, which is often related to better opportunities. However, economic integration processes tend to change inequalities between regions, mainly through a reorientation of the structure and flows within transportation networks at the transnational level.

The efficiency of a network can be measured through graph theory and network analysis. These methods rest on the principle that the efficiency of a network depends partially on the layout of nodes and links. Obviously some network structures have a higher degree of accessibility than others, but careful consideration must be given to the basic relationship between the revenue and costs of specific transport networks. Rates thus tend to be influenced by the structure of transportation networks; the hub-and-spoke structure, particularly, has a notable impact on transport costs, namely through economies of scale.

The topology and typology of networks

Transportation networks, like many networks, are generally embodied as a set of locations and a set of links representing connections between those locations. The arrangement and connectivity of a network is known as its topology, with each transport network having its own. The most fundamental elements of such a structure are the network geometry and the level of connectivity. Transport networks can be classified in specific categories depending on a set of topological attributes that describe them. It is thus possible to establish a basic typology of a transport network that relates to its geographical setting as well as its modal and structural characteristics.

There are many criteria that can be used to classify transportation networks (Figure 1.10). The level of abstraction can be considered with tangible network representations closely matching the reality (such as a road map) while conversely an abstract network would only be a symbolization of the nodes and flows (such as the network of an airline). Since transportation networks have a geographical setting, they can be defined according to their relative location to main elements of a territory. Networks also have an orientation and an extent that approximates their geographical coverage or their market area. The number of nodes and edges is relevant to express the complexity and structure of transportation networks with a branch of mathematics, graph theory, developed to infer structural properties from these numbers.

Since networks are the support of movements they can be considered from a modal perspective, their edges being an abstraction of routes (roads, rail links, maritime routes) and their nodes an abstraction of terminals (ports, railyards). Specific modes can further be classified in terms types of road (highway, road, street, etc.) and level of control (speed limit, vehicle restrictions, etc.). Flows on a network have a volume and a direction, enabling links to be ranked by their importance and the general direction of flows evaluated (e.g. centripetal or centrifugal). Each segment and network has a physical capacity related to the volume it can support under normal conditions. The load (or volume to capacity) is the relation between the existing volume and the capacity. The closer a network is to its full load (a ratio of 1), the more congested it is. The structure of some networks imposes a hierarchy reflecting the importance of each of its nodes and a pattern reflecting their spatial arrangement. Finally, networks have a dynamic where both their nodes and links can change due to new circumstances.

The physical grounding of a network varies in relevance depending on the transport mode considered. Roads and railways are composed of track infrastructure while

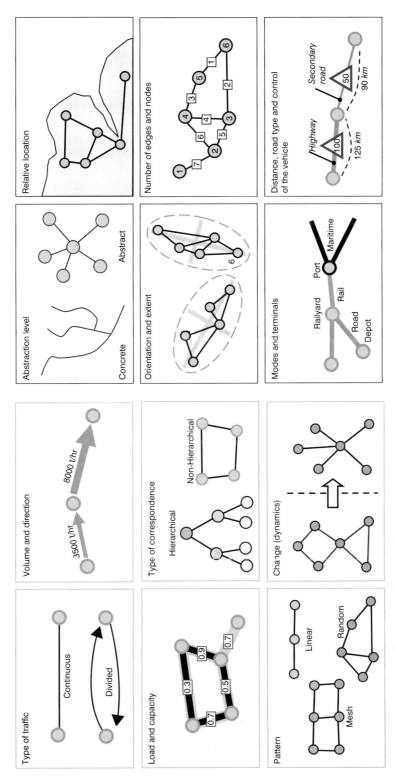

Figure 1.10 A typology of transportation networks

maritime and air transports remain vaguely defined due to their higher spatial flexibility except for their terminals, whereas maritime networks remain more constrained than airline networks due to the necessity of bypassing coastlines. River networks typically form basins and can be classified as trees or dendrograms. Therefore, there are three types of physical spaces on which transport networks are set and where each represents a specific mode of territorial occupation:

- **Clearly defined and delimited**. The space occupied by the transport network is strictly reserved for its exclusive usage and can be identified on a map. Ownership can also be clearly established. Major examples include road, canal and railway networks.
- **Vaguely defined and delimited**. The space of these networks may be shared with other modes and is not the object of any particular ownership, only of rights of way. Examples include air and maritime transportation networks.
- **Without definition**. The space has no tangible meaning, except for the distance it imposes. Little control and ownership are possible, but agreements must be reached for common usage. Examples are radio, television, WiFi and cellular networks, which rely on specific wave frequencies granted by regulatory agencies.

Networks provide a level of transport service which is related to their costs. An optimal network would be a network servicing all possible locations but such a service would have high capital and operational costs. Transport infrastructures are established over discontinuous networks since many were not built at the same time, by the same entity or with the same technology. Therefore, operational networks rarely service all parts of the territory directly. Some compromise must often be found among a set of alternatives considering a variety of route combinations and level of service. Networks are also labeled depending on their overall properties:

- **Regular network**. A network where all nodes have the same number of edges. In the same vein, a random network is a network that is formed by random processes. While regular networks tend to be linked with high levels of spatial organization (e.g. a city grid), random networks tend to be linked with opportunistic development opportunities such as accessing a resource.
- **Small-world network**. A network with dense connections among close neighbors and few but crucial connections among distant neighbors. Such networks are particularly vulnerable to catastrophic failures around large hubs.
- **Scale-free network**. A network having a strong hierarchical dimension, with few vertices having many connections and many vertices having few connections. Such networks evolve through the dynamic of preferential attachment by which new nodes added to the network will primarily connect larger nodes instead of being connected randomly.

Investigating the interdependencies among different transport networks, notably when they are of a different nature and structure, is challenging. Some crucial aspects and problems related with inter-network relations may be as follows:

- **Coevolution**. Different transport networks might follow similar or different paths based on spatial proximity and path-dependence of economic development, with a wider variety of networks at core regions than at remote regions.

- **Complementarity**. Some locations may be central in one network but peripheral in another, depending on their specialization and function and on the scale of analysis (terminal, city, region, country); the complementarity between networks can be measured based on the number of common nodes and links.
- **Interoperability**. Typically, cargo flows from a maritime network to a road network shift from a scale-free structure to a regular structure, thus following different topologies that are not easily combined; air and sea terminals remain few in the world due to the difficulty of combining and integrating technically air and sea networks physically at the same locations.
- **Vulnerability**. How do changes in one network affect the other network, on a global level (entire network) or local level (single node or region)? This is particularly important for two networks sharing common nodes, such as global cities, logistics platforms and multilayered hubs. In the case of abrupt conjunctures (e.g. natural disasters, targeted attacks, labor disputes, security and geopolitical tensions), thus posing the problem of rerouting flows through alternative routes and locations.

Networks and space

Transportation networks underline the territorial organization of economic activities and the efforts incurred to overcome distance. These efforts can be measured in absolute (distance) or relative terms (time) and are proportional to the efficiency and the structure of the networks they represent. The relationships transportation networks establish with space are related to their continuity, their topographic space and the spatial cohesion they establish. The territory is a topological space having two or three dimensions depending on the transport mode considered (roads are roughly set over a two-dimensional space while air transport is set over a three-dimensional space). However, flows and infrastructures are linear; having one dimension since they conceptually link two points. The establishment of a network is thus a logical outcome for a one-dimensional feature to service a territory by forming a lattice of nodes and links. In order to have such a spatial continuity in a transport network, three conditions are necessary:

Ubiquity. The possibility to reach any location from any other location on the network thus providing a general access. Access can be a simple matter of vehicle ownership or bidding on the market to purchase a thoroughfare from one location to another. Some networks are continuous, implying that they can be accessed at any location they service. Roads are the most salient example of a continuous network. Other networks are discrete, implying that they can only be accessed at specific locations, commonly at a terminal. Rail, maritime and rail networks are considered discrete networks since they can only be accessed through their terminals.

Fractionalization. The possibility for a traveler or a unit of freight to be transported without depending on a group. It becomes a balance between the price advantages of economies of scale and the convenience of a dedicated service.

Instantaneity. The possibility to undertake transportation at the desired or most convenient moment. There is a direct relationship between fractionalization and instantaneity since the more fractionalized a transport system is, the more likely time convenience can be accommodated.

These three conditions are never perfectly met, and some transport modes fulfill them better than others. For instance, the automobile is the most flexible and ubiquitous mode for passenger transportation, but has important constraints such as low capacity and high levels of space and energy consumption. In comparison, public transit is more limited in the spatial coverage of its service, implies batch movements (bus loads, train loads, etc.) and follows specific schedules (limited instantaneity), but is more cost and energy efficient. Freight transportation also varies in its spatial continuity, ranging from massive loads of raw materials (oil and ores) that can be handled only in a limited number of ports to highly flexible parcels movements. Containerization has been a remarkable attempt to address the issue of ubiquity (the system permits intermodal movements), fractionalization (each container is a load unit) and instantaneity (units can be loaded by trucks at any time of the day and container ships make frequent port calls).

An important cause of discontinuity is linked to the spatial distribution of economic activities, notably industrial and urban, which tend to agglomerate. Congestion may also alter these conditions. Road congestion in a metropolitan area may impair ubiquity as some locations may be very difficult to reach since their accessibility is reduced. Fractionalization may also be reduced under such circumstances as people would consider public transit and carpooling and would thus move as batches. Further, as commuters cope with increasing congestion, several trips may be delayed or canceled altogether reducing instantaneity.

Transportation networks have always been a tool for spatial cohesion and occupation. The Roman and Chinese empires relied on transportation networks to control their respective territories, mainly to collect taxes and move commodities and military forces. During the colonial era, maritime networks became a significant tool of trade, exploitation and political control, and this was later on expanded by the development of modern transportation networks within colonies. In the nineteenth century, transportation networks also became a tool of nation building and political control. For instance, the extension of railways in the American hinterland had the purpose to organize the territory, extend settlements and distribute resources to new markets. In the twentieth century, road and highways systems (such as the Interstate system in the United States and the autobahn in Germany) were built to reinforce this purpose. In the later part of the twentieth century, air transportation networks played a significant role in weaving the global economy. For the early twenty-first century, telecommunication networks have become means of spatial cohesion and interactions abiding well to the requirements of global supply chains.

Network expansion

The co-evolution of roads, canals and ports during the industrial revolution in England reveals noticeable interdependencies among the different nodes and networks over time, based on spatial and functional proximity. Initial network developments are often done to support and then compete with an existing network by expanding geographically and topologically in ways unavailable to the prior network. As transport networks expand, existing transport infrastructures are being upgraded to cope with spatial changes. Airports and ports are being transformed, expanded or relocated. In the air transport sector, emphasis is being given to integrate airports within fully fledged multimodal transport systems, networking air with rail and road transport. In maritime transport, networks are also being modified with increasing attention being paid to the expansion of the Panama and Suez canals. There is increasing traffic on inland waterways and the creation of new inland passages between semi-enclosed or enclosed seas.

The growing competition between the sea and land corridors are not only reducing transport costs and encouraging international trade but prompting many governments to reassess their land-based connections and seek shorter transit routes. Existing land routes are also being extended. Passages through difficult terrain are being investigated with a view to creating fully fledged land-based continental connections, notably through railways. These land network expansions are driven by economic globalization and inter-regional cooperation and eventually become multimodal transcontinental corridors for rail, road, pipelines and trunk telecommunications routes. But the impact of increasing world trade on land network expansion, notably over railways, is scale specific. The expansion of railways has permitted inter and intra-continental connections, namely landbridges (Figure 1.11).

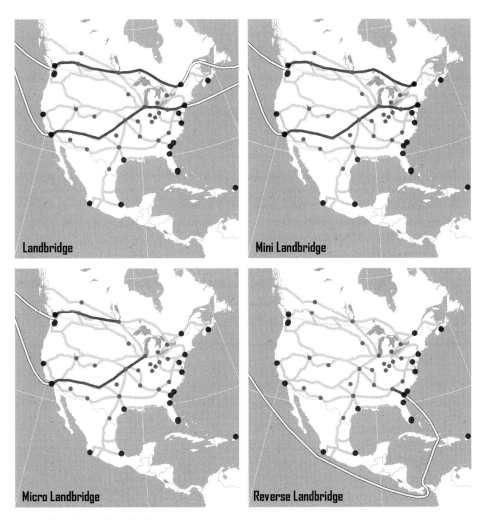

Figure 1.11 Types of landbridges

Landbridges provide a level of continuity between maritime and long distance inland transport networks. There are four main types of landbridges depending on the destination of the transoceanic cargo:

- **Landbridge**. Using a landmass as a link in a maritime transport chain involving a foreign origin and destination. This type of link is not highly used in the North American context as it is more convenient for Europe-bound cargo from Asia to use the maritime route instead or the emerging Eurasian landbridge for niche cargo.
- **Mini landbridge**. Using a landmass as a link in a transport chain involving a foreign origin and a destination at the end of the landmass.
- **Micro landbridge**. Using a landmass as a link in a transport chain involving a foreign origin and an inland destination. A common type of service for Asian cargo bound to an inland North American market.
- **Reverse landbridge**. Through a maritime detour reaching an inland destination by using the closest maritime facade instead of the landbridge. This type of landbridge has experienced remarkable growth with the usage of the Panama Canal to reach East Coast ports by an all-water route from Asia. The expansion of the Panama Canal is expected to increase the significance of this type of landbridge service.

In recent years new rail routes in North America, Eurasia, Latin America and Africa have been developed or are being considered. There is scope for shippers to increase their trade through these new routes, particularly if rising insurance premiums, charter rates and shipping risks prompt them to opt for a land route instead of the sea route through the Suez or Panama canals. These developments linked to the integration of regional economies to the world market are part of a rationalization and specialization process of rail traffic presently occurring around the world. But the success of these rail network expansions depends on the speed of movement and the unitization of general cargo by containerization. Railways servicing ports tend to consolidate container flows, which allows an increase in capacity and the establishment of door-to-door services through a better distribution of goods among different transport modes. New links are establishing and reshaping new trade flows underpinning outward cargo movements and the distribution of goods. As some coastal gateways are now emerging as critical logistics services centers that rationalize distribution systems to fit new trading patterns, the land network development and cross-border crossings throughout the world have far-reaching geopolitical implications.

CASE STUDY   Strategic maritime passages

Authors: Jean-Paul Rodrigue and Theo Notteboom

Global maritime routes and chokepoints

Maritime transportation is the dominant purveyor of international freight distribution and evolves over a global maritime space. This space has its own constraints such as the profile of continental masses and the imperatives it creates in terms of detours and passages. Maritime routes are spaces of a few kilometers wide trying to avoid the discontinuities of land transport. They are a function of obligatory points of passage, which are almost all strategic places, physical constraints such as coasts, winds, marine currents, depth, reefs or ice and political boundaries where sovereignty may impede

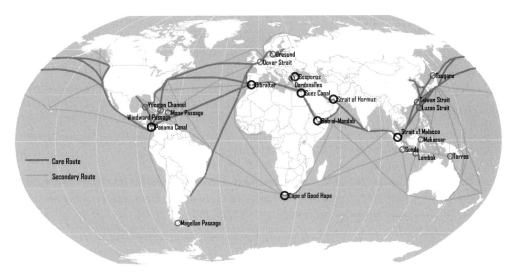

Figure 1.12 Main maritime bottlenecks and shipping routes

circulation. The majority of the maritime circulation takes place along coasts and three continents have limited fluvial trade (Africa, Australia and Asia, except China).

International maritime routes are thus forced to pass through specific locations corresponding to passages, capes and straits (Figure 1.12). These routes are generally located between major industrial regions such as Western Europe, North America and East Asia where an active system of commercial containerized trade is in place. The importance of these large manufacturing regions and their consumption markets are structuring exchanges of semi-finished and finished goods. Also, major routes involve flows of raw materials, namely minerals, grains, some food products (coffee, cocoa and sugar) and, most importantly, petroleum. The location of strategic oil and mineral resources shapes maritime routes for bulks since they represent the most transported commodities. For instance, over 30 Mb/d (million barrels per day) are being shipped around the world.

Maritime routes are a function of obligatory points of passage, which are strategic places, of physical constraints and of political borders. As a result, maritime routes draw arcs on the earth water surface as intercontinental maritime transportation tries to follow the great circle distance. Main shipping lanes are those supporting the most important commercial shipping flows servicing major markets. Secondary shipping lanes are mostly connectors between smaller markets. Figure 1.12 provides a simple taxonomy of the main strategic passages.

- **Primary passages** are the most important passages since without them there would be limited cost-effective maritime shipping alternatives which would seriously impair global trade.
- **Secondary passages** have alternatives, but would still involve a notable detour.

The most important strategic maritime passages are known as chokepoints (or bottlenecks) due to:

- **Capacity constraints**. Chokepoints tend to be shallow and narrow, impairing navigation and imposing capacity limits on ships. For canals such as the Panama and Suez, the capacity must effectively be managed with appointment and pricing systems.

- **Potential for disruptions or closure**. Disruption of trade flows through any of these export routes could have a significant impact on the world economy. Many chokepoints are next to politically unstable countries, increasing the risk of compromising their access and use, such as with piracy. Closures are a rare instance that have only taken place in situations of war as one proponent prevented another from accessing and using the chokepoint (e.g. Gibraltar and Suez during the Second World War). A closure of a maritime chokepoint in the current global economy, even if temporary, would have important economic consequences with the disruption of trade flows and even the interruption of some supply chains (e.g. oil).

The Panama Canal, the Suez Canal, the Strait of Malacca and the Strait of Hormuz are the world's four most important strategic maritime passages in part because they are chokepoints that affect global freight circulation and in part because of the economic activities and resources they grant more efficient access to. Their continuous availability for global maritime circulation is challenging.

The Panama Canal

The Panama Canal joins the Atlantic and Pacific oceans across the Isthmus of Panama, running from Cristobal on Limon Bay, an arm of the Caribbean Sea, to Balboa, on the Gulf of Panama. Its operational characteristics involve a length of 82 kilometers, a depth of 12.5 meters (39.5 feet), a width of 32 meters (106 feet) and a length of 294 meters (965 feet). Its construction ranks as one of the greatest engineering works of all time as it serves a long detour around South America, thus supporting the maritime flows of world trade (Figure 1.13). The Panama Canal is of strategic importance to the United States as it serves to link the East and the West coast more quickly, saving about 13,000 km (from 21,000 km to 8,000 km) for a maritime journey. It is composed of three main elements, the Gatun Locks (Atlantic Ocean access), the Gaillard Cut (continental divide; Photo 1.1) and the Miraflores/Pedro Miguel Locks (Pacific Ocean access).

The Panama Canal is obviously the shortest operational route between the Atlantic and Pacific oceans. The maritime alternatives are the Magellan Route circumnavigating South America and the potential use of the Northwest Passage through the Arctic Ocean. The Magellan Route imposes a substantial detour but offers the opportunity to pick up or drop off cargo along the way. The Northwest Passage is the shortest route between the North Pacific and the North Atlantic, but remains hazardous to navigation and does not offer any significant opportunity to pick up or drop off cargo. The landbridge alternatives are more numerous with the North American landbridges composed of the Canadian, American and Mexican landbridges being operational realities. Still, their role is not necessarily to offer a substantial alternative to the Panama Canal, but options to shippers servicing North American supply chains. Other landbridges in Central or South America are simply projects of unknown market potential.

In its 96 years of existence (as of 2010) since its completion in 1914, more than one million vessels have transited the canal, carrying 8.1 billion tons of cargo. About 13,000 ships transit the canal every year, with an average of 35 ships per day. However, the canal has the capacity to handle 50 ships per day with an average transit time of about 16.5 hours if the passage has been reserved in advance and about 35 hours if no reservations have been made, for an average crossing time of 23 hours. Containers, grains and petroleum account for the dominant share of the cargo transited. The

Figure 1.13   Main routing alternatives between the Pacific and Atlantic

introduction of super-tankers at the beginning of the 1950s forced the reconsideration of its strategic importance as economies of scale in petroleum shipping are limited by the size of the canal. It is synonymous of a standard in maritime transport related to capacity, the Panamax standard, which equates to 65,000 deadweight tons, a draft of 12 meters and a capacity of about 4,500 TEUs depending on the load configuration.

The canal handles about 5 percent of the global seaborne trade and about 12 percent of the American international seaborne trade. Under the control of the United States until 1979, its administration was entrusted to the State of Panama by the Panama Canal Treaty of 1977. In December 1999, the canal reverted to Panama under the jurisdiction of the Panama Canal Authority. The authority generates revenue by collecting tolls on all ships crossing the canal and is responsible for the operation and maintenance of the facility. A loaded ship pays about $2.57 per net ton and the average toll is about $45,000. For container ships the toll (as of 2011) is $74 per TEU of capacity on laden containers and $65.60 per TEU of capacity on ships with empty containers. In 2008, $1.32 billion in tolls were collected, of which 54 percent was generated by container shipping.

Photo 1.1 Container ship exiting the Gaillard Cut, Panama Canal

A container ship of the Hamburg Sud shipping line completes the transit through the Gaillard Cut, the part of the Panama Canal that slices through the continental divide. The ship mostly carries empty containers likely originating from the American East Coast and bound back to Asia. In the background is the Centennial Bridge, which opened in 2004 and which offers a crossing over the canal in addition to the Bridge of the Americas that was opened in 1962. Originally named Culebra (snake) Cut for the nature of its curves, it was renamed for Col. David DuBose Gaillard, the engineer who supervised its construction. The Cut is approximately 12.8 kilometers long and, during the 1960s was widened from its original 91.5 meters to 152.4 meters. The currently ongoing Gaillard Cut widening program is widening it still further to increase the canal capacity and to facilitate vessel transit scheduling.

In 1999, the Hong Kong port terminal operator, Hutchison Port Holdings (HPH), took control of the operation of port terminals on both the Atlantic (Port of Colon) and Pacific (Port of Balboa) sides of the canal with a 25-year lease. The company also became involved in the improvement of the rail line between the two ports to handle the growing containerized traffic. This rail line, the Panama Canal Railway Company (owned by KCS and Lanco Group), is important as it offers an alternative to the size limitations of the canal and supports transshipment activities between the Atlantic and the Pacific sides. The same rationale applies to oil circulation with the trans-Panama pipeline which resumed its operations in 2003, but the additional capacity this pipeline conveys is only about 1 Mb/d.

In spite of being close to a century old, the Panama Canal remains a critical bottleneck in global trade. The continuous growth of global trade since the 1990s has placed additional pressures on the Panama Canal to handle a growing number of ships in a timely and predictable manner. This raised concerns that the existing canal would reach capacity by the second decade of the twenty-first century. Because of these capacity limits, many shipping companies have changed the configuration of their routes. This became increasingly apparent as a growing share of the global container ship fleet reached a size beyond the capacity of the Panama Canal, which came to be known as "post-Panamax" container ships. Through economies of scale, they offer significant operational cost advantages that cannot be exploited by the existing canal. The increasing usage of these ships along the Pacific Asia/Suez Canal/Mediterranean routes as well as the development of the North American rail landbridge have created a substantial competition to the Panama Canal as an intermediate location in global maritime shipping. There are thus a range of alternatives to the Panama Canal route, with the North American landbridges the most important. Yet, concerns about the reliability of the landbridge connection have encouraged the setting of "all-water routes" linking directly Pacific Asia and the American East Coast, particularly in light of the booming China–United States trade relations.

A decision to expand the Panama Canal was finally reached in 2006 by the Panamanian government. The expansion is a 5.25 billion US dollars project that involves building a new set of locks on both the Atlantic and Pacific sides of the canal to support a depth of 60 feet, a width of 190 feet and a length of 1,400 feet, which would accommodate ships up to 12,000 TEU depending on their load configuration. The dredging of access channels as well as the widening of several sections of the existing canal will also be required. This would allow Aframax and Suezmax vessels to pass through the canal, thus permitting new opportunities for container shipping services such as the re-emergence of round-the-world services. The new locks will complement the existing lock systems, creating a two-tier service: one for the very large ships and the other for the Panamax ships. It is expected that the new infrastructures will be available by 2014 or 2015 depending on unforeseen construction problems.

The Suez Canal

The Suez Canal is an artificial waterway of about 190 km in length running across the Isthmus of Suez in northeastern Egypt which connects the Mediterranean Sea with the Gulf of Suez, an arm of the Red Sea. It has no locks, because the Mediterranean Sea and the Gulf of Suez have roughly the same water level and is thus the world's longest canal without locks. It acts as a shortcut for ships between both European and American ports and ports located in southern Asia, eastern Africa, and Oceania. The maritime route from Europe to the Indian and Pacific oceans otherwise must go round the Cape of Good Hope at the southernmost point of the African continent. The minimum width of the channel is 60 meters and ships of 18 meters (62 feet) draft can make the transit. The canal can accommodate ships as large as 220,000 deadweight tons fully loaded.

The opening of the Suez Canal in 1869 brought forward a new era of European influence in Pacific Asia. The journey from Asia to Europe was considerably reduced by saving 6,500 km from the circum Africa route. In 1874, Great Britain bought the shares of the Suez Canal Company and became its sole owner. According to the Convention of Constantinople signed in 1888, the canal was to be open to vessels of all nations in time of peace or in war. However, Great Britain claimed the need to control the area to maintain its maritime power and colonial interests (namely in South Asia). In 1936, it acquired the right to maintain defense forces along the Suez Canal, which turned out to be of strategic importance during the Second World War to uphold Asia–Europe supply routes for the Allies.

The second half of the twentieth century saw renewed geopolitical instability in the region with the end of colonialism and the emergence of Middle Eastern nationalisms.

In 1954 Egypt and Great Britain signed an agreement that superseded the 1936 treaty and provided for the gradual withdrawal of British troops from the zone. All the British troops were gone by June 1956 as the canal was nationalized by Egypt. This triggered problems with Israel, as Israeli ships were not permitted to cross the canal. This threat was also extended to France and Britain, the former owners of the canal, because they refused to help finance the Aswan High Dam project, as initially promised. Israel, France and Britain thus invaded Egypt in 1956. Egypt responded by sinking ships in the canal effectively closing it between 1956 and 1957. An agreement about the usage of the canal was then reached.

However, geopolitical problems persisted as tensions between Israel and Arab nations increased in the 1960s. The Six Day War between Israel and Egypt and the invasion of the Sinai Peninsula by Israel caused the closure of the Suez Canal between 1967 and 1975. This event significantly destabilized international transportation and favored the development of ever larger tankers to use the long circum Africa route. The canal was finally reopened in 1975 as Egypt agreed to let Israel use it. Significant improvements were made between 1976 and 1980, mainly the widening of the canal to accommodate very large crude carriers (VLCC) of about 200,000 tons supporting the oil trade between Europe and the Middle East. However, ultra large crude carriers (ULCC; tankers of more than 300,000 tons) cannot pass through the canal when fully loaded. A common practice is to unload part-loads of Mediterranean-bound ships and use the Sumed pipeline. With additional deepening and widening projects, the depth of the canal reached 22.5 meters in 2001.

The canal has the capacity to accommodate up to 25,000 ships per year (about 78 per day), but handles about 20,000, on average 55 ships per day, which roughly account for 15 percent of the global maritime trade. Since the canal can only handle unidirectional traffic, crossings must be organized into convoys of about 10 to 15 ships. Three convoys per day, two southbound and one northbound, are organized. The transit time is about 10 hours northbound and 12 hours southbound. Missing a convoy involves supplementary delays to the point that many maritime shipping companies (particularly for containers) will skip a port call to ensure that their ships arrive on time at the Suez Canal to be part of a specific convoy. A rail line also runs parallel to the canal.

The transit rates are established by the Suez Canal Authority (SCA). They are computed to keep the canal transit fees attractive to shippers. In the fiscal year 2008, Egypt earned 5 billion USD in canal fees making it Egypt's third largest revenue generator after tourism and remittances from expatriate workers. Container ships account for just under half of the canal's traffic and a slightly higher percentage of its net tonnage and revenues. The average canal transit fee per TEU (at 90 percent vessel utilization) amounts to 102 USD for a vessel of 1,000 TEU down to 56 USD for the largest container vessels. In early 2009, a number of shipowners started to boycott the Suez Canal because of the high transit fees. There are a number of alternatives to the Suez Canal (Figure 1.14).

The Suez route is the main commercial artery between Asia and Europe and transits through locations such as the Strait of Malacca, Bab el-Mandab, the Suez Canal and the Strait of Gibraltar. It offers opportunities to pick up and drop cargo at major transshipment hubs such as Singapore, Colombo, Salalah, Suez, Gioa Tauro and Algeciras. These hubs are collecting points for their respective regional cargoes as well as connectors to north–south routes. Because of the level of trade and the draft of the Suez Canal maritime shipping companies have been able to allocate their largest ships along this route, with services including container ships with a capacity above 10,000 TEU. The Cape

Figure 1.14 Main routing alternatives between East Asia and Northern Europe

Source: Adapted from Notteboom, T. and J-P. Rodrigue (2011) "Emerging global networks in the container terminal operating industry", in T. Notteboom (ed.) *Current Issues in Shipping, Ports and Logistics*, Brussels: Academic & Scientific Publishers.

route is a shipping alternative involving a long detour past the southern tip of Africa, but represents an opportunity to handle African cargo as well as connect with routes bound to South America. Next to the Cape route, a number of other routing alternatives are being planned or are in operation to accommodate part of the trade volumes between Europe and Asia, but their market shares are expected to remain low compared to the Suez route, which offers substantial capacity, low costs and reliability.

First there is the **Northern Sea Route**, a set of all-water shipping lanes between the Atlantic Ocean and the Pacific Ocean along the Russian coast of Siberia and the Far East. Future polar ice cap reductions would open new possibilities for commercial shipping on this route. In cost terms the route today is still less favorable due to the need for ice-classed ships and ice-breaker assistance, non-regularity of the liner services, slower sailing speeds, navigation difficulties and Russian transit fees.

Second, **North–South land corridors** could develop as landbridges from the Persian Gulf via Iran to Russia. However, geopolitical and infrastructure issues forbid any serious consideration of this alternative in the medium term.

Third, the **East–West rail corridors**, a set of railway lines connecting East Asia and the western part of Russia with the eastern part of Russia, are becoming more commercially interesting. One of the main arteries is the Trans-Siberian Railway which connects St. Petersburg with the port of Vladivostok. Other primary rail connections are the Trans-Manchurian Railway, the Trans-Mongolian Railway and the Baikal Amur Mainline (BAM – opened in 1991). The "Trans-Siberian in Seven Days" program sets a target speed of 1,500 km a day by 2015, but is facing the problem of different rail gauges between the Chinese, Russian and Western European systems. Rail landbridges in principle offer lead time advantages to shippers, but capacities remain low compared to container liner services. They offer a niche potential for time-sensitive cargo.

The Strait of Malacca

The Strait of Malacca is one of the most important strategic passages in the world because it supports the bulk of the maritime trade between Europe and Pacific Asia, which accounts for 50,000 ships per year. About 30 percent of the world's trade and 80 percent of Japan's, South Korea's and Taiwan's imports of petroleum transits through the strait, which involved approximately 11.7 Mb/d in 2004. It is the main passage between the Pacific and the Indian oceans with the strait of Sunda (Indonesia) being the closest alternative (Figure 1.15). It measures about 800 km in length, has a width between 50 and 320 km (2.5 km at its narrowest point) and a minimal channel depth of 23 meters (about 70 feet). It represents the longest strait in the world used for international navigation and can be transited in about 20 hours.

Traditionally, the Strait of Malacca was an important passage point between the Chinese and the Indian worlds and was controlled at different times by Javanese and Malaysian kingdoms. From the fourteenth century, the region came under the control of Arab merchants who established several fortified trading towns, Malacca being the most important commercial center in Southeast Asia. Again, the control of the trade route shifted as the era of European expansion began in the sixteenth century. In 1511, Malacca fell to the Portuguese and this event marked the beginning of European control over the strait. In 1867, England took control of the passage with Singapore as a main harbor and other important centers, such as Malacca and Penang, forming the Strait Settlements. This control lasted until the Second World War and the independence of Malaysia in 1957. As Pacific trade increased considerably after the Second World War, so did the importance of the passage. Singapore, located at the southern end of the Strait of Malacca, is one of the most important ports in the world and a major oil refining center.

One of the main problems about the Strait of Malacca is that at some points it requires dredging, since it is barely deep enough to accommodate ships of about 300,000 deadweight tons. The strait runs between Malaysia, Indonesia and Singapore, and it has been difficult for these countries to reach agreement about how the dredging costs should be shared and how fees for its usage should be levied. Political stability and piracy along it are also major issues for the safety of maritime circulation, especially on the Indonesian side with the province of Aceh in a state of civil unrest.

The Strait of Malacca ends up in the South China Sea, another extremely important shipping lane and a region subject to contention since oil and natural gas resources are

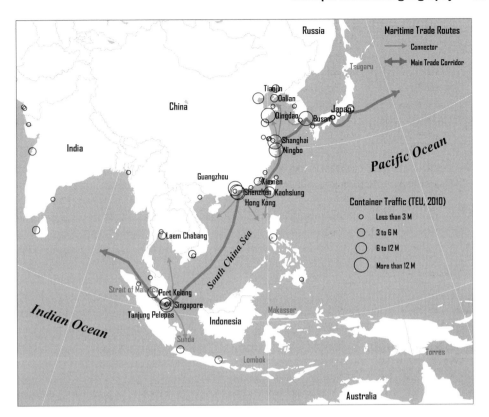

Figure 1.15 Shipping lanes and strategic passages in Pacific Asia

present. The Spratly and Paracel groups of islands are claimed in whole or in part by China, Vietnam, Malaysia, Indonesia, Brunei and the Philippines. The region has proven oil reserves estimated at about 7.0 Bb with oil production accounting for 2.5 Mb/d. With the substantial economic growth taking place in the region large flows of oil, liquefied natural gas and other raw materials (iron ore, coal) are transiting towards East Asia. About 25 percent of the global shipping fleet transits through the region each year, underlining the importance of the South China Sea as an extension of the Malacca chokepoint. Economic development in Pacific Asia has been accompanied by a strong growth of container ports and the setting of a major maritime trade corridor and its connectors. This corridor is structured along a north–south maritime axis that runs from a cluster of Japanese ports (Tokyo/Yokohama, Nagoya, Osaka/Kobe) to Singapore, including along the way major gateways such as the port of Busan in South Korea, the Chinese ports of the Yangtze River and the Pearl River Delta, and the port of Kaohsiung, Taiwan.

The Strait of Hormuz

The Strait of Hormuz forms a strategic link between the oil fields of the Persian Gulf, which is a maritime dead end, the Gulf of Oman and the Indian Ocean. It has a width between 48 and 80 km, but navigation is limited to two 3 km wide channels, each exclusively used for inbound or outbound traffic. Circulation in and out of the Persian Gulf is thus highly constrained, namely because the sizable amount of tanker and container ship traffic makes navigation difficult along the narrow channels. In addition, islands that ensure the control of the strait are contested by Iran and the United Arab Emirates.

The security of the strait has been often compromised and its commercial usage has been the object of contentions. Between 1984 and 1987 a "Tanker War" took place between Iran and Iraq, during the Iran–Iraq War (1980–88), where each belligerent began firing on tankers, even neutrals, bound for their respective ports. Shipping in the Persian Gulf dropped by 25 percent, forcing the intervention of the United States to secure the oil shipping lanes. About 88 percent of all the petroleum exported from the Persian Gulf transits through the Strait of Hormuz, bound for Asia, Western Europe and the United States. Its importance for global oil circulation cannot be overstated. For instance, 75 percent of all Japanese oil imports transit through the strait. There are thus very few alternative outlets to oil exports if the traffic of about 14 Mb/d going through Hormuz were compromised. In addition to geopolitics, the geography of the Middle East creates challenges for petroleum distribution. The first is the concentration of reserves around the Persian Gulf (Saudi Arabia, Kuwait, Iraq, Iran and the United Arab Emirates) leaving the Strait of Hormuz as one of the only viable outlets to export oil using large oil tankers (Figure 1.16). The second is the constrained access to the main

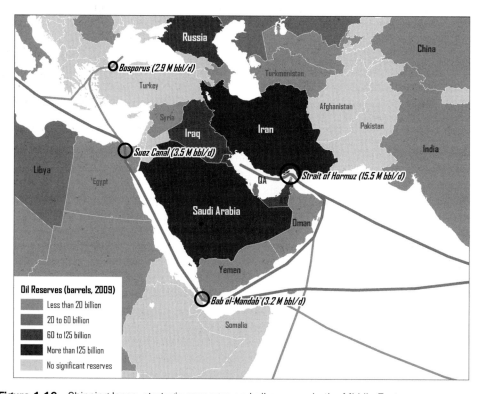

Figure 1.16 Shipping lanes, strategic passages and oil reserves in the Middle East

Sources: US Energy Information Administration, *World Oil Transit Chokepoints* and BP, *Statistical Review of World Energy*.

European markets through the Suez Canal, whereby a fully laden VLCC cannot transit through the canal and must round Africa.

While the Persian Gulf has conventionally been centered on oil production and distribution, the growth of container shipping has also expanded its commercial importance. For instance, Dubai ranked in 2010 as the world's ninth largest container port with traffic above 11.6 million TEU but it can only be accessed through the Strait of Hormuz. It has become a major transshipment hub linking the main Asian, Middle Eastern and East African trade routes. Consequently, compromising circulation through the Strait of Hormuz would impair global oil trade as well as commercial trade along Europe/Asia routes.

Bibliography

Banister, D. (2002) *Transport Planning*, 2nd edition, London: Spon Press.

Black, W. (2003) *Transportation: A Geographical Analysis*, New York: Guilford.

Briggs, K. (1972) *Introducing Transportation Networks*, London: University of London Press.

Dalton, R., J. Garlick, R. Minshull and A. Robinson (1978) *Networks in Geography*, London: George Philip.

Goetz, A.R., T.M. Vowles and S. Tierney (2009) "Bridging the qualitative–quantitative divide in transport geography", *Professional Geographer*, 61(3): 323–35.

Haggett, P. (2001) *Geography: A Modern Synthesis*, 4th edition, New York: Prentice Hall.

Haggett, P. and R.J. Chorley (1969) *Network Analysis in Geography*, London: Edward Arnold.

Harrington, R. (1999) "Transport: then, now, and tomorrow", *Royal Society of Arts Journal*, 146(5488).

Hoover, E.M. (1948) *The Location of Economic Activity*, New York: McGraw-Hill.

Hoyle, B. and R. Knowles (eds) (1998) *Modern Transport Geography*, 2nd edition, London: Wiley.

Hugill, P.J. (1995) *World Trade since 1431*, Baltimore, MD: Johns Hopkins University Press.

Keeling, D.J. (2007) "Transportation geography: new directions on well-worn trails", *Progress in Human Geography*, 31(2): 217–25.

Keeling, D.J. (2008) "Transportation geography – new regional mobilities", *Progress in Human Geography*, 32(2): 275–83.

Knowles, R.D. (2006) "Transport shaping space: the differential collapse of time/space", *Journal of Transport Geography*, 14(6): 407–25.

Knowles, R., J. Shaw and I. Docherty (eds) (2008) *Transport Geographies: Mobilities, Flows and Spaces*, Malden, MA: Blackwell.

Leinbach, T. (1976) "Networks and flows", *Progress in Human Geography*, 8: 179–207.

Merlin, P. (1992) *Géographie des Transports, Que sais-je?* Paris: Presses Universitaires de France.

Rimmer, P. (1985) "Transport geography", *Progress in Human Geography*, 10: 271–77.

Rodrigue, J-P (2003) "Teaching transport geography: conference report and viewpoint", *Journal of Transport Geography*, 11(1): 73–5.

Taaffe, E.J., H.L. Gauthier and M.E. O'Kelly (1996) *Geography of Transportation*, 2nd edition, Upper Saddle River, NJ: Prentice Hall.

Tolley, R. and B. Turton (1995) *Transport Systems, Policy and Planning: A Geographical Approach*, Harlow, Essex: Longman.

② Transportation and the spatial structure

Transportation has a strong influence on the spatial structure at the local, regional and global levels. A historical perspective on the evolution of transport systems underlines the impacts of technological innovations and how improvements in transportation were interdependent with economic, social and spatial changes. The current transport systems are thus the outcome of a long evolution marked by periods of rapid changes where new transport technologies were adopted. Following the industrial revolution in the nineteenth century, transportation systems were mechanized with the development of steam engine technology, which permitted the setting of networks servicing regions. This process was further expanded in the twentieth century with the setting of global air transport, container shipping and telecommunication networks. The impacts of transport on the spatial structure became multiscalar. Transportation systems are composed of a complex set of relationships between the demand, the locations they service and the networks that support movements. Such conditions are closely related to the development of transportation networks, both in capacity and in spatial extent. Future transportation systems will likely be shaped by the same forces as in the past but it remains to be seen which technologies will prevail and what will be their impacts on the structure.

Concept 1 – Historical geography of transportation: the emergence of mechanized systems

Transportation in the pre-industrial era (pre 1800s)

Transportation is closely linked with the process of globalization. Efficiently distributing freight and moving people has always been an important factor for maintaining the cohesion of economic systems from empires to modern nation states and economic blocs. With technological and economic developments, the means to achieve such a goal have evolved considerably with a series of historical revolutions and evolutions. This process is very complex and is related to the spatial evolution of economic systems and associated technical developments. It is possible to summarize this evolution, from the pre-industrial era to transportation in the early twenty-first century, in four major stages, each linked with specific technological innovations in the transport sector: the pre-industrial era, the industrial revolution, Fordism and post-Fordism (globalization).

Before the major technical transformations brought forward by the industrial revolution at the end of the eighteenth century, no forms of motorized transportation existed. Transport technology was mainly limited to harnessing animal labor for land transport and wind for maritime transport. The transported quantities were very limited and so was the speed at which people and freight were moving. The average overland speed by horse, which was domesticated around 2,000 BC, was between 8 to

15 kilometers per hour and maritime speeds were barely above these figures. Also, a horse can only carry a load of about 125 kg while a camel can carry about 200 kg. Waterways were the most efficient transport systems available and cities next to rivers were able to trade over longer distances and maintain political, economic and cultural cohesion over a larger territory. It is not surprising to find that the first civilizations emerged along river systems for agricultural but also for trading purposes (Tigris–Euphrates, Nile, Indus, Ganges, Huang He).

Because the efficiency of the land transport system of this era was poor, the overwhelming majority of trade was local in scope. Economies based on autonomy and basic subsistence could not generate much trade. From the perspective of regional economic organization, the provision of cities in perishable agricultural commodities was limited to a radius of about 50 kilometers, at most. The size of cities also remained constant in time. Since people can walk about 5 km/hr and that they are not willing to spend more than one hour per day walking, the daily space of interaction would be constrained by a 2.5 km radius, or about 20 square kilometers. Thus, most rural areas centered around a village and cities rarely exceeded a 5 km diameter. The largest cities prior to the industrial revolution, such as Rome, Beijing, Constantinople or Venice, never surpassed an area of 20 square kilometers. International trade did exist, but traded commodities were high value (luxury) goods such as spices, silk, wine and perfume, notably along the Silk Road (Figure 2.1). Around the Mediterranean, the amphora permitted a form of intermodalism as an effective standard transport product of olive oil, grain or wine.

The Silk Road was the most enduring trade route in human history, being used for about 1,500 years. Its name is taken from the prized Chinese textile that flowed from Asia to the Middle East and Europe, although many other commodities were traded along the route. The Silk Road consisted of a succession of trails followed by caravans

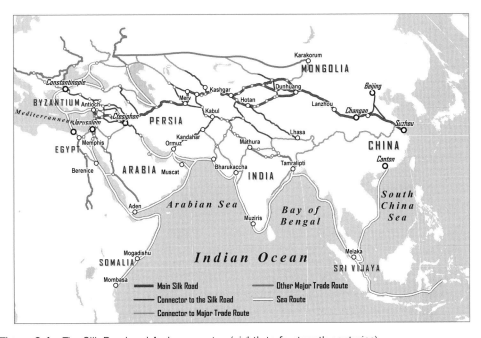

Figure 2.1 The Silk Road and Arab sea routes (eighth to fourteenth centuries)

through Central Asia, about 6,400 km in length. Travel was favored by the presence of steppes, although several arid zones had to be bypassed such as the Gobi and Takla Makan deserts. Economies of scale, harsh conditions and security considerations required the organization of trade into caravans slowly trekking from one stage (town and/or oasis) to the next.

Although it is suspected that significant trade occurred for about 1,000 years beforehand, the Silk Road opened around 139 BC once China was unified under the Han dynasty. It started at Changan (Xian) and ended at Antioch or Constantinople (Istanbul), passing by commercial cities such as Samarkand and Kashgar. It was very rare that caravans traveled for the whole distance since the trade system functioned as a chain. Merchants with their caravans were shipping goods back and forth from one trade center to the other. Since the transport capacity was limited, over long distances and often unsafe, luxury goods were the only commodities that could be traded. The Silk Road also served as a vector for the diffusion of ideas and religions (initially Buddhism and then Islam), enabling civilizations from Europe, the Middle East and Asia to interact.

The initial use of the sea route linking the Mediterranean basin and India took place during the Roman era. Between the first and sixth centuries, ships were sailing between the Red Sea and India, aided by summer monsoon winds. Goods were transshipped at the town of Berenike along the Red Sea and moved by camels inland to the Nile. From that point, river boats moved the goods to Alexandria, from which trade could be undertaken with the Roman Empire. From the ninth century, maritime routes controlled by the Arab traders emerged and gradually undermined the importance of the Silk Road. Since ships were much less constraining than caravans in terms of capacity, larger quantities of goods could be traded. The main maritime route started at Canton (Guangzhou), passed through Southeast Asia, the Indian Ocean, the Red Sea and then reached Alexandria. A significant feeder went to the Spice Islands (Moluccas) in today's Indonesia. The diffusion of Islam was also favored through trade as many rules of ethics and commerce are embedded in the religion.

The Silk Road reached its peak during the Mongolian Empire (thirteenth century) when China and Central Asia were controlled by Mongol Khans, who were strong proponents of trade even if they were ruthless conquerors. At the same time relationships between Europe and China were renewed, notably after the voyages of Marco Polo (1271–92). During the Middle Ages, the Venetians and Genoese controlled the bulk of the Mediterranean trade which connected to the major trading centers of Constantinople, Antioch and Alexandria. As European powers developed their maritime technologies from the fifteenth century, they successfully overthrew the Arab control of this lucrative trade route to replace it with their own. Ships being able to transport commodities faster and cheaper marked the downfall of the Silk Road by the sixteenth century.

Prior to the industrial revolution, it was difficult to speak of an urban system, but rather of a set of relatively self-sufficient economic systems with very limited trade. The preponderance of city-states during this period can a priori be explained by transportation, in particular the difficulties of shipping goods (therefore to trade) from one place to another. Among the most notable exceptions to this were the Roman and Chinese empires, which committed extraordinary efforts at building transportation networks and consequently maintained control over an extensive territory for a long period.

The transport system of the Roman Empire was a reflection of the geographical characteristics and constraints of the Mediterranean basin. The Mediterranean provided a central role to support trade between a network of coastal cities, the most important of

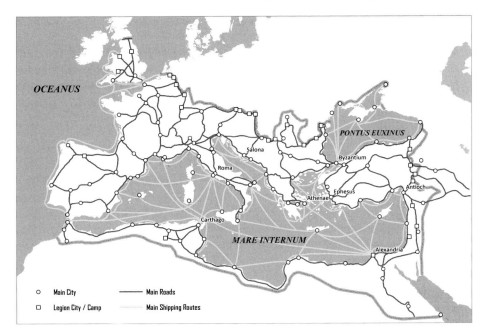

Figure 2.2 Roman Empire, c. AD 125

the empire (Rome, Constantinople, Alexandria, Carthage, etc.). These cities were serviced by a road network permitting trade within their respective hinterlands. Little fluvial transportation took place since the major pan-European rivers, the Rhine and the Danube, were military frontiers, not the core of the empire. The road served numerous functions, such as military movements, political control, cultural and economic (trade). To improve the traveling speed posthouses with fresh horses were laid every 15 kilometers along the route and lodgings for travelers could be found about every 40 kilometers. This 40 km corresponded to the average distance a traveler could cross each day. Courier services could travel twice that daily distance. The Appian Way (Via Appia), about 560 kilometers in length, was one of the first Roman roads (Via) to be constructed (around 312 BC) under the initiative of emperor Appius Claudius Caecus. It linked Rome to Brundisium (Brindisi).

As the empire grew, this system was expanded to cover 80,000 kilometers of first-class roads at the height of the Roman Empire (around AD 200). Most of the roads were constructed by soldiers, prisoners of war and slaves. The minimum requirement of a first-class road was a width of 5 meters and a drained stone surface. The Romans also built the world's first dual-carriageway, Via Portuensis, between Rome and its port Ostia at the mouth of the Tiber. The Roman road network covered most of the conquered provinces, with Rome as the focal point (thus the saying "All roads lead to Rome"). At the center of Rome was located the milliareum aureum (the golden milestone), from which the Roman roads radiated. Maintenance was the responsibility of the inhabitants of the district through which the road passed by, but access was public. Way stations where travelers could rest and eat dotted the network. The system collapsed during the Middle Ages because of the lack of maintenance and plundering for construction material, but still, the remains of the Roman network provided transportation in Europe for a thousand years. Only small segments of this system are left today.

The economic importance and the geopolitics of transportation were recognized very early, notably for maritime transportation since before the industrial revolution; it was the most convenient way to move freight and passengers around. Great commercial empires were established with maritime transportation. Initially, ships were propelled by rowers and sails were added around 2,500 BC as a complementary form of propulsion. By medieval times, an extensive maritime trade network, the highways of the time, was established, centered along the navigable rivers, canals, and coastal waters of Europe (and also China). Shipping was extensive and sophisticated using the English Channel, the North Sea, the Baltic and the Mediterranean, and the most important cities were coastal or inland ports (London, Norwich, Königsberg, Hamburg, Bruges, Bordeaux, Lyon, Lisbon, Barcelona and Venice). Trade of bulk goods, such as grain, salt, wine, wool, timber and stone, was taking place. By the fourteenth century galleys were finally replaced by full-fledged sailships (the caravel and then the galleon) that were faster and required smaller crews. The year 1431 marked the beginning of European expansion with the discovery by the Portuguese of the North Atlantic circular wind pattern, better known as the trade winds. A similar pattern was also found on the Indian and Pacific oceans with the monsoon winds.

The fall of Constantinople, the capital of the Byzantium Empire (Eastern Roman Empire), to the Turks in 1453 disrupted the traditional land trade route from Europe to Asia. Europe was forced to find alternate maritime routes. One alternative, followed by Columbus in 1492, was to sail to the west, and the other alternative, followed by Vasco de Gama in 1497, was to sail to the east. Columbus stumbled upon the American continent, while Gama found a maritime route to India using the Cape of Good Hope. These events were quickly followed by a wave of European exploration and colonization, initially by Spain and Portugal, the early maritime powers, then by Britain, France and the Netherlands. The traditional trade route to Asia no longer involved Italy (Venice) and Arabia, but involved direct maritime connections from ports such as Lisbon and Amsterdam. European powers were able to master the seas with larger, better armed and more efficient sailing ships and thus were able to control international trade and colonization. Private charter companies, such as the Dutch East India Company, were agents initially used to establish maritime trading networks that spanned the world. By the early eighteenth century, most of the world's territories were controlled by Europe, providing wealth and markets to their thriving metropolises through a system of colonial trade.

Figure 2.3 plots the density of a large sample of ship logs, which is indicative of the global trade network structure of the eighteenth century. Due to differences in the number of log entries geocoded, such as only 7,698 entries for French flagged ships as opposed to 83,036 entries for UK flagged ships, there are density variations that are not necessarily reflective of the size of nationally flagged fleets and their level of commercial activity. Limited shipping is observed over the Pacific Ocean, mostly due to the yet limited scale of European involvement in this part of the world at that time. The second half of the nineteenth century will see an intensification of maritime trade to and across the Pacific with the First Opium War of 1838–42 being the turning point. Prior to that, the China trade was mostly serviced by Portuguese flagged ships, which are not covered by the sample. The following observations can be made regarding the respective trade patterns of national flagged ships for the 1750 to 1810 period:

Spain. The Spanish trade was mainly focused on their American colonies, particularly Mexico (which was linked to the transpacific trade through an overland route),

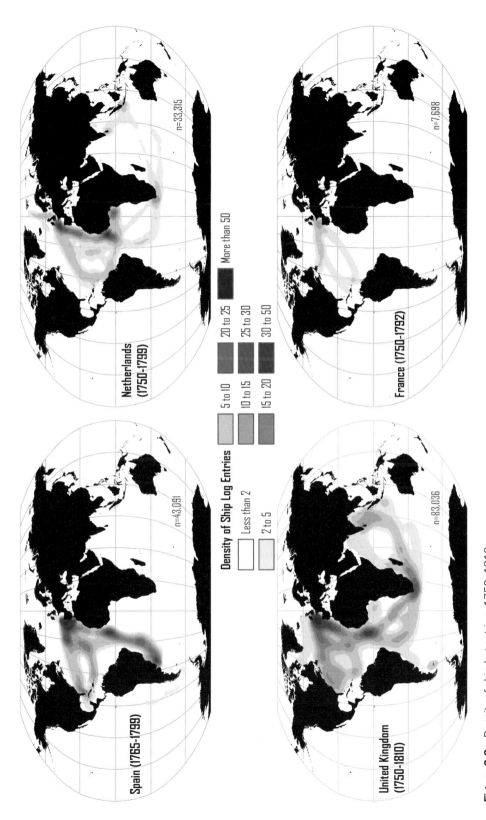

Figure 2.3 Density of ship log entries, 1750–1810

Source: Ship log data from CLIWOC Project. Data geocoded by David Hopp.

Cuba and Hispaniola. The transatlantic triangular trade system follows the pattern of wind and sea currents that benefit the southern part of the North Atlantic for westbound travel and the northern part of the North Atlantic for eastbound travel. The second main component of the Spanish maritime trade network concerns the Argentinian trade. Less visible are the connections to Chile and Peru through the Strait of Magellan.

Netherlands. The Dutch maritime trade network is mostly associated with the activities of the Dutch East India Company (VOC) with a long distance network heading towards VOC's main trade ports of Batavia (Indonesia) and Galle (Sri Lanka) and using Cape Town as an intermediary staging point (a Dutch colony between 1652 and 1806). Dutch interests were also involved in the triangular transatlantic trade with their colonies of the Caribbean and Guyana (Dutch colony between 1616 and 1814).

United Kingdom. Its trade network is the most extensive as it reflects the emergence of the UK during that period as the world's dominant maritime power with colonial interests in the Americas and South Asia. The Arctic trade segment is mainly related to the activities of the Hudson Bay Company established in 1670 for the fur trade around Rupert's Land.

France. The trade network depicted by French flagged ships is the simplest and limited to the triangular trade structure between France, its Caribbean colonies (Haiti, Guadeloupe and Martinique) and New France (St. Pierre and Miquelon after 1763).

Prior to the industrial revolution, the quantity of freight transported between nations was negligible by contemporary standards. For instance, during the Middle Ages, French imports via the Saint-Gothard Passage (between Italy and Switzerland) would not fill a freight train. The total amount of freight transported by the Venetian fleet, which dominated Mediterranean trade for centuries, would not fill a modern cargo ship. The volume, but not the speed, of trade improved under mercantilism (fifteenth to eighteenth century), notably with maritime transportation. In spite of all, distribution capacities were very limited and speeds slow. For example, a stagecoach going through the English countryside in the sixteenth century had an average speed of two miles per hour; moving one ton of cargo 30 miles (50 km) inland in the United States by the late eighteenth century was as costly as moving it across the Atlantic. The inland transportation system was thus very limited, both for passengers and freight. By the late eighteenth century, canal systems started to emerge in Europe, initially in the Netherlands and England. They permitted the beginning of large movements of bulk freight inland and expanded regional trade. Maritime and fluvial transportation were consequently the dominant modes of the pre-industrial era.

The industrial revolution and transportation (1800–70)

The factors that have led to the remarkable economic, technological and social changes brought by the industrial revolution are subject to debate in terms of their role and importance in the emergence of capitalism. Four of them appear to be prevalent and interdependent:

● **The scientific method**. Mostly the outcome of changes that took place in the seventeenth century, often dubbed the "Age of Reason", it triggered a rational approach to the laws of nature and formalized technical professions (physics, chemistry, engineering, etc.).

- **Property rights**. The strengthening of democratic institutions leaning on the rule of law that guarantees and protects private ownership.
- **Capital markets**. Institutions such as banks able to gather capital pools and invest them in economic ventures. The process of capital accumulation and allocation became increasingly rational.
- **Communications and transport infrastructure**. The setting and development of mechanized transport systems supporting the distribution of resources and the setting of comparative advantages.

It was during the industrial revolution that massive modifications of transport systems occurred in two major phases, the first centered on the development of canal systems and the second on railways. This period marked the development of the steam engine that converted thermal energy into mechanical energy, providing an important territorial expansion for maritime and railway transport systems. Much of the credit of developing the first efficient steam engine in 1765 is attributed to the British engineer Watt, although the first steam engines were used to pump water out of mines. It was then only a matter of time to see the adaptation of the steam engine to locomotion. In 1769, the French engineer Cugnot built the first self-propelled steam vehicle, and was also responsible for the first automobile accident ever recorded. The first mechanically propelled maritime vehicle was tested in 1790 by the American inventor Fitch as a mode of fluvial transportation on the Delaware River. By 1807, commercial steam boat services were inaugurated. This marked a new era in the mechanization of land and maritime transport systems alike.

From the perspective of land transportation, the early industrial revolution faced problems over bottlenecks, as inland distribution was unable to carry the growing quantities of raw materials and finished goods. Roads were commonly unpaved and could not be used to effectively carry heavy loads. Although improvements were made on road transport systems in the early seventeenth century, such as the Turnpike Trusts in Britain (1706) and the development of stagecoaches, this was not sufficient to accommodate the growing demands on freight transportation. The first coach services had speeds of about 5.5 miles per hour in the 1750s. By the 1820s turnpikes greatly improved overland transportation but roads were not profitable if used to haul anything except compact and valuable goods. In a horse-drawn era, road economics were clearly disadvantageous. Bulk products could be transported for about 100 miles, but in a slow, costly and inefficient manner. For instance, four horses could pull a wagon weight of one ton 12 miles a day over an ordinary road and one-and-a-half tons 18 miles a day over a well-maintained turnpike. Comparatively, four horses could draw a barge of 100 tons 24 miles a day on a canal.

From the 1760s a set of freight shipping canals were slowly built in emerging industrial cores such as England (e.g. Bridgewater Canal, 1761) and the United States (e.g. Erie Canal, 1825). These projects relied on a system of locks to overcome changes in elevation, and thus link different segments of fluvial systems into a comprehensive waterway system. Barges became increasingly used to move goods at a scale and a cost that were not previously possible. Economies of scale and specialization, the foundation of modern industrial production systems, became increasingly applicable through fluvial canals. Physical obstacles made canal construction expensive, however, and the network was constrained in its geographical coverage. In 1830 there were about 2,000 miles of canals in Britain and by 1850, there were 4,250 miles of navigable waterways.

In the late eighteenth and particularly in the early nineteenth centuries the construction of canals was undertaken to improve inland transportation in North America which was limited to trails and coastal navigation (see Figure 2.4). The setting of such canals was constrained as no navigable river system from the East Coast reached far inland, with the exception of the St. Lawrence that was navigable up to Montreal. The Appalachian mountains limited the inland reach to just a few hundred miles, with navigation often blocked by rapids or waterfalls (the Fall Line). The Great Lakes offered a significant agricultural potential but their access was blocked by the Lachine Rapids and the Niagara Escarpment. Segments between navigable waterways involved a portage where freight was carried by horses.

Canal construction closely followed the course of rivers and some river segments were essentially canalized (replaced by the canal). Rivers provided water supply to be used in locks as well as a path of minimal impendence. The problem was that canals never followed a direct path. The exception was when a "cut" was needed, which was done with the straightest path possible, but at great expense. As the first major canals were being constructed in the 1820s and 1830s they provided the first significant economies of scale for North American inland transportation. While a horse could carry one-eighth of a ton, a canal barge could carry 30 tons. Two canal systems emerged, one east of the Appalachians along the East Coast and one west of the Appalachians in the Midwest:

The first canal system was an attempt to connect the interior from a set of coastal cities and go as far inland as technically possible. It took place from two main corridors. The first went from Montreal and along the St. Lawrence to Lake Erie with the completion of the Lachine Canal in 1825 and the Welland Canal in 1829, which

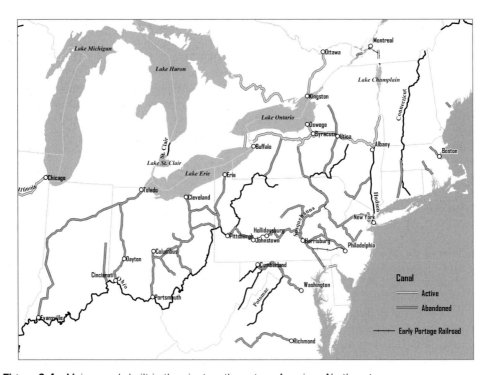

Figure 2.4 Major canals built in the nineteenth century, American Northeast

overcame the Niagara Escarpment between Lake Ontario and Lake Erie. The second was the Erie Canal system, completed in 1825 and connecting Albany, Syracuse and Buffalo. A connection to Lake Ontario was provided to the Oswego branch completed in 1828. Many other branch canals were built to carry coal from the Appalachians to the cities of the East Coast.

The second canal system in the Midwest was mostly connecting the Ohio River to Lake Erie, enabling access to the agricultural resources of the region and for them to be carried to the East Coast through the Erie Canal. The two most important canals were the Ohio and Erie Canal completed in 1833 linking Cleveland, Columbus and the Ohio River, and the Wabash and Erie Canal completed in 1853 linking Toledo to Evansville. A canal completed in 1848 between Chicago and the Illinois River was an important factor in the subsequent role of the city as the most important transportation hub in North America.

The first canals were constrained by several technical limitations related to their draft (4 to 10 feet) and the lift that locks could provide. Early locks could elevate a barge only by about 8 to 10 feet, so that a climb of 100 feet required about 10 to 15 locks. For instance, the Chesapeake and Ohio Canal that linked Washington, DC, to Cumberland, Maryland, climbed 605 feet and required 74 locks (average of 8.2 feet per lock). By the late nineteenth century, improvement in lock technology permitted a single lift of 30 to 40 feet. For instance, a lock system at the town of Lockport climbing the Niagara Escarpment along the Erie Canal was modernized into one lock offering a lift of 40 feet instead of five locks lifting 8 feet each. The first barges were propelled manually by pushing a pike and using a rudder (mostly downstream) or hauled by horses along the towpath (mostly upstream). Later, barges were motorized, but tended to be larger and used for canals with deeper drafts.

Paradoxically, the setting of canals also induced the construction of the first rail lines to compete with an existing transport market or for portage between unserviced segments. For instance, one of the first rail lines to be established in the United States in 1834, the Philadelphia and Columbia Railroad, was built to complement the Schuylkill and Union canals between Harrisburg and Philadelphia. Another important rail line completed the same year was the Allegheny Portage Railroad, which was the first railroad constructed through the Allegheny Mountains (part of the Appalachian Range), linking two canal cities: Johnstown (east of Pittsburgh) and Hollidaysburg (west of Harrisburg). The first railroad in Canada, the Champlain and St. Lawrence Railroad, completed in 1838 between La Prairie and St. Jean-sur-Richelieu, was also built with the same rationale: a portage between the St. Lawrence and Lake Champlain.

By the late nineteenth century the great majority of the canals were abandoned as they lost their commercial utility. Many had limited draft, had high upkeep and were no longer able to compete effectively with railways. Those that are left today, such as the Erie Canal, the Rideau Canal and the Champlain Canal, are used for recreational purposes and managed by state or federal governments as parks. Portions of some canals have been restored, again for recreational purposes. The only commercial exceptions are the Welland Canal, upgraded several times, which is now part of the St. Lawrence Seaway that was completed in 1959, and the Illinois and Michigan Canal that links Chicago to the Illinois River and was supplemented by the Chicago Sanitary and Ship Canal in 1900. The canal era was short-lived as a new mode that would revolutionize and transform inland transportation emerged in the second half of the nineteenth century.

Steam railway technology initially appeared in 1814 to haul coal. It was found that using a steam engine on smooth rails required less power and could handle heavier loads. The first commercial rail line linked Manchester to Liverpool in 1830 (distance of 65 km). Shortly after, rail lines began to be laid throughout developed countries, leading to the setting of national systems. Speed improvements were significant as the first rail networks ran between 30 and 50 km/hr, three times faster than stagecoach services. The capital costs to build railway networks were enormous and often left to the private sector but with significant public involvements in terms of loans and land grants. They included rights of way, building, maintenance and operating costs. This was accompanied by a few railway manias (and their subsequent busts) with capital pouring in to a sector that was perceived, at least by the general public, as limitless in possibilities. By the 1850s, railroad towns were being established and the railways were giving access to resources and markets of vast territories. Some 10,000 km of railways were then operating in England and railways were quickly being constructed in Western Europe and North America (Figure 2.5). The need to organize and schedule rail services instigated the adoption of standard time (often labeled standard railroad time). England was the first to implement a standard time system in 1855, Greenwich Mean Time, which became the global reference time.

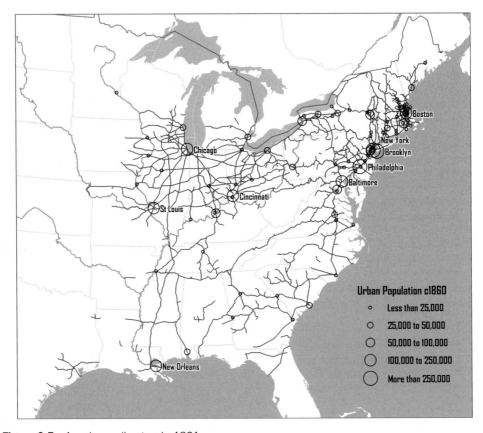

Figure 2.5 American rail network, 1861

Sources: Railroads and the Making of Modern America, University of Nebraska, Lincoln (Rail Network). US Census Bureau (Urban Population).

In only 30 years after its introduction the American rail network totaled about 28,900 miles (46,500 km) on the eve of the Civil War (1861–65; Figure 2.5). Yet, the American rail network was composed of two systems reflecting the sociopolitical division between the North (Union States) and the South (Confederate States). Outside the connection through Washington, the networks were not connected and serviced rather different economic systems. This lack of connectivity was compounded by the fact that railways servicing the same city were often not connected, requiring cargo to be ferried from one terminal to the other and for passengers to spend a night to catch the train the next day (schedules were not effectively coordinated). The dominantly rural society of the South was mainly serviced by penetration lines seeking to connect the agricultural hinterland to ports where surpluses were exported (e.g. New Orleans and Charleston). As such, the network was not very cohesive. The more urbanized North developed a network based on interconnecting its main urban centers and agricultural regions in the Midwest in a complex lattice. At the end of the Civil War the expansion of the network resumed as well as its level of integration.

Railroads represented an inland transport system that was at the same time flexible in its spatial coverage and could carry heavy loads. As a result many canals fell into disrepair and were closed as they were no longer able to compete with rail services. In their initial phase of development, railways were a point-to-point process where major cities were linked one at a time by independent companies. Thus, the first railroad companies bore the name of the city pairs or the region they were servicing (e.g. the Camden and Amboy Railroad Company chartered in 1830). From the 1860s, integrated railway systems started to cohesively service whole nations with standard gauges (made mandatory in the United States by the Interstate Commerce Act of 1887) and passenger and freight services. The journey between New York and Chicago was reduced from three weeks by stagecoach in 1830 to 72 hours by train in 1850. Many cities thus became closely interconnected. The transcontinental line between New York and San Francisco, completed in 1869, represented a remarkable achievement in territorial integration made only possible by rail. It reduced the journey across the continent (New York to San Francisco) from six months to one week, thus opening for the Eastern part of the United States a vast pool of resources and new agricultural regions. Canada followed in 1886 (Trans-Canada railway) and Russia in 1904 (Trans-Siberian railway).

In terms of international transportation, the beginning of the nineteenth century saw the establishment of the first regular maritime routes linking harbors worldwide, especially over the North Atlantic between Europe and North America. Many of these long distance routes were navigated by fast Clipper ships, which dominated ocean trade until the late 1850s. Another significant improvement resided in the elaboration of accurate navigation charts where prevailing winds and sea currents could be used to the advantage of navigation (Figure 2.6).

The navigator Matthew Fontaine Maury between 1842 and 1861 collected an extensive array of ship logs which enabled him to chart prevailing winds and sea currents, as well as their seasonal variations. While mariners were well aware of specific regional conditions such as the Gulf Stream, Maury was the first to comprehensively tally oceanographic conditions at the global level. Much of this information appeared in *The Physical Geography of the Sea*, published in 1855, which many consider to be the first significant oceanography textbook. Such knowledge enabled sailing times to be considerably shortened, especially over long distance travel. For instance, a journey from New York to Rio de Janeiro was reduced by 32 days, without any technical improvements to the ship, but by simply taking better advantage of prevailing winds and sea currents (Figure 2.6).

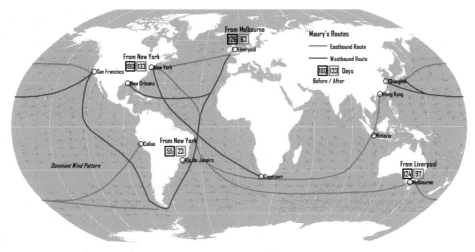

Figure 2.6 Impacts of Maury's navigation charts on sailing time, 1850s

Source: Leighly, J. (ed.) (1963) *The Physical Geography of the Sea and its Meteorology by Matthew Fontaine Maury*, 8th edition, Cambridge, MA: Belknap Press. Cited by Knowles, R.D. (2006) "Transport shaping space: the differential collapse of time/space", *Journal of Transport Geography*, 14(6): 407–25.

Maury was also able to demonstrate that using the longer eastbound sailing route rounding Cape Horn from Australia to Europe or North America was actually faster than the shorter westbound sailing route rounding the Cape of Good Hope. The outcome was the creation of relatively well-defined navigation routes that followed dominant wind patterns. It represented a close to optimal use of routing for sailing which remained until steamships replaced sailships in the late nineteenth century.

Composite ships (a mixture of wood and iron armature) then took over a large portion of the trade until about 1900, but they could not compete with steamships which had been continually improved since they were first introduced a hundred years before. Regarding steamship technology, 1807 marks the first successful use of a steamship, Fulton's *North River/Clermont*, on the Hudson servicing New York and Albany. In 1820, the *Savannah* was the first steamship (used as auxiliary power) to cross the Atlantic, taking 29 days to link Liverpool to New York. The first regular service for transatlantic passengers transported by steamships was inaugurated in 1838, followed by the usage of the helix, instead of the paddle wheel as a more efficient propeller in 1840. The gradual improvement of steam engine technology slowly but surely permitted longer and safer voyages, enabling steamships to become the dominant mode of maritime transportation by the late nineteenth century. Shipbuilding was also revolutionized by the usage of steel armatures (1860), which overcame the structural constraints of wood and iron armatures in terms of ship size. Iron armature ships were 30 to 40 percent lighter and had 15 percent more cargo capacity compared with wood armature ships of the same size.

The main consequence of the industrial revolution was a specialization of transportation services and the establishment of large distribution networks of raw materials and energy.

Emergence of modern transportation systems (1870–1920)

By the end of the nineteenth century, international transportation entered a new growth phase, especially with improvements in engine propulsion technology of the steamship and a gradual shift from coal to oil in the 1870s. Although oil had been known for centuries for its combustion properties, its commercial use was only applied in the early nineteenth century. Inventors started experimenting with engines that could use the cheap new fuel. Oil increased the speed and the capacity of maritime transport. It also permitted the energy consumption of ships to be reduced by a factor of 90 percent relatively to coal, the main source of energy for steam engines prior to this innovation. An equal size oil-powered ship could transport more freight than a coal-powered ship, reducing operation costs considerably and extending range. Also, coal refueling stages along trade routes could be bypassed. Global maritime circulation was also dramatically improved when infrastructures to reduce intercontinental distances, such as the Suez (1869) and the Panama (1914) canals, were constructed (Figure 2.7). With the Suez Canal, the far reaches of Asia and Australia became more accessible.

The Suez Canal represents, along with the Panama Canal, one of the most significant maritime "shortcuts" ever built. It brought a new era of European influence in Pacific Asia by reducing the journey from Asia to Europe by about 6,000 km by avoiding the route around the Cape of Good Hope. Asia became more commercially accessible and colonial trade expanded as a result of increased interactions because of a reduced friction of distance. Great Britain, the maritime power of the time, benefited substantially from this improved access. For instance, the Suez Canal shortened the distance on a maritime journey from London to Bombay by 41 percent and shortened the distance on a journey from London to Shanghai by 32 percent. The Panama Canal considerably shortened the maritime distances between the American East and West coasts by some 13,000 km.

The increasing size of ships, the outcome of advances in shipbuilding, imposed massive investments in port infrastructures such as piers and docks to accommodate them. Ship size grew dramatically, from the largest tonnage of 3,800 gross registered tons (revenue-making cargo space) in 1871 to 47,000 tons in 1914. Accordingly, ocean freight rates dropped by 70 percent between 1840 and 1910. The commercial demise of sailships took place during that period as trade shifted to steamships and expanded substantially. While sailships accounted for 85 percent of the total maritime tonnage in 1870, this share plummeted to 14 percent in 1910: 1878 was about the last year when

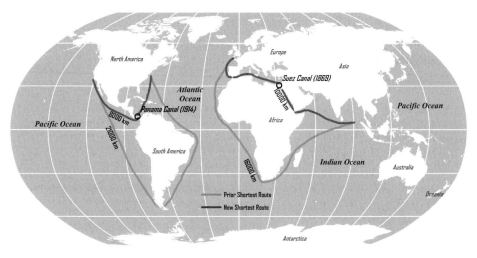

Figure 2.7 Geographical impacts of the Suez and Panama canals

sailships could compete effectively with steamships for the China trade. Harbors, while integrating production and transshipping activities, became industrial complexes around which agglomerated activities using ponderous raw materials. From the 1880s, liner services linked major ports of the world, supporting the first regular international passenger transport services, until the 1950s when air transportation became the dominant mode.

The period between 1890 and 1925 underlines the dominance of the steamship as a support to global trade with its diffusion to ports across the world (Figure 2.8). The year 1890 marks the threshold when steamship services became competitive for distances above 16,000 km, placing Asia within reach, particularly after the opening of the Suez Canal in 1869. Still, steamship services show for both periods a high level of concentration around industrialized clusters. While only two notable clusters are active around 1890, the North Sea and the American Northeast, by 1925 several additional clusters had emerged. One concerns the American West Coast that saw its industrialization expanded by the opening of the Panama Canal in 1914 and by an oil extraction boom. Another concerns Argentina, which at that time was going through a period of prosperity, including a phase of European immigration, and accounted for one of the world's highest GDP per capita. Buenos Aires rivaled other metropolises such as Paris, London and New York. The activity in South Asia was linked with the British colonial trade. This also included Australia, which at that time was becoming an important exporter of agricultural goods. The fast industrialization of Japan as an outcome of the Meiji Revolution was also becoming apparent.

This period also marked the golden era of the development of the railway transport system as railway networks expanded tremendously and became the dominant land transport mode both for passengers and freight. As the speed and power of locomotives improved above 100 km/hr and as the market expanded, rail services became increasingly specialized with trains entirely devoted to passengers or freight. Japan, the first Asian country to undertake its industrial revolution, saw its first train service in 1872. Rail systems reached a phase of maturity by the early twentieth century as in most developed economies the rail network reached its maximum extent in terms of total length.

Many European countries were undergoing a demographic transition, implying a rapid growth of their population and related urbanization and migration pressures. In such a context a significant technological change of this era involved urban transportation, which until then solely relied on walking and different types of carriages (mainly horse-drawn). The fast growth of the urban population favored the construction of the first public urban transport systems. Electric energy became widely used in the 1880s and considerably changed urban transport systems with the introduction of tramways (streetcars), notably in Western Europe and in the United States. They enabled the first forms of urban sprawl and the specialization of economic functions, through a wider separation between place of work and residence. In large agglomerations, underground metro systems were constructed, London being the first in 1863. The bicycle, first shown at the Paris Exhibition of 1867, was also an important innovation which changed commuting in the late nineteenth century. Initially, the rich used it as a form of leisure, but it was rapidly adopted by the working class as a mode of transportation to the workplace. Today, the bicycle is much less used in developed countries (outside recreational purposes), but it is still a major mode of transportation in developing countries, especially China.

This era also marked the first significant developments in telecommunications. The telegraph is considered to be the first efficient telecommunication device gaining wide

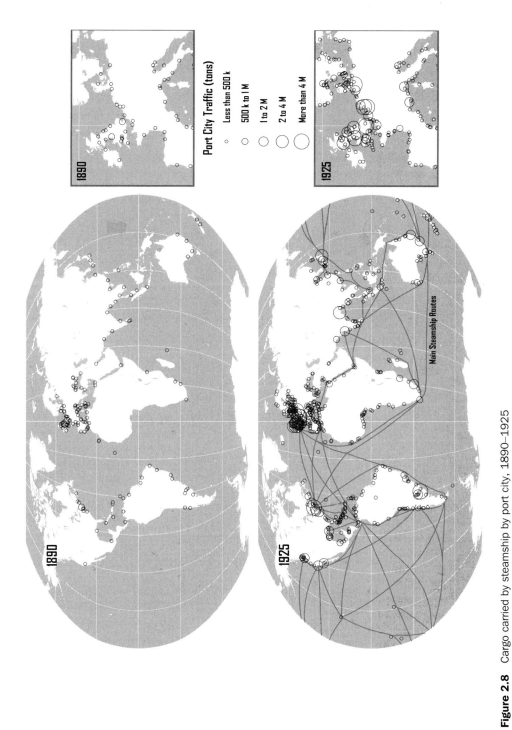

Figure 2.8 Cargo carried by steamship by port city, 1890–1925

Source: Lloyd's List. Adapted from Ducruet, C. (2012) "Ports et routes maritimes dans le monde (1890–1925)", *Mappemonde*, 106.

market coverage. In 1844, Samuel Morse built the first experimental telegraph line in the United States between Washington and Baltimore, opening a new era in the transmission of information. By 1852, more than 40,000 km of telegraph lines were in service in the United States. In 1866, the first successful transatlantic telegraph line marked the inauguration of an intercontinental telegraphic network that was dubbed the "Victorian Internet". The growth of telecommunications is thus closely associated with the growth of railways and international shipping. Managing a rail transport system, especially at the continental level, became more efficient with telegraphic communication. In fact, continental rail and telegraphic networks were often laid concomitantly. Telecommunications were also a dominant factor behind the creation of standard times zones in 1884. From a multiplicity of local times, zones of constant time with Greenwich (England) as the reference were created. This improved the scheduling of passenger and freight transportation at national levels. By 1895, every continent was linked by telegraph lines, a precursor of the global information network that would emerge in the late twentieth century. Business transactions became more efficient as production, management and consumption centers could interact with delays that were in hours instead of weeks and even months.

Concept 2 – Historical geography of transportation: the setting of global systems

Transportation in the Fordist era (1920–70)

The Fordist era was epitomized by the adoption of the assembly line as the dominant form of industrial production, an innovation that benefited transportation substantially. The internal combustion engine, or four-stroke engine by Daimler (1889), which was a modified version of the Diesel engine (1885), together with the pneumatic tire (1885) by Dunlop made road vehicles operations faster and more comfortable. Compared with steam engines, internal combustion engines have a much higher efficiency and use a lighter fuel: petrol. Petrol, previously perceived as an unwanted by-product of the oil refining process, which was obtaining kerosene for illumination, became a convenient fuel. Initially, diesel engines were bulky, limiting their use to industrial and maritime propulsion, a purpose which they still fulfill today. The internal combustion engine permitted an extended flexibility of movements with fast, inexpensive and ubiquitous (door-to-door) transport modes such as automobiles, buses and trucks. Mass producing these vehicles changed considerably the industrial production system, notably by 1913 when Ford began the production of the Model T car using an assembly line. From 1913 to 1927, about 15 million Ford Model T were built, making it the second most produced car in history, behind the Volkswagen Beetle. Economies of scale realized along the assembly line were passed on to the consumer which made the automobile even more affordable and popular. The rapid diffusion of the automobile marked an increased demand for oil products and other raw materials such as steel and rubber.

Economies of scale also improved transportation in terms of capacity, which enabled low-cost bulk commodities such as minerals and grain to be moved over long distances. The process was however slow to start as cargo ships require large amounts of labor to be loaded and unloaded. This informally imposed a limit of 10,000 deadweight tons to break-bulk cargo ships that remained in place until containerization began in the late 1950s. Still, the gradual growth of international trade and the Second World War gave a strong impetus for shipbuilding. The end of the war left an ample supply of military cargo ships, namely Liberty Ships, which could be cheaply used for commercial purposes and became the workhorses of global trade until the 1960s. Oil tankers are a good example of the application of the principle of economies of scale to transport larger quantities of oil at a lower cost, especially in the postwar period when global demand surged. Maritime routes were thus expanded to include tanker routes, notably from the Middle East, the dominant global producer of oil. The very long distances concerned in the oil trade favored the construction of larger tankers. In the 1960s, tanker ships of 100,000 tons became available, to be supplanted by VLCCs (Very Large Crude Carrier) of 250,000 tons in the 1970s and by the ULCCs (Ultra Large Crude Carrier) of 550,000 tons at the end of the 1970s. A ship of 550,000 tons is able to transport 3.5 million tons of oil annually between the Persian Gulf and Western Europe.

Economies of scale were also favored by unitization where batches of break-bulk cargo could be combined into one handling unit. In ancient times amphorae were used as load units, but had limitations for carrying goods. The pallet became the first effective load unit, particularly after the invention of the forklift in 1937. By the early 1930s about three days were required to unload a rail boxcar containing 13,000 cases of unpalletized canned goods. With pallets and forklifts, a similar task could be done in about four hours. However, it was during the Second World War that pallets as the standard supply unit load became widely adopted by the US military, as it permitted fast handling of goods and turnaround of transport assets.

Although the first balloon flight took place in 1783, due to the lack of propulsion no practical applications for air travel were realized until the twentieth century. The first propelled flight was made in 1903 by the Wright brothers and inaugurated the era of air transportation. The initial air transport services were targeted at mail since it was a type of freight that could be easily transported and due to technical limitations in carrying capacity initially proved to be more profitable than transporting passengers. The first commercial air transport service between England and France began in 1919, but air transport suffered from limitations in terms of capacity and range. Several attempts were made at developing dirigible services, with the Atlantic crossed by a Zeppelin dirigible in 1924. The 1920s and 1930s saw the expansion of regional and national air transport services in Europe and the United States with mass produced propeller aircrafts such as the Douglas DC-3.

Through the first half of the twentieth century the Atlantic remained an important technical challenge for non-stop long distance transportation modes since it linked large markets in Europe and North America (Figure 2.9). Both sides of the North Atlantic have substantial levels of economic activity so that a mode able to safely cross the North Atlantic non-stop would be a commercial success. The steamship was the first powered mode to set regular passenger services from the late 1830s, initially taking more than 10 days, but technical improvements reduced the transit time to about 6 days by the 1870s. By the time transatlantic liner services started to be abandoned in the 1950s transit times were reduced to 4 days.

Dirigibles were seriously considered in the 1930s with the setting of the first transatlantic air passenger flights, but their slow speeds and the *Hindenburg* disaster during a landing accident in 1937 marked the end of dirigibles as modes for passenger transportation. Both propeller and sea plane ("flying boat") transatlantic services began in the mid-1930s. The sea plane initially had an advantage with its capacity to land on water, but this advantage turned out to be an impediment as technical improvements were

Figure 2.9 Powered transatlantic passenger modes

made to propeller planes, which left the sea plane with limited range. It was later abandoned as a long distance transport mode. The introduction of pressurized propeller
planes (e.g. Constellation) in 1945 permitted the first regular transatlantic services with
two technical stops, Gander (Newfoundland) and Shannon (Ireland), and a flight time of
about 11 hours.

The postwar period was the turning point for air transportation as the range, capacity
and speed of aircrafts increased as well as the average income of passengers. A growing
number of people were thus able to afford the speed and convenience of air transportation. The application of the gas turbine principle led to the development of jet engines,
and 1952 marks the beginning of commercial jet services with the Comet, although a
design flaw grounded the plane the following year. In 1958, the first successful commercial jet plane, the Boeing 707, entered in service and revolutionized international movements of passengers, marking the end of passenger transoceanic ships (liners) and
replacing propeller planes for long distance services. The jet plane enabled the setting of
time dependent trade relations between producers across the world (such as electronics),
created a long distance market for perishables (fruits and vegetables) and supported the
development of mass tourism.

Basic telecommunications infrastructures, such as the telephone and the radio, were
mass marketed during the Fordist era. However, the major change was the large
diffusion of the automobile, especially from the 1950s as it became a truly
mass consumption product, and when the first major highway systems, such as the
American Interstate, began to be built. No other modes of transportation have so
drastically changed lifestyles and the structure of cities. It created suburbanization
and expanded cities to areas larger than 100 km in diameter in some instances. In
dense and productive regions, such as the Northeast of the United States, Japan and
Western Europe, the urban system became structured and interconnected by transport

networks to the point that it could be considered as one vast urban region: the Megalopolis.

A new context for transportation: the post-Fordist era (1970–)

Among the major changes in international transportation from the 1970s are the massive development of telecommunications, the globalization of trade, more efficient distribution systems, and the considerable development of air transportation. Telecommunications enabled growing information exchanges, especially for the financial and service sectors. After 1970 telecommunications successfully merged with information technologies. As such, telecommunication also became a medium of doing business in its own right, in addition to supporting and enhancing other transportation modes.

The setting of the first submarine cables took place in the second half of the nineteenth century, notably with the laying of the first successful transatlantic cable in 1866. By 1900 a global telegraphic cable network was established with transpacific connections completed in 1902. By 1956 the telegraphic system started to switch to telephonic cables with the first transatlantic telephone line (TAT-1). However, since their inception submarine cables were facing a bandwidth problem which made transoceanic communication expensive and mostly used for business or government transactions. As was the case in the nineteenth century, submarine cables are laid by ships and are thus capital-intensive projects. The information highway became a reality as fiber optic cables gradually replaced copper wires, multiplying the capacity to transmit information between computers. Global submarine cable networks, which have existed since the setting of telegraph networks in the nineteenth century, were overhauled with fiber optics to become the backbone of the global telecommunication system, particularly the Internet (Figure 2.10).

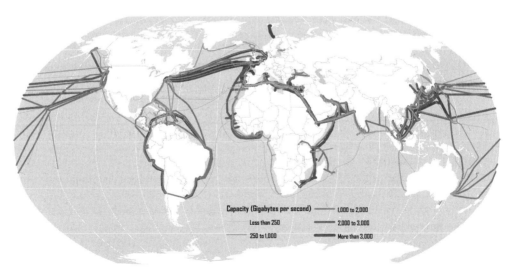

Figure 2.10 Global submarine cable network

Source: Dataset encoded by Greg Mahlknecht.

The development of fiber optic transmission technology provided a substantial impetus in the setting of a global telecommunication network since it permitted significantly higher bandwidth and less signal degradation. Throughputs of hundreds of gigabytes of information per second became possible. The first transatlantic fiber optic cable (TAT-8) was laid in 1988 and over the years fiber optic cables were laid across the world, connecting economies and societies increasingly dependent on telecommunications (see Figure 2.10). The Internet could not have existed otherwise. While initial submarine cables were laid on a point-to-point basis, technical advances permitted branching so that one cable could service a sequence of hubs (e.g. Africa and Latin America). The global network is designed for redundancy as several cables are laid in parallel for major connections (transatlantic and transpacific), which means that a failure in one cable can be mitigated by rerouting traffic to the others. In recent years, Pacific Asia has seen significant submarine cable laying activities, in support of its economic development. In 2012, a new route for submarine cables was being established with the setting of the first Arctic cables between London and Tokyo through the Northwest Passage. As well as providing an additional redundancy to the global telecommunication network, it will shave off about 60 milliseconds in the connection speed. Through lower latency (delay) levels, this would thus improve bandwidth-intensive telecommunications between Europe and Pacific Asia, such as financial transactions and videoconferencing.

This growth was, however, dwarfed by the tremendous growth in processing power of personal computing devices, which are now fundamental components of economic and social activities across the world. A network of satellite communication was also created to support the growing exchanges of information, especially for television images, but remains of marginal use because of lower bandwidth. Out of this wireless technology emerged local cellular networks which expanded and merged to cover whole cities, countries, regions and then continents. Telecommunications have reached the era of individual access, portability and global coverage.

In a post-Fordist system, the fragmentation of production, organizing an international division of labor, as well as the principle of "just-in-time" increased the quantity of freight moving at the local, regional and international levels. This in turn required increasing efforts to manage freight and reinforced the development of logistics, the science of physical distribution systems. Containers, main agents of the modern international transport system, enabled an increased flexibility of freight transport, mainly by reducing transshipment costs and delays; handling a container requires about 25 times less labor than its equivalent in bulk freight. This increased the unitization trend brought forward by the pallet, particularly since pallets could be loaded into containers. Containers were introduced by the American entrepreneur Malcolm McLean who initially applied containerization to land transport but saw the opportunity of using container shipping as an alternative to acute road congestion in the early 1950s before the construction of the first Interstate highways. The initial attempts at containerization thus aimed at reducing maritime transshipment costs and time. Before containerization, a cargo ship could spend as much time in a port being loaded or unloaded than it did at sea. Later on, the true potential of containerization became clear when interfacing with other modes became an operational reality, mainly between maritime, rail and road transportation.

The first container ship (the *Ideal-X*, a converted T2 oil tanker) set sail in 1956 from New York to Houston and marked the beginning of the era of containerization. In 1960, the Port Authority of New York/New Jersey foreseeing the potential in container trade constructed the first specialized container terminal next to Port Newark: the Port Elizabeth Marine Terminal. The Sea-Land Company established the first regular maritime container line in 1965 over the Atlantic between North America and Western Europe. By the early 1980s, container services with specialized ships (cellular container ships, first introduced in 1967) became a dominant aspect of international and regional transport systems, transforming the maritime industry. However, the size of those ships remained for 20 years constrained by the size of the Panama Canal, which de facto became the Panamax standard. In 1988, the first post-Panamax container ship was introduced, an indication of the will to further expand economies of scale in maritime container shipping. The container revolution was concomitant with globalization by supporting an increasingly complex system of trade involving parts, manufactured goods and even commodities. Few other transport innovations had such an impact on the global economic landscape.

Air and rail transportation experienced remarkable improvements in the late 1960s and early 1970s through massification and network developments. The first commercial flight of the Boeing 747 between New York and London in 1969 marked an important landmark for international transportation (mainly for passengers, but freight became a significant function in the 1980s). This giant plane could transport around 400 passengers, depending on the configuration. It permitted a considerable reduction of air fares through economies of scale and opened intercontinental air transportation to the mass market. Attempts were also undertaken to establish faster than sound commercial services with the Concorde (1976; flying at 2,200 km/hr). However, such services proved to be unprofitable and no new supersonic commercial planes have been built since the 1970s. The Concorde was finally retired in 2003. At the regional level, the emergence of high speed train systems provided fast and efficient inter-urban services, notably in France (1981; TGV; speeds up to 300 km/hr) and in Japan (1964; Shinkansen; speeds up to 275 km/hr). More recently, high speed train systems were constructed in China, Korea and Taiwan.

Major corporations making transportation equipment, such as car manufacturers, have become dominant players in the global economy. Even if the car is not an international transport mode, its diffusion has expanded global trade of vehicles, parts, raw materials and fuel (mainly oil). Car production, which used to be mainly concentrated in the United States, Japan and Germany, has become a global industry with a few key players being part of well-integrated groups such as Ford, General Motors and Toyota. Along with oil conglomerates, they have pursued strategies aimed at the diffusion of the automobile as the main mode of individual transportation. This has led to growing mobility but also to congestion and waste of energy. As the twenty-first century begins, the automobile accounts for about 80 percent of the total oil consumption in developed countries.

Since the beginning of the modern era in the fifteenth century, transportation has played a significant role in the development of the world economy (Figure 2.11). The current period is also one of transport constraints and the search for alternatives, mainly because of a dual dependency. First, transportation modes have a heavy dependence on fossil fuels and second, road transportation has assumed dominance. The first oil shock of the early 1970s, which saw a significant increase in fuel prices, induced innovations in transport modes, the reduction of energy consumption and the search for alternative sources of energy (electric cars, adding ethanol to gasoline and fuel cells). However, from the mid-1980s to the end of the 1990s, oil prices declined and attenuated the importance of these initiatives. Again, oil prices surged in the beginning of the

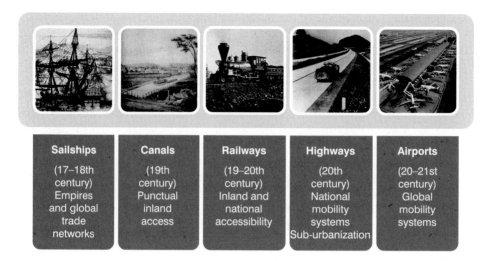

Sailships	Canals	Railways	Highways	Airports
(17–18th century) Empires and global trade networks	(19th century) Punctual inland access	(19–20th century) Inland and national accessibility	(20th century) National mobility systems Sub-urbanization	(20–21st century) Global mobility systems

Figure 2.11 Cumulative waves of transport development

twenty-first century, placing alternative sources of energy on the agenda. Still, the reliance on fossil fuels continues with a particularly strong growth of motorization in developing countries.

Transport development is a cumulative process as each new transport technology adds to the capacity and mobility potential of the previous technologies. It may also lead to the obsolescence and decline of prior technologies when they service similar markets. Since the modern era, five major waves of transport developments can be identified:

First wave: sailships. The mastery of high sea navigation in the mid-sixteenth and early seventeenth centuries led to the gradual setting of a global trade network supported by the emergence of colonial empires. In the late nineteenth century, the steamship would mark the demise of the sailship, but not of commercial maritime shipping networks that continued to expand. Subsequently in the late twentieth century, the container ship would strengthen global commercial relations to an unparalleled level.

Second wave: canals. The early stages of the industrial revolution in the nineteenth century were accompanied by the setting of canals complementing existing rivers or linking them. They provided the first level of inland access with the economies of scale they were able to confer; such accessibility was highly punctual, where canals could be built. Although canal systems were historically set in other parts of the world (China being the most salient example), it is in Western Europe and North America that their impacts on economic development were the most significant. Even if canal systems were supplemented by railways for many commercial relations, they remained active transport modes, particularly in Europe, China and North America.

Third wave: railways. The setting of rail systems in the second half of the nineteenth century permitted the first effective forms of inland accessibility and concomitantly of cohesive national transport systems, although interconnecting different rail systems took time. At the beginning of the twentieth century, rail systems were the dominant mode supporting passengers and freight flows. Although their relative importance has declined with the setting of highways, railways are far from being an obsolete technology with the setting of high speed rail systems around the world as well as their conversion to intermodalism.

Fourth wave: highways. The diffusion of the internal combustion engine and the availability of cheap oil supplies permitted an effective setting of individual or small load (truckload) mobility (national mobility systems). This, however, could not take place without the construction of national highway systems, such as the Interstate in the United States. Another important impact of the highways was lower density forms of urbanization, namely suburbs.

Fifth wave: airports. The introduction of jet services in the late 1950s permitted for the first time the setting of true global mobility systems where locations can be reached within hours. Airports became important nodes in the national and global systems of passenger flows as well as freight flows.

Concept 3 – Transport and spatial organization

The spatial organization of transportation

Geography imposes an organization to activities and consequently a spatial structure. The opposite also applies as the spatial structure influences geography. Spatial organization relies on two dimensions. The first relates to spatial differentiation where attributes such as location, size and density are illustrative of the distribution inequalities of a feature. This differentiation is the outcome of a cumulative process as several elements of the spatial structure, such as urban areas, are the outcome of a long process of accumulation, which tends to change slowly. The second relates to spatial interactions where attributes such as origins, destinations and flows are also illustrative of inequalities. Transportation not only favors economic development but also has an impact on spatial organization. Throughout history, transport networks have structured space at different scales. The fragmentation of production and consumption, the locational specificities of resources, labor and markets generate a wide array of flows of people, goods and information. The structure of these flows in terms of origin, destination and routing is closely related to spatial organization. Space shapes transport as much as transport shapes space, which is a salient example of the reciprocity of transport and its geography. This reciprocity can be articulated over two points:

Reciprocity in location. This relationship concerns the transport system itself. Since the transport system is composed of nodes and links as well as the flows they are supporting, the spatial organization of this system is a core defining component of the spatial structure. Even if streets are not the city, they are shaping its organization in terms of locations and relations. The same applies to maritime shipping networks, which are not international trade, but reflect the spatial organization of the global economy.

Reciprocity in mobility. This relationship concerns activities that are all dependent on transportation at one level or another. Since every single activity is based on a level of mobility, the relationship they have with transportation is reflected in their spatial organization. While a small retail activity is conditioned by local accessibility for its customers, a large manufacturing plant relies on accessibility to global freight distribution for its inputs as well as its outputs.

The more interdependent an economy is, the more important transportation becomes as a support and a factor shaping this interdependence. It is also a reason why the importance of transportation can be neglected as the interdependence will be noticed but its structural support less so. The relationship between transport and spatial organization can be considered from three major geographical scales: the global, the regional and the local (Figure 2.12).

While the major nodes structuring spatial organization at the global level are gateways supported by port, airport and telecommunication activities, at the local level, employment and commercial activities, which tend to be agglomerated, are the main structuring elements. Each of these scales is also characterized by specific links and relations ranging from locally based commuting to global trade flows.

Global spatial organization

At the global level, transportation supports and shapes economic specialization and productivity through international trade. Improvements in transport are expanding markets and development opportunities, but not uniformly. The inequalities of the global economy are reflected in its spatial organization and the structure of international transport systems. The patterns of globalization have created a growth in spatial flows (trade) and increased interdependencies. Telecommunications, maritime transport and

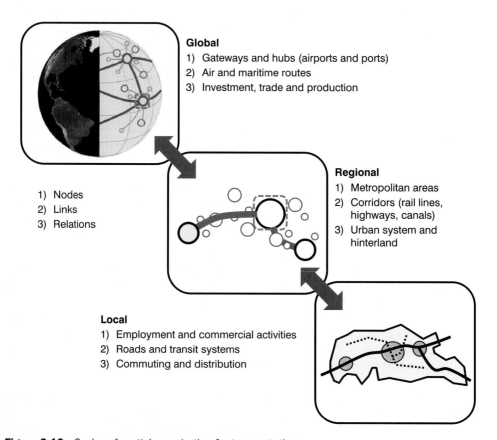

Global
1) Gateways and hubs (airports and ports)
2) Air and maritime routes
3) Investment, trade and production

1) Nodes
2) Links
3) Relations

Regional
1) Metropolitan areas
2) Corridors (rail lines, highways, canals)
3) Urban system and hinterland

Local
1) Employment and commercial activities
2) Roads and transit systems
3) Commuting and distribution

Figure 2.12 Scales of spatial organization for transportation

air transport, because of their scale of service, support the majority of global flows. The nature and spatial structure of these flows can be considered from two major perspectives that seek to explain global differences in growth and accessibility:

Core/periphery. This basic representation assumes that the global spatial organization favors a few core areas that grow faster than the periphery. Differential growth creates acute inequalities in levels of development. For instance, global migration flows are illustrative of different levels of economic development with flows from locations with lower development levels to higher development levels dominating. Transportation is thus perceived as a factor of polarization and unequal development. From this perspective, parts of the global economy are gaining, because they are more accessible, while others are marginalized and bound to dependency. However, this trend can be reversed if international transport costs are significantly reduced. This is evidenced by the substantial growth of many Pacific Asian countries that have opted for an export-oriented strategy which requires good access to global freight distribution. Consequently, the core/periphery relationship is flexible and relative.

Poles. Transportation is perceived as a factor of articulation in the global economy where the circulation of passengers and freight is regulated by poles corresponding to a high level of accumulation of transport infrastructures, distribution and economic activities. These poles are subject to centrifugal and centripetal forces that have favored geographical concentration of some activities and the dispersion of others. The global economy is thus based on the backbone of freight distribution, which in turn relies on networks established to support its flows and on nodes that are regulating the flows within networks. Networks, particularly those concerning maritime shipping and air transportation, are flexible entities that change with the ebb and flows of commerce while nodes are locations fixed within their own regional geography.

The global spatial organization is a priori conditioned by its nodality. Global flows are handled by gateways and hubs, each of which account for a significant share of the flows of people, freight and information.

Gateway. A location offering accessibility to a large system of circulation of freight and passengers. Gateways reap the advantage of a favorable physical location such as highway junctions, the confluence of rivers, a good port site, and have been the object of a significant accumulation of transport infrastructures such as terminals and their links. A gateway is commonly an origin, a destination and a point of transit. It generally commands the entrance to and the exit from its catchment area. In other words, it is a pivotal point for the entrance and the exit in a region, a country, or a continent and often requires intermodal transfers.

Hub. A central point for the collection, sorting, transshipment and distribution of goods for a particular area. This concept comes from a term used in air transport for passengers as well as for freight and describes collection and distribution through a single point such as the "hub-and-spoke" concept.

Gateways and hubs refer to rather similar elements of the spatial structure of flows (Figure 2.13). However, the concept of gateway tends to be more restrictive in its definition. While a hub is a central location in a transport system with many inbound and outbound connections of the same mode, a gateway commonly implies a shift from one

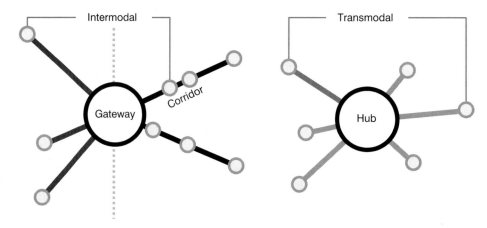

Figure 2.13 Gateways and hubs

mode to the other (such as maritime/land). A gateway is performing an intermodal function (between modes) while a hub is mostly transmodal (within a mode). Transport corridors are commonly linking gateways to the hinterland. Gateways also tend to be most stable in time as they often have emerged at the convergence of inland transport systems while the importance of a hub can change. For instance, a transport company such as a maritime shipping line or an air carrier may decide to use another hub. If this is the case, then the hub can lose its status. Also, a location can at the same time be a gateway and a hub.

Often through the principle of economies of agglomeration and notable accessibility advantages a region can accumulate several major nodal infrastructures, namely port and airport terminals. When these nodes act as an intermodal interface they can be characterized as gateway systems (or regions) that play a substantial role in the global distribution of freight, connecting major systems of circulation. Gateways also act as bottlenecks in global freight distribution imposing capacity constraints because of capacity, infrastructure or supply chain management.

The global system of freight circulation is articulated by major gateway regions (or gateway systems), often composed of a cluster of ports and airports within a metropolitan area (Figure 2.14). This does not mean that ports and airports are functionally integrated (they are not), but that the region they service is a major load center serviced by a variety of globally oriented supply chains, some using the ports and others using the airports. The 39 largest gateway regions account for 90 percent of the containerized and air freight traffic, with the most important being the Pearl River Delta (Hong Kong, Shenzhen and Guangzhou) with a share of 16.7 percent. Other significant gateway systems concern the Rhine/Scheldt delta for Western Europe (e.g. Antwerp, Rotterdam) and Los Angeles/Long Beach for the American West Coast. Intermediary locations such as Singapore and Dubai are also relevant.

Services are following a spatial trend which appears to be increasingly different than of production. As production disperses worldwide to lower cost locations, high level services increasingly concentrate into a relatively few large metropolitan areas, labeled as world cities. They are centers for financial services (banking, insurance), head offices of major multinational corporations, nexuses for the arts and the seats of major governments. Thus, gateways and world cities may not necessarily correspond as locations,

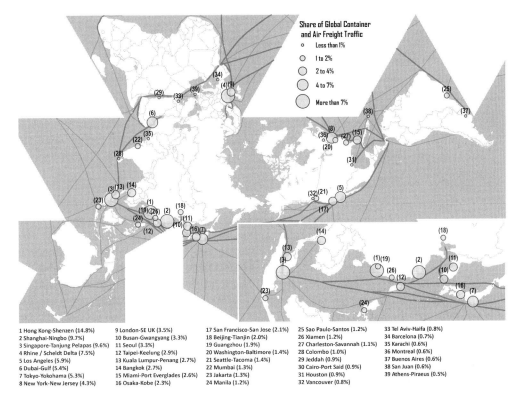

Figure 2.14 World's major gateway systems, 2006

Source: O'Connor, K. (2010) "Global city regions and the location of logistics activity", *Journal of Transport Geography*, 18(3): 354–62.

underlining the ongoing dichotomy between central places and transport places. This is particularly the case for containerized traffic which is linked with new manufacturing clusters and the usage of intermediary hubs.

Regional spatial organization

Regions are commonly organized along an interdependent set of cities forming what is often referred to as an urban system. The key spatial foundation of an urban system is based on a series of market areas, which are a function of the level of activity of each center in relation with the friction of distance. The spatial structure of most regions can be subdivided into three basic components:

● A set of **locations of specialized industries** such as manufacturing and mining, which tend to group into agglomerations according to location factors such a raw materials, labor, markets, etc. They are often export-oriented industries from which a region derives the bulk of its basic growth.
● A set of **service industry locations**, including administration, finance, retail, wholesale and other similar services, which tend to agglomerate in a system of central places (cities) providing optimal accessibility to labor or potential customers.

● A pattern of **transport nodes and links**, such as roads, railways, ports and airports, which services major centers of economic activity.

Jointly, these components define the spatial order of a region, mostly its organization in a hierarchy of relationships involving flows of people, freight and information. More or less well defined urban systems spatially translate such development. Many conceptual models have been proposed to explain the relationships between transport, urban systems and regional development, the core/periphery stages of development and the network expansion being among those. Three conceptual categories of regional spatial organization can be observed:

> **Central places/urban systems** models try to find the relationships between the size, the number and the geographic distribution of cities in a region. Many variations of the regional spatial structure have been investigated by central place theory. The great majority of urban systems have a well-established hierarchy where a few centers dominate. Transportation is particularly important in such a representation as the organization of central places is based on minimizing the friction of distance. The territorial structure depicted by central place theory is the outcome of a region seeking the provision of services in a (transport) cost-effective way.
>
> **Growth poles** where economic development is the structural change caused by the growth of new propulsive industries that are the poles of growth. The location of these activities is the catalyst of regional spatial organization. Growth poles first initiate then diffuse development. It attempts to be a general theory of the initiation and diffusion of development models. Growth gets distributed spatially within a regional urban system, but this process is uneven with the core benefiting first and the periphery eventually becoming integrated in a system of flows. In the growth poles theory transportation is a factor of accessibility which reinforces the importance of poles.
>
> **Transport corridors** represent an accumulation of flows and infrastructures of various modes and their development is linked with economic, infrastructural and technological processes. Where these processes involve urban development, urbanization corridors are a system of cities oriented along an axis, commonly fluvial or a coastline since historically they permitted cities to maintain transport and commercial relations. Many urban regions such as BosWash (Boston–Washington) or Tokaido (Tokyo–Osaka) share this spatial commonality. Corridors are also structured along articulation points that regulate the flows at the local, regional and global levels either as hubs or gateways. The development of high speed train systems around the world takes place along major urban corridors and reinforces the existing regional spatial structure.

Local spatial organization

Although transport is an important element in rural spatial organization, it is at the urban level that transportation has the most significant local spatial impact. Urbanization and transport are interrelated concepts (see Chapter 6 for a detailed perspective concerning urban transportation). Every city relies on a need for mobility of passengers (residence, work, purchases and leisure) and freight (consumption goods, food, energy, construction materials and waste disposal) and where the main nodes are employment zones. Urban demographic and spatial evolution is translated in space by the breadth and amplitude of movements. Employment and attraction zones are the most important elements shaping the local urban spatial organization:

Employment zones. The growing dissociation between the workplace and the residence is largely due to the success of motorized transport, notably the private automobile. Employment zones being located away from residential zones have contributed to an increase in the number and length of commuting trips. Before suburbanization, public transit was wholly responsible for commuting. Today, the automobile supports the majority of these trips. This trend is particularly prevalent in highly populated, industrialized and urbanized zones, notably in North America and Western Europe, but motorization is also a dominant trend in developing countries.

Attraction zones. Attraction zones linked to transport modes are areas to which a majority of the population travels for varied reasons such as shopping, professional services, education and leisure. As with central place theory, there is a certain hierarchy of services within an urban area ranging from the central business district offering a wide variety of specialized services to small local centers offering basic services such as groceries and personal banking.

The development of cities is conditioned by transport and several modes, from urban transit to the automobile, have contributed to the creation of urban landscapes. Three distinct phases can be noted:

Conventional/classic city. Constructed for pedestrian interactions and constrained by them, the historic city was compact and limited in size. The emergence of the first urban transit systems in the nineteenth century permitted the extension of the city into new neighborhoods. However, pedestrian movements still accounted for the great majority of movements and the local spatial organization remained compact. Many European and Asian cities still have a significant level of compactness today where urban transit remains a defining element of the spatial organization.

Suburbanization. The advent of more efficient urban transit systems and later of the automobile permitted an increased separation between basic urban functions (residential, industrial and commercial) and their spatial specialization. The resulting rapid expansion of urban areas, especially in North America, created a new spatial organization, less cohesive than before but still relatively adjacent to the existing urban fabric. Although this process started in the early twentieth century, it accelerated after the Second World War.

Exurbanization. Additional improvements in mobility favored urban expansion in the countryside where urban and rural activities are somewhat intermixed. Many cities became extended metropolitan regions, with a wide array of specialized functions including residential areas, commercial centers, industrial parks, logistics centers, recreational areas and high tech zones. These exurban developments have also been called "edge cities".

The automobile has clearly influenced contemporary spatial organization but other socioeconomic factors have also shaped urban development such as gentrification and differential changes in land values. The diffusion of the automobile has led to an urban expansion relying on the mobility of individuals and permitting a disorderly growth and an allocation of space between often conflicting urban functions (residential, industrial, commercial). Still, distance decay remains a force shaping urban spatial organization since suburban and exurban developments tend to occur along concentric rings within large metropolitan areas. Transport thus contributes to the local spatial organization, however, it must also adapt to urban morphologies. Transport networks and urban centers complement and condition each other.

Concept 4 – Transport and location

The importance of transport in location

In addition to being a factor of spatial organization, transportation is linked with the location of socioeconomic activities, including retail, manufacturing and services. In a market economy, location is the outcome of a constrained choice where many issues are being considered, transportation being one of them. The goal is to find a suitable location that would maximize the economic returns for this activity. There is a long tradition within economic geography in developing location theories which aim to explain and predict the locational logic of economic activities by incorporating market, institutional and behavioral considerations.

The majority of location theories have an explicit or implicit role attributed to transport since accessibility is an important factor in the location preferences of firms and individuals. As there are no absolute rules dictating locational choices, the importance of transport can only be evaluated with varying degrees of accuracy. At best, the following observations concerning transportation modes and terminals and their importance for location can be made:

- **Ports and airports**. Convergence of related activities around terminals, particularly for ports since inland distribution costs tend to be high. The location and the level of activity of ports and airports are reflective of global trade patterns.
- **Roads and railroads**. A structuring and convergence effect that varies according to the level of accessibility. For rail transport, terminals also have a convergence effect.
- **Telecommunications**. No specific local influence, but the quality of regional and national telecommunication systems tends to ease transactions.

Globalization has been associated with significant changes in business operations and markets. Managing operations in such an environment has become increasingly complex, especially with the globalization of production and consumption. Manufacturing strategies tend to use different locations for each component of a product in order to optimize respective comparative advantages and reduce input costs. Transport requirements have proportionally increased to support and organize the related flows. The requirement of faster long distance transport services has increased the importance of air transport, especially for freight. Air terminals have thus become a significant location factor for globally oriented activities, which tend to agglomerate in the vicinity. Additionally, the surge in long distance trade has put logistical activities, namely transport terminals and distribution centers, at the forefront of locational considerations. Technological changes have also been linked with the relocation of industrial and even service activities. Global telecommunication facilities can favor the outsourcing of several services to lower cost locations, such as the case of call centers in India indicates.

Location factors

The location of economic activities is a priori dependent on the nature of the activity itself and on certain location factors such as the attributes of the site, the level of accessibility and the socioeconomic environment:

- **Site**. Specific micro-geographical (local) characteristics of the site, including the availability of land, basic utilities, the visibility (for activities related to prestige such as head offices), amenities (quality of life) and the nature and level of access to local transportation (such as the proximity to a highway). These factors have an important effect on the costs associated with a location.
- **Accessibility**. Include a number of opportunity factors related to a location, mainly labor (wages, availability, level of qualification), materials (mainly for raw materials-dependent activities), energy, markets (local, regional and global) and accessibility to suppliers and customers (important for intermediate activities). These factors tend to have a meso (regional) connotation.
- **Socioeconomic environment**. Specific macro-geographical characteristics that can to apply to jurisdictional units (nation, region, municipality). They consider the availability of capital (investment, venture capital), various subsidies and incentives, regulations, taxation and technology.

The role and importance of each factor depends on the nature of the activity and its locational behavior. Although each type of economic activity has its own set of location factors, some general factors can be identified by major economic sectors:

Primary economic activities. Their dominant location factor is related to environmental endowments, such as natural resources. For instance, mining takes place where economically recoverable mineral deposits are found and agriculture is

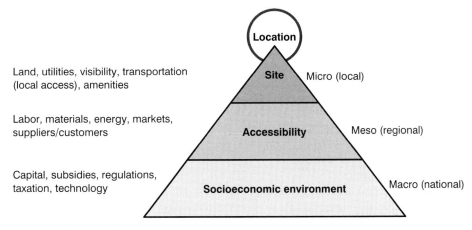

Figure 2.15 Basic location factors

subject to environmental constraints such as soil fertility, precipitation and temperature. Primary activities are thus characterized by the most basic location factors but have a strong reliance on transportation since their locations rarely are close to centers of demand and they usually concern ponderous goods. Substantial investments in extraction and distribution infrastructures must thus be made before resources can be brought to markets. The capacity to transport raw materials plays a significant role in the possible development of extractive activities at a location.

Secondary economic activities. Imply a complex web of location factors which, depending upon the industrial sector, relate to labor (cost and/or skill level), energy costs, capital, land, markets and/or proximity of suppliers. Location is thus an important cost factor and the general purpose is usually to minimize it. Considering the wide variety of industrial and manufacturing activities, understanding the rationale of each sector is a difficult task that has been subject to many investigations in economic geography. Globalization, recent developments in supply chain management and global production networks have made the situation even more complex with the presence of many intermediaries and significant locational changes. The industrialization of China has been supported by several strategies trying to multiply locational advantages, such as the setting of export-oriented special economic zones and large investment in transport infrastructure.

Tertiary economic activities. Involve activities that are most bound to market proximity, since the capacity to distribute a product or service is their most important location requirement. As many of these activities are retail-oriented, consumer proximity (as well as their level of income) is essential and is directly related to sale levels. The main focus is to maximize sales revenues. Location is thus an important revenue factor. The retail industry has significantly changed with the emergence of large retail stores that maximize sales through economies of scale and local road accessibility. E-commerce also provides a new dynamic where information can easily be traded and where niche retailing markets can be developed in a situation of high product diversity.

Quaternary economic activities. Imply activities not linked to environmental endowments or access to a market, but to high level services such as banking, insurance, education, research and development. This often relates to the high technology sector where innovation is a key commercial factor. With improvements in telecommunications, many of these activities can be located almost anywhere as demonstrated by the trend to locate call centers offshore. There are still some strong locational requirements for high technology activities that include proximity to large universities and research centers and to a pool of highly qualified workers (as well as cheap labor for supporting services), availability of venture capital, a high quality of life (cultural amenities) and access to excellent transportation and telecommunication facilities. The sector has shown a propensity at clustering as close inter-firm relations are a factor of innovation.

Each of these sectors thus has its own set of economies related to its relations with production, distribution and consumption. However, basic location strategies appear to be primarily cost-minimization or revenue-maximization endeavors. Understanding location factors enables a better overview of the dynamics of the global economy and the associated territorial changes at the global, regional and local levels.

Accessibility and location economies

Since accessibility is predominantly the outcome of transportation activities, namely the capacity of infrastructures to support mobility, it presents the most significant influence of transportation on location. Hence, location (accessibility) and economic activities are interrelated. Accessibility plays an important role by offering more customers through an expanded market area, by making distribution more efficient (in terms of capacity, costs and time), or by enabling more people to reach workplaces (labor cost and qualification). While some transport systems have favored the dispersion of socioeconomic activities (e.g. automobiles and suburbanization), others have favored their concentration (e.g. airports and container terminals). All transport systems are bearers of spatial specialization and configuration.

Activities involved in production (manufacturing), distribution (transportation) and consumption (retail) are constantly seeking economies to improve their competitiveness and increase their market share (Figure 2.16). This is commonly done by opting for a specific location and type of facility:

Economies of transportation. Relate to the benefits that lower transport costs may grant to specific activity sectors. For production, it relates to a location that minimizes total transport costs (accessibility to suppliers and customers) and thus lowers production unit costs. These considerations are at the core of classic industrial location theories where transport-dependent activities seek to minimize total transport costs, such as Weber's location triangles. Economies of transportation in distribution consider the management of transport chains, often of several modes, to reduce total transport costs (modal and intermodal). Some are elements of transport costs in

	Production	Distribution	Consumption
Economies of transportation	Lower unit costs through accessibility to suppliers and customers	Lower unit distribution costs through transport chains management	Lower unit output costs through accessibility to suppliers and customers
Economies of scale	Lower unit costs with larger plants	Lower unit transport costs through larger modes and terminals	Lower unit costs with larger retail outlets
Economies of scope	Lower unit output costs with more product types	Lower transport costs with bundling of different loads	Product diversification attracts more customers
Economies of agglomeration	Industrial and service linkages with manufacturing clusters	Lower input costs with clustering of distribution activities	Lower input costs with clustering of retail activities
Economies of density	Increased accessibility to labor (skills) with higher densities	Lower unit distribution costs with higher densities	Increased accessibility to goods and services with higher densities

Figure 2.16 Main types of economies in production, distribution and consumption

production while others are elements of transport costs in consumption. Transport costs must be considered in a wider context where the quality and reliability of transport is of growing importance. It has been demonstrated that travel time, instead of distance, is the determining factor behind commuting ranges. For freight distribution, while cost factors are significant, there is a growing importance of the concept of reliability. Economies of transportation in consumption can be derived from economies in distribution as well as accessibility (proximity) to customers.

Economies of scale. Relate to the benefits that scale may offer to activity sectors. For production, the larger the production plant, the lower the unit costs since fixed costs (e.g. the factory) are spread over a larger quantity of units. For transportation, the principle involves unit cost reductions derived from larger modes (e.g. megaships), terminals and distribution centers: the massification of transportation. For consumption, larger retail outlets tend to reduce input costs, underlining the success of large megastore chains such as Wal-Mart. This concept also implies that diseconomies of scale can be reached after a certain size, particularly through growing complexity and management costs. Diseconomies vary substantially by the type of activity; steel production is prone to large economies of scale while restoration much less so.

Economies of scope. Relate to the benefits derived by expanding the range of goods and services. For production, they are commonly based on product diversification and flexible manufacturing systems able to produce a variety of products in view of changes in demand. For distribution, economies of scope are very important and commonly achieved when a transporter is able to bundle several different loads into fewer loads. For instance, a container ship is able to bundle the loads (and offer economies of scale) for several customers in often completely different sectors of activity that are simply sharing a similar origin and destination. For consumption, activities offering a wider range of goods or services are usually able to attract more customers since they have more choices. Economies of scale and economies of scope are highly related.

Agglomeration economies. The benefits derived from locating in proximity to other activities, even if the location is suboptimal. Often referred to as the clustering effect and involves the sharing of common infrastructures such as roads and utilities. Clustering continues to be a powerful force in location as the reduction in transport costs favors the agglomeration of retail, manufacturing and distribution activities at specific locations. For production, industrial and service linkages are offered as respective suppliers and customers benefit from close proximity and interactions. The outcome is a manufacturing cluster. A similar trend is observed in freight distribution as logistics activities tend to cluster and often co-locate next to an intermodal terminal. The distribution centers of parcels companies are commonly located directly adjacent to runways so that their air freight services can be tightly synchronized with the consolidation and deconsolidation of air parcels. For intermodal rail terminals, inland port facilities are built on the principle of co-location. For consumption, commercial districts or shopping malls are common expressions of the benefits of agglomeration.

Economies of density. The benefits derived from the increasing density of features on the costs of accessing them. For production, this could involve access to a larger labor pool (and skills) or resources (e.g. mining, agriculture). Higher market densities reduce distribution costs as shorter distances service the same number of customers and the same freight volume. A similar rationale applies to consumption where higher market densities involve higher accessibility levels to goods and services. For instance, if the customer density is sufficient, a retailer can achieve several types of cost savings by locating its stores in proximity to one another. Such a structure reduces logistics and delivery costs by sharing a distribution center. High densities can also lead to diseconomies, particularly with congestion.

Because of the level of accessibility they provide, new transport infrastructures influence the setting of economic activities. It becomes a particularly strong effect when new infrastructures are added to an undeveloped (or underdeveloped) site and thus locational decisions tend to be simpler and unhindered by the existing spatial structure. The locational effects on activities are not always automatic or evident. They are important, however, when infrastructure is accompanied by social, economic and urban transformations of space. New infrastructures therefore play a catalytic role, because they are able to transform space through land use and mobility changes.

Concept 5 – Future transportation

Past trends and uncertain future

In 200 years of history since the introduction of forms of mechanized transportation, the capacity, speed, efficiency and geographical coverage of transport systems have improved dramatically. Modes, terminals and networks alike have been subject to remarkable changes that have two functional aspects:

> **Revolutionary changes**. Concern a completely new technology that creates new markets and growth opportunities for transportation and the economy as well. It often marks the obsolescence of an existing transport mode as the new mode has substantial cost, capacity or time benefits. Revolutionary changes tend to be rare but profound since they commonly involve the setting of entirely new networks. They commonly cannot be predicted, but once they occur it is possible to assess their potential impacts.
> **Incremental (evolutionary) changes**. Concern the stepwise improvement of an existing transport technology and operations. This leads to increases in productivity with more capacity, lower costs and better performance of the involved mode or terminal. Incremental changes are possible to extrapolate but the rate of change they bring is difficult to assess.

Considering those changes, the following observations can be made:

> Each mode, due to its geographical and technical specificities, was characterized by different technologies and different rates of innovation and diffusion. A transport innovation can thus be an additive/competitive force where a new technology expands or makes an existing mode more efficient and competitive. It can also be a destructive force when a new technology marks the obsolescence and the demise of an existing mode often through a paradigm shift. Still, in many cases an older technology will endure because of its wide level of adoption, utilization and accumulated capital investment. This is commonly known as path dependency. Vested interests in an existing mode, particularly if publicly owned, may also delay or even prevent an innovation taking place.
> Technological innovation was linked with faster and more efficient transport systems. This process implied a space/time convergence where a greater amount of space could be exchanged with lesser amount of time. The comparative advantages of space could thus be more efficiently used.

Technological evolution in the transport sector has been linked with the phases of economic development of the world economy. Transportation and economic development are consequently interlinked as one cannot occur without the other.

One of the pitfalls in discussing future trends resides at looking at the future as an extrapolation of the past. It is assumed that the future will involve a technology that already exists, but simply operating at an extended scale beyond what is currently possible. It can be seen as an incremental change bias. The parameters of such an extrapolation commonly involve a greater speed, mass availability, a higher capacity and/or a better accessibility, all of which imply similar or lower costs. Popular literature (such as *Popular Mechanics* or *Popular Science*) of the first half of the twentieth century is abundant with extrapolations and speculations, some spectacular, about how transportation technology would look like in the (their) future. Looking at such perspectives is labeled "paleo-futurology": how the past perceived the future.

At the start, the prediction of future outcomes must consider what is within the realm of forecasting, scenario building or speculation. Forecasting tries to evaluate near-term outcomes by considering that parameters do not change much, while scenario building tries to assess a series of possible outcomes based upon expected fluctuation in key parameters. A common failure about predictions is their incapacity at anticipating paradigm shifts brought by new technologies as well as economic and social conditions. Another failure relates to the expectation of a massive diffusion of a new technology with profound economic and social impacts, and this over a short period of time (the "silver bullet effect"). This rarely takes place as most innovations go through a cycle of introduction, adoption, growth, peak and then obsolescence, which can take several years, if not decades. Even in the telecommunication sector, which accounts for the fastest diffusion levels, the adoption of a technology takes place over a decade.

Any discussion about the future of transportation must start with the realization that much of what is being presented as plausible is unlikely to become a reality, more so if the extrapolation goes several decades into the future. Thus, as much as someone would have been unable at the beginning of the twentieth century to even dream of what transportation would look like half a century later (e.g. air transportation and the automobile), we may be facing the same limitations at the beginning of the twenty-first century. However, since substantial technological innovations took place in the twentieth century and that the laws of physics are much better understood, we are likely better placed to evaluate which technological trends will emerge in the near future. Still, the socioeconomic impacts of new transport technologies and systems remain complex to assess.

Technological trends

Since the introduction of commercial jet planes, high speed train networks and the container in the late 1960s, no significant technological changes have impacted passengers and freight transport systems, at least from a paradigm shift perspective. The early twenty-first century is an era of car and truck dependency, which tends to constrain the development of alternative modes of transportation, as most of the technical improvements aim at ensuring the dominance of oil as a source of energy. However, with dwindling oil reserves, there is strong evidence that the end of the dominance of the internal combustion engine is approaching. As oil production is expected to peak within a decade and then gradually decline, energy prices are expected to continue their upward trend, triggering the most important technological transition in transportation since the automobile. In such an environment the most promising technologies are:

Automated/Intelligent transport systems. Refer to the development of a set of information and communication technologies (ICT) to improve the speed, efficiency, safety and reliability of movements, by relying upon complete or partial automation of the vehicle, transshipment and control. These systems could involve the improvement of existing modes such as automated highway systems, or the creation of new modes and new transshipment systems such as for public transit and freight transportation (automated terminals). The goal of such initiatives is mainly to efficiently use existing infrastructures through information technologies. Many gains still remain to be achieved through the better management of existing infrastructures and vehicles. Yet, the diffusion of ICT is influenced by the business models of the transport sectors it takes place in.

Alternative modes. There is a range of modes that could replace but more likely complement existing modes, particularly for the transportation of passengers. One such technology is for maglev (short for magnetic levitation) trains, which have the advantage of having no friction (except air friction), enabling operational speeds of 500–600 km/hr to be reached (higher speeds are possible if the train circulates in a low pressure tube). This represents an alternative for passengers and freight land movements in the range of 75 to 1,000 km. Maglev improves from the existing technology of high speed train networks which are limited to speeds of 300 km/hr. In fact, maglev is the first fundamental innovation in railway transportation since the industrial revolution. The first commercial maglev system opened in Shanghai in 2003 and has an operational speed of about 440 km/hr. On the other side of the mobility spectrum urban transportation shows some potential for a more effective use of alternative modes, particularly in car-dependent cities and this would be for passengers and freight transportation alike.

Alternative fuels. These mainly concern existing modes but the sources of fuel, or the engine technology, are modified. For instance, hybrid vehicles involve the use of two types of motor technologies, commonly an internal combustion engine and an electric motor. Simplistically, braking is used to recharge a battery, which then can be used to power the electric motor. Although gasoline appears to be the most prevalent fuel choice, diesel has a high potential since it can also be made from coal or organic fuels. Diesel can thus be part of a lower petroleum dependency energy strategy. Hybrid engines have often been perceived as a transitional technology to cope with higher energy prices. There is also a possibility of greater reliance on biofuels as an additive (and possibly a supplement) to petroleum, but their impacts on food production must be carefully assessed. More far reaching in terms of energy transition are fuel cells, which involve an electric generator using the catalytic conversion of hydrogen and oxygen. The electricity generated can be used for many purposes, such as supplying an electric motor. Current technological prospects do not foresee high output fuel cells, indicating they are applicable only to light vehicles, notably cars, or to small power systems. Nevertheless, fuel cells represent a low environmental impact alternative to generate energy and fuel cell cars are expected to reach mass production by 2015. Additional challenges in the use of fuel cells involve hydrogen storage (especially in a vehicle) as well as establishing a distribution system to supply consumers.

Still, anticipating future transport trends is very hazardous since technology is a factor that historically has created paradigm shifts and is likely to do so again in the future with unforeseen consequences. For instance, one of the major concerns about future transportation for London, England, in the late nineteenth century, was that by the mid-twentieth century the amount of horse manure generated by transport activities would become unmanageable . . .

Economic and regulatory trends

Through recent history, there are few, if any, cases where a revolutionary transport technology was the outcome of a public endeavor. Still, the public sector came to play a growing role as transport innovations became more complex and required a concerted approach in infrastructure, management or regulation. For instance, the massive diffusion of the automobile in the twentieth century was associated with regulations concerning operations (e.g. speed limits), safety (e.g. seatbelts), emissions, as well as public investments in road infrastructures. While vehicle production came to be predominantly private, road infrastructures were perceived as a public good and provided as such. Similar processes took place for maritime transportation (port authorities), air transportation (national carriers), rail (national carriers), public transit (transit agencies) and telecommunications (frequencies). The complexity of transport systems, particularly with intermodalism, is likely to rise in the future – will this complexity be linked with additional public sector involvement?

Future transportation systems are also facing growing concerns related to energy, the environment, safety and security. Transport systems are either going to be developed to accommodate additional demands for mobility or to offer alternatives (or a transition) to existing demand. An important challenge relies in the balance between market forces and public policy, as both have a role to play in the transition. Since transportation is a derived demand, a core aspect of future transportation pertains to the level of economic activity and to what extent this level will be linked with specific passenger and freight volumes. In recent years, economic development and globalization have been important factors behind the surge in mobility. It remains to be seen to what extent this process will endure and if the global transportation system will become more globalized or regionalized:

> **Globalization**. Assumes affordable energy prices, growing accessibility and an enduring openness to trade. The exploitation of comparative advantages continues, leading to a more complex lattice of trade and transportation systems. In addition to active networks of regional transportation are superposed various transnational relations.
>
> **Regionalization**. Assumes higher energy prices and a commercial environment that is more prone to protectionism, all of which conveys more friction to long distance interactions. The exploitation of comparative advantages is thus done on a more regional foundation. This environment does not forbid international trade, but the latter mostly concerns goods and services that cannot be effectively substituted. It is also prone to the setting of more effective regional transport systems.

A fundamental component of future transport systems, freight and passengers alike, is that they must provide increased flexibility and adaptability to changing market circumstances (origins, destinations, costs, speed, etc.), some of which are unforeseen, while complying to an array of environmental, safety and security regulations. This cannot be effectively planned and governments have consistently been poor managers and slow to understand technological changes, often impeding them through regulations and preferences to specific modes or to specific technologies. Regulations have the tendency of preventing technological innovations and their potential positive impacts. This is often referred to as the status quo bias where the dominant strategy of a public agency is to maintain existing conditions. Also, if a new mode or technology competes with a nationalized transport system, then it is likely that the government will intervene to prevent its emergence with regulations (e.g. permits) and delays (e.g. public safety hearings). Recent history indicates that it was when deregulation took place that the most significant changes and innovations resulted for transportation. One of the most salient examples is the Staggers Act (1980) in American rail transportation, which was linked with substantial productivity improvements and new investments.

The drivers of change for the transportation system come in six major categories: policy, demography and society, energy and environment, technology, economics and finance (Figure 2.17). Each plays a role individually and in conjunction. It is therefore virtually impossible to establish outcomes accurately as there are too many interrelationships and uncertainties, particularly if a longer time frame is considered. It is, however, possible to identify trends that may impact each driver individually and try to assess how these trends will shape different components of the transport system:

Policy. The inherent scale and complexity of transportation systems, particularly when they span multiple jurisdictions, will require novel approaches in governance. The role and impact of government policy is commonly subject to cycles of increasing commitments followed by different forms of retrenchment (e.g. privatization) as regulations lead to unproductive practices and unintended consequences. In spite of

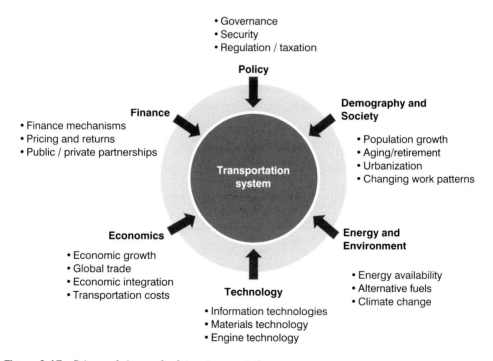

Figure 2.17 Drivers of change for future transportation

Source: Adapted from ICF International (2008) "Long range strategic issues facing the transportation industry", Final Future-focused Research Framework, National Cooperative Highway Research Program, Project 20–80, Task 2.

deregulation, transportation is subject to many forms of regulations pertaining to safety, security and the environment. These regulations, as well as the taxation of transport activities, add to the management complexity and the cost burden.

Demography and society. Population growth is expected to endure in many parts of the world until the mid-twenty-first century, a process which will be linked with demands in mobility and increased consumption. Yet, in other parts of the world, such as in Western Europe, North America and Japan, the rapid aging of the population and more people in retirement age will be associated with changes in mobility and lower levels of consumption per capita. Urbanization is expected to continue in many developing countries, underlining issues linked with the urban mobility of passengers and freight. As a greater share of the global population lives in urban areas, additional pressures are felt on terminal facilities, such as airports, railyards and ports, that find themselves with limited room for expansion. New sites are therefore located further away from existing activity centers. It also remains to be seen how changes in work patterns, such as a greater share of the population in the service sector, will be reflected in mobility.

Energy and environment. Issues related to the availability of energy and raw materials, particularly fossil fuels, are likely to endure and become more acute. This will be reflected in higher energy prices and since each mode has a different elasticity, the comparative advantages of modal options will change towards the most energy-efficient transport chains. A whole range of alternative fuels will be brought forward and transportation activities will increasingly be considered within a sustainability framework. Climate change is also an issue that may add to the sustainability of transport systems, particularly in terms of a more stringent regulatory framework.

Technology. Technological innovation is a very difficult process to anticipate and its impacts even more complex to assess. For transportation, technological innovations either concern the management, the mode (or infrastructure) or the motion (engine). It is expected that information technologies (IT) are likely to transform mobility with an improved command of flows and supply chain management practices. This is commonly linked with a better utilization of existing assets and derived productivity gains. IT also has a high potential to contribute to trade facilitation through more efficient custom procedures as cargo information is standardized and exchangeable. Improvements in materials and engines are also highly possible with the expected benefits on modes and terminals, namely in terms of performance.

Economics. Economic development and global trade have been significant vectors for the growth of mobility. Yet this process is subject to cycles of growth and recession and limits in credit-based consumption. The level of activity and the structure of national economies, as well as their trade patterns, are important influences on national and global transport systems. Economic integration is likely to endure, which will favor more comprehensive and seamless regional transport systems. The relative price of transportation is also linked with the viability of several supply chains and the comparative advantages they extract value from. As transportation costs are expected to rise in the medium term, largely due to fossil fuels, transport demand, from commuting to global supply chains, will be readjusted accordingly in volume but also in the locations they concern.

Finance. Transportation projects, due to their size and technological complexity, are getting increasingly capital-intensive. In several cases, only the largest financial institutions, often in partnership with the public sector, can provide an adequate level of capitalization. The value of transportation assets and the revenue they generate are likely to be important factors behind their financing. Thus, a transport innovation cannot be adopted effectively if financing cannot be secured. Financial considerations are also linked with demographic issues, namely aging. An aging population tends to be more wealth consuming as opposed to wealth producing, which may undermine the availability of capital.

CASE STUDY High speed rail systems

Existing high speed networks

High speed rail (HSR) refers to passenger rail systems running at operational speeds between 200 and 300 km/hr. The high speed train passenger system era originates from Japan with the Tokaido line, bridging Tokyo and Osaka, which entered into service in 1964 in time for the Tokyo Olympics. Today, HSR is perceived as an efficient alternative to highway and airport congestion. Evidence underlines that rail travel time is cut by about a half when a high speed service is established between two cities. The setting of high speed rail systems has accelerated around the world over the last two decades, particularly in China where since 2000 several high speed rail corridors have been rapidly set to reach 13,000 km in 2012. Several countries, including the United States, are also planning for high speed rail corridors, but these projects tend to take decades to implement in part due to funding issues, the limited importance of existing passenger rail services as well as the dominance of air and road. Dedicated high speed postal trains are used in Europe (e.g. France and Sweden) on a daily basis, but the relative decline of postal use leaves such endeavors with questionable growth potential. High speed rail currently functions under two discrete technologies:

> **Improvement of conventional rail**. The first type uses existing conventional rail systems and its great velocity is primarily the fact of considerable improvements in locomotive performance and train design. They may not be considered as pure high speed trains per se. England (London–Edinburgh), Sweden (Stockholm–Gothenburg), Italy (Rome–Florence and Rome–Milan), and the United States (Boston–Washington) are examples of this type of technology. Trains can reach peak speeds of approximately 200 km/hr in most cases and up to 250 km/hr in Italy. The principal drawback from using this system, however, is that it must share existing lines with regular freight services.
>
> **Exclusive high speed networks**. In contrast, the second category runs on its own exclusive and independent tracks. In Japan, trains can attain speeds of 240 km/h, but ongoing projects to raise peak speeds to 300 km/hr aim at maintaining the competitiveness of rail passenger transport versus air. In France, the TGV Sud-Est (Trains à Grande Vitesse) reaches speeds of 270 km/hr while the TGV Atlantique can cruise at speeds of 300 km/hr. One of the key advantages of such a system is since passenger trains have their exclusive tracks, the efficiency of rail freight transport increases as it inherits the almost exclusive use of the conventional rail system.

HSR systems are built to reinforce the accessibility and functional relations within well-developed regional urban systems by filling a gap between short-range mobility provided by cars and buses and medium-range mobility provided by air transport. They were initially set as corridors between city-pairs and their growth eventually led to an integrated system spanning an extended region. While services remain designed along corridors, nodes where it is possible to effectively switch from one corridor to the other are emerging. Like air transport, a good frequency of high speed services enables these interconnections to occur effectively.

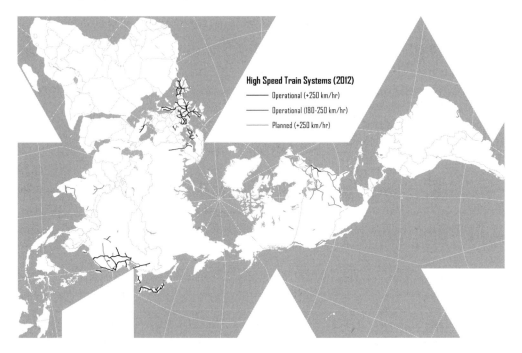

Figure 2.18 World high speed rail systems, 2011

Source: Adapted from International Union of Railways.

HSR systems are mainly found in three regions of the world:

In **Asia**, Japan was the innovator and from 1964 a national integrated HSR system was built to service the major cities and three of the four main islands of the Japanese territory. Both South Korea and Taiwan have built HSR to service a corridor between their two largest cities. China has embarked on an ambitious plan to build a national HSR system that is showing a growing level of integration. With a 8,000 km network in 2010, China's HSR system is already the world's largest and it is expected that by 2020, China's HSR system will total 16,000 km. India has several high speed corridors planned, but high capital requirements and securing rights of way are serious challenges, like in many other parts of the world.

In **Europe**, the setting of HSR systems has gradually permitted a growing level of integration, particularly between France, Belgium, the Netherlands and Germany. The completion of the Eurotunnel in 1994 enabled a link between London and the European HSR system. Northern European countries (Norway, Sweden and Finland) have mostly developed their HSR system through the reconversion of existing lines, a strategy that reflects the relatively short distances involved and lower population densities.

In **North America**, only one high speed rail corridor is in operation between Boston and Washington, but technical requirements limit the speed along several segments of the corridor. The setting of HSR corridors linking regional urban systems has been debated for more than two decades, with many corridor projects clearly identified (e.g. Quebec–Toronto, Miami–Orlando–Tampa or Vancouver–Seattle–Eugene). Yet, the prominence and relatively low cost of road and air transport have been factors playing against the development of HSR systems.

The setting of high speed rail networks consequently must take into consideration the following constraints:

- **Distance between stations**. A distance of 50 km is often considered a minimum, leaving enough for trains to accelerate and reach cruising speed. Servicing too many stations undermines the rationale of high speed systems, which is to service large urban agglomerations in a fast and continuous manner.
- **Separation from other rail systems**. This is mainly the case in and out of metropolitan areas where high speed trains are forced to use the standard rail network so that they may connect to central rail stations.
- **Availability of land**, both for terminals and high speed lines. This problem can be mitigated by using existing central rail stations.

Modal impacts and challenges

High speed rail systems can have substantial impacts on other transport modes, even freight transport systems. One of the most apparent is on air transportation services between cities of the high speed rail corridor, particularly the most distant ones. High speed is able to compete successfully with short to medium distance air transport services as it conveys the advantage of servicing downtown areas and has much lower terminal time, mainly because of fewer security constraints. Another emerging trend concerns a complementarity between HSR and air transportation, which involves cooperation between a national air and rail carrier. For instance, Lufthansa and Deutsche Bahn as well as Air France and SNCF offer single fares and tickets for selected routes where a high speed rail segment is offered instead of a flight. There is thus a balance between competition and complementarity for HSR and air transportation services, particularly when there is congestion in the air transport system. In this situation the complementarity may help release airport gate slots that can be used to support more revenue-generating (longer distance) flights or to reduce congestion.

Rail stations with high speed rail services are also increasingly becoming transport hubs with the associated demands on urban transport systems, particularly public transit. Regarding high speed rail stations, two dynamics have emerged:

- **The reconversion and usage of central railway stations**. Such facilities benefit from high accessibility levels due to their central locations and can thus grant a significant customer base for HSR services (Photo 2.1).
- **The setting of new facilities in suburbia**. In this case, the HSR station represents an opportunity to create a new node of activity within a metropolitan area.

For freight transportation, there are several potential impacts, mostly indirect. The most straightforward is that since high speed rail uses its own right of way, the separation between passenger and freight systems promotes the efficiency and reliability of both networks. The main reason is that passengers and freight have different operational characteristics, namely in terms of speed and frequency of service. The setting of high speed networks may also invite additional investments in rail freight infrastructure, particularly in metropolitan areas, better signaling technologies and cost-sharing initiatives. Although there have been discussions about the potential of using high speed rail to move freight, these have not yet led to concrete realizations. There are plans to have a high speed rail

Photo 2.1   Antwerp Central train station

Antwerp train station, completed in 1905, is considered to be among the world's most architecturally impressive. However, due to increased ridership as well as the setting of a high speed train system along the Paris–Brussels–Amsterdam corridor (Thalys), the station was increasingly inadequate. The renovations of the train station, completed in 2007, addressed two issues. The first was a shortage of boarding platforms, which was expanded by providing three levels (two underground) with a total of 16 quays. The second was that Antwerp was a dead-end spur with bay platforms, implying that trains had to reverse back to the mainline to continue intercity services, which created additional delays. A tunnel servicing the lowest level was constructed with through platforms enabling a continuous intercity service along the high speed rail corridor.

cargo network in Europe by 2015, which would link major air cargo hubs such as Paris, Liège, Amsterdam, London and Frankfurt. The goal is to provide an alternative to short haul air cargo routes as well as the possibility to move cargo between the hubs.

Yet, HSR does not have the far-reaching impacts on passenger mobility that its proponents suggest, at least in the medium term. Although HSR in Europe is considered to be successful, its implementation required massive subsidies and its profitability remains difficult to achieve. The case of China is illustrative: despite the massive potential of HSR in a context of existing high usage levels of passenger rail and a dense urban system, the rush to construct the system in China has raised technical and security issues and has been associated with low ridership. China remains a developing country where low fares are the dominant factor in mode selection, implying that HSR is not affordable for the great majority of the population. The location of stations remains a salient issue as suburban locations are advantageous from an availability of land perspective. However, suburban locations tend to be not well connected to the local transport system and are remote from central areas, which is commonly the destination for most passenger traffic. The impacts of new HSR stations as poles for urban growth and development remain so far elusive.

In addition to present technologies, an entirely new technological paradigm has been under development in Japan and Germany since the late 1970s. This technology is known as maglev (magnetic levitation) (mentioned above); it utilizes magnetic forces to uplift trains, guide them laterally and to propel them, relying upon highly efficient electromagnetic systems. The first commercial maglev rail system was inaugurated in Shanghai in 2003. Maglev systems have experienced some constraints on widespread commercialization, however, such as difficulties with integration in established rail corridors and perceptions of high construction costs.

Bibliography

Ausubel, J.H. and C. Marchetti (2001) "The evolution of transportation", *Industrial Physicist*, April/May, pp. 20–4.

Bernstein, W.J. (2008) *A Splendid Exchange: How Trade Shaped the World*, New York: Atlantic Monthly Press.

Bogart, D. (2009) "Inter-modal network externalities and transport development: evidence from roads, canals, and ports during the English industrial revolution", *Networks and Spatial Economics*, 9(3): 309–38.

Clark, G. (2008) *A Farewell to Alms: A Brief Economic History of the World*, Princeton, NJ: Princeton University Press.

Givoni, M. (2006) "Development and Impact of the Modern High-speed Train: a review", *Transport Reviews*, 26(5): 593–611.

Gottmann, J. (1961) *Megalopolis: The Urbanized Northeastern Seaboard of the United States*, New York: Twentieth Century Fund.

Harris, C. and E. Ullman (1945) "The nature of cities", *Annals of the the American Academy of Political Science*, 242: 7–17.

Henderson, J.V., Z. Shalizi and A.J. Venables (2000) "Geography and development", *Journal of Economic Geography*, 1: 81–106.

Lundgren, N.G. (1996) "Bulk trade and maritime transport costs: the evolution of global markets", *Resources Policy*, 22(1): 5–32.

McQuaid, R.W., G. Malcolm, A. Smyth and J. Cooper (2004) *The Importance of Transport in Business' Location Decisions*, London: Department for Transport.

Rioux, J-P. (1989) *La révolution industrielle, 1780–1880*, Paris: Éditions du Seuil.

Sletmo, G.K. (1989) "Shipping's fourth wave: ship management and Vernon's trade cycles", *Maritime Policy and Management*, 16(4): 293–303.

Smith, R.A. (2003) "The Japanese Shinkansen", *Journal of Transport History*, 24(2): 222–37.

Taaffe, E.J., H.L. Gauthier and M.E. O'Kelly (1996) *Geography of Transportation*, 2nd edition, Upper Saddle River, NJ: Prentice Hall.

Weber, A. 1909 (1929 Translation) *Alfred Weber's Theory of the Location of Industries*, Chicago, IL: University of Chicago Press.

Williams, A. (1992) "Transport and the future", in B.S. Hoyle and R.D. Knowles (eds) *Modern Transport Geography*, London: Belhaven Press, pp. 257–70.

Woxenius, J. (2006) "Temporal elements in the spatial extension of production networks", *Growth and Change*, 37(4): 526–49.

③ Transportation modes

Transportation modes are an essential component of transport systems since they are the means by which mobility is supported. Geographers consider a wide range of modes that may be grouped into three broad categories based on the medium they exploit: land, water and air. Each mode has its own requirements and features, and is adapted to serve the specific demands of freight and passenger traffic. This gives rise to marked differences in the ways the modes are deployed and utilized in different parts of the world. More recently, there is a trend towards integrating the modes through intermodality and linking the modes ever more closely into production and distribution activities. At the same time, however, passenger and freight activity is becoming increasingly separated across most modes.

Concept 1 – A diversity of modes

Transport modes are the means by which people and freight achieve mobility. They fall into one of three basic types, depending on over what surface they travel – land (road, rail and pipelines), water (shipping) and air. Each mode is characterized by a set of technical, operational and commercial characteristics (see Figures 3.1 and 3.2).

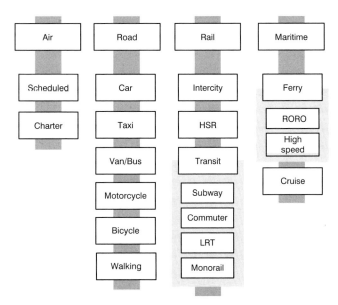

Figure 3.1 Main passenger modal options

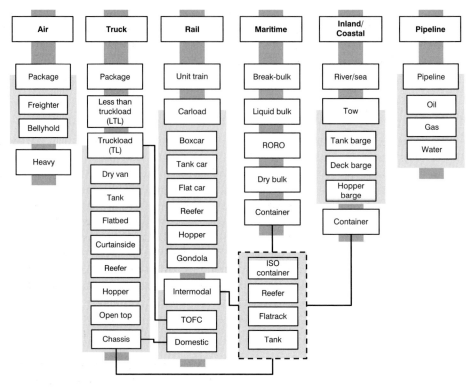

Figure 3.2 Main freight modal options

Source: Adapted from W.J. DeWitt, Freight Transport and Modes in Global Logistics and Supply Chains.

Telecommunications are a special case as they cover a gray area in terms of if they can be considered a transport mode since unlike true transportation, telecommunications often does not have a physicality. Yet, they are structured as networks with a practically unlimited capacity with very low constraints, which may include the physiography and oceanic masses that may impair the setting of cables. They provide for the instantaneous movement of information (speed of light in theory). Wave transmissions, because of their limited coverage, often require substations, such as for cellular phone and wireless data networks. Satellites often use a geostationary orbit which is getting crowded. High network costs and low distribution costs characterize many telecommunication networks, which are linked to the tertiary and quaternary sectors (stock markets, business-to-business information networks, etc.). Telecommunications can provide a substitution for personal movements in some economic sectors.

Road transportation

Road transportation and its infrastructures are large consumers of space with the lowest level of physical constraints among transportation modes. However, physiographical constraints are significant in road construction with substantial additional costs to overcome features such as rivers or rugged terrain. While historically road transportation was developed to support non-motorized forms of transportation (walking, domestication of animals and cycling at the end of the nineteenth century), it is motorization that has shaped its development the most since the beginning of the twentieth century. Road transportation has an average operational flexibility as vehicles can serve several purposes but are rarely able to move outside roads. It offers a range of motorized and non-motorized options for short distance movements that a user may opt for depending on affordability, convenience, availability and comfort. The automobile has emerged as a preferred form of passenger transportation as it offers flexibility and convenience, but also contributes to the dominant share of the congestion of roads, particularly in urban areas.

Road transport systems have high maintenance costs, both for the vehicles and infra-structures. They are mainly linked to light industries where rapid movements of freight in small batches are the norm. Yet, with containerization, road transportation has become a crucial link in freight distribution. Trucks are highly flexible vehicles able to carry almost every type of cargo over short to medium distances. Package trucks are commonly used in urban freight distribution since they carry a variety of cargo (in boxes or pallets) servicing a fluctuating demand. Less-than-truckload (LTL) carriers usually consolidate and deconsolidate loads coming from different customers, which is common in the parcel-carrying business. Truckload (TL) transportation carries large volumes that have been broken down into the largest possible truck load unit; several truckloads are required to fulfill an order. The variety of modal options is related to the technical requirements to carry specific cargoes such as bulk, liquids or containers. Trucks using chassis are able to carry domestic (usually 53 feet in North America) and ISO containers (20 and 40 feet).

Road development accelerated in the first half of the twentieth century. By the 1920s, the first all-weather transcontinental highway, the Lincoln Highway, spanned over 5,300 km between New York and San Francisco. The Germans were, however, the first to build the modern highway (autobahn) in 1932 with specifications such as restricted access, overpasses and road separation that would eventually become common charac-teristics of highway systems. The post-Second World War era represented a period of rapid expansion of road transportation networks worldwide. The most remarkable achievement is without doubt the American Interstate highway system initiated in 1956. Its strategic purpose was to provide a national road system servicing the American economy and also able to support troop movements and act as air strips in case of an emergency. About 56,000 km was built from the 1950s to the 1970s, but between 1975 and 2006 only 15,000 km were added to the system, underlining growing construction costs and diminishing returns. Overall, about 70,000 km of four-lane and six-lane high-ways were constructed, linking all major American cities, coast to coast. A similar project took place in Canada with the Trans-Canada highway completed in 1962. By the 1970s, every modern nation had constructed a national highway system, which in the case of Western Europe resulted in a pan-European system. This trend now takes place in many industrializing countries. For instance, China is building a national highway system that expanded to 80,000 km in 2011, with construction taking place at a pace of about 2,000 km per year.

An overview of the growth of the American and the Chinese highway systems under-lines different sequences and rates (Figure 3.3). From its inception, the American Inter-state highway system expanded substantially, but at a declining rate as the system neared its planned size (46,000 miles; 74,000 km). By 1991, after more than three decades of construction, the system was considered completed, with a total cost of about 129 billion dollars. Between 1954 and 2001, 370 billion dollars were invested by the federal govern-ment in the construction and the maintenance of the system. Close to three-quarters of

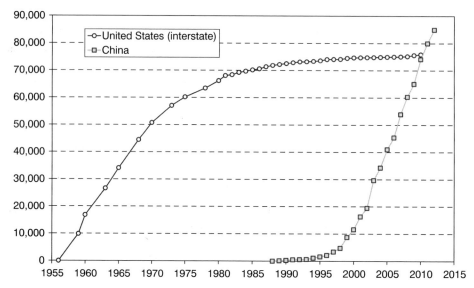

Figure 3.3 Length of the Interstate highway system and of the Chinese expressway system, 1959–2012 (in km)

Note: Includes Puerto Rico.

Source: Federal Highway Administration and National Bureau of Statistics of China.

the financing came from fuel taxes, which created a positive feedback loop as the more Interstate roads were available, the more fuel consumption and thus tax collection. However, the Interstate is facing diminishing returns due to high construction and maintenance costs, which is forcing many state governments to consider privatization of several highway segments. Construction costs went from four million dollars per mile in 1959 to 20 million dollars in 1979. Still, the system has returned more than six dollars in economic productivity for each dollar it cost, placing it at the core of American economic productivity gains in the second half of the twentieth century.

The Chinese expressway system was developed later but at a much faster rate. Prior to 1989, there were no highways in China, but as the economy opened up, the development of a national system of expressways was seen as a priority. To facilitate the fast construction of the system almost all expressways are toll roads financed by private companies under contract from provincial governments, commonly as public–private partnerships. Debt contracted for expressway construction is expected to be recovered through toll collection. Unlike the United States, China did not implement a national fuel tax road financing mechanism. This mode of financing thus differs from the publicly funded highway systems built in Europe and North America. A significant landmark was achieved in 2011, when the length of the Chinese expressway system surpassed that of the American Interstate system. The planned length of the expressway system was set at 85,000 km. The construction of new expressways will likely slow down afterwards, underlining that at this point China will have achieved an important step in its motorization transition. It remains to be assessed in light of the significant urban population, raising levels on national production and consumption and the fast growth of car ownership, to what extent the national expressway network will be sufficient to effectively support China's socioeconomic mobility needs.

All road transport modes have limited potential to achieve economies of scale. This is due to size and weight constraints imposed by governments and also by the technical and economic limits of engines. In most jurisdictions, trucks and buses have specific weight and length restrictions which are imposed for safety reasons. In the United States, the maximum gross vehicle weight is 36 metric tons (80,000 pounds), while in Europe and China these figures are 40 (88,000 pounds) and 49 (100,000 pounds) metric tons respectively. In addition, there are serious limits on the traction capacities of cars, buses and trucks because of the considerable growth in energy consumption that accompany increases in the vehicle weight. For these reasons the carrying capacities of individual road vehicles are limited.

Road transport, however, possesses significant advantages over other modes:

The **capital cost** of vehicles is relatively small, which makes it comparatively easy for new users to gain entry. This helps ensure that the trucking industry, for example, is highly competitive. Low capital costs also ensure that innovations and new technologies can diffuse quickly through the industry.

The **high relative speed** of vehicles, the major constraint being government-imposed speed limits.

Flexibility of route choice, once a network of roads is provided. Road transport has the unique opportunity of providing door-to-door service for both passengers and freight.

These multiple advantages have made cars and trucks the modes of choice for a great number of trip purposes, and have led to their market dominance for short distance trips. The success of cars and trucks has given rise to a number of serious problems. Road congestion has become a feature of most urban areas around the world. In addition, the mode is behind many of the major environmental externalities linked to transportation. Addressing these issues is becoming an important policy challenge at all levels of jurisdiction, from the local to the global.

Rail transportation

Rail transportation is composed of a traced path on which vehicles are bound. They have an average level of physical constraints linked to the types of locomotives and a low gradient is required, particularly for freight. Rail transportation is characterized by a high level of economic and territorial control since most rail companies are operating in a monopoly, as in Europe, or oligopoly, as in North America where seven large rail freight carriers control and operate large networks. Operating a rail system involves using regular (scheduled), but rigid, services. Rail transportation, like roads, has an important relationship with space, since it is the transport mode the most constrained by the physiography. Gauges, however, vary around the world, often complicating the integration of rail systems. The global rail network, which is mostly a collection of unlinked national rail systems, is an overlay of three main types of rail lines (Figure 3.4):

Penetration Lines. Their main purpose is to link a port city with its hinterland, particularly in order to access natural resources such as minerals, agricultural products and wood products. They also represented one of the initial stages of rail development, notably in the United States, which later became regional networks linked by transcontinental lines. This type of system is today mainly found in developing countries (Africa and Latin America) and was partially the result of the colonial era. Transporting freight is the dominant function of this type of network, although passenger traffic can be significant.

Regional Networks. They service high density population areas of developed countries with the goal to support massive shipment of freight and passengers. Regions with the highest rail density are Western Europe, the Northeastern part of North America, coastal China and Japan.

Transcontinental Lines. These lines were mainly established for territorial conquest and the establishment of national sovereignty. The most relevant examples are in the United States, Canada, Russia and Australia, which have built rail systems of this scale. Today, transcontinental rail lines are being established, such as the North American landbridge, to attenuate the discontinuity of maritime transportation by transporting containers. They are a chain in the global intermodal transport system.

The standard intercity passenger services that have been active in many parts of the world for a century and a half are being expanded by the setting of high speed rail (HSR) services between high density city-pairs. Another salient form of rail services concerns urban transit systems that rely on specific applications of rail technology. Subway systems are those supporting the densest forms of mobility in large metropolitan areas. Such systems are usually supported by commuter rail linking a central station to a network of satellite cities. Light rail transit (LRT) systems are also set in lower density situations.

Heavy industries are traditionally linked with rail transport systems, although containerization has improved the flexibility of rail transportation by linking it with road and maritime modes. Rail is by far the land transportation mode offering the highest capacity with a 23,000 tons fully loaded coal unit train being the heaviest load ever

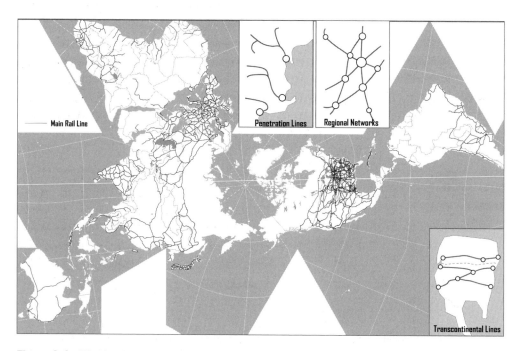

Figure 3.4 World rail network and rail systems

carried. A unit train carries the same cargo between one origin and one destination, with several carload configurations possible depending on what is being carried. There can be unit trains for coal, grain, cars or containers. Trains can also be assembled with different carloads servicing different customers, origins and destinations. This is, however, more costly and time-consuming. Containerization had significant impacts on rail transportation and spurred the development of intermodal rail services that are specific to rail. The first concerns trailers on flatcars (TOFC) where a complete truckload is loaded on a purposely designed flatcar. The second involves carrying domestic containers on well cars that are also designed to carry ISO containers.

Rail freight services are also facing the challenge of improving their reliability, which leads to a fragmentation of the types of services being offered. For conventional rail freight markets such as coal, grain, forest products or chemicals, the priority has consistently been the provision of high capacity and low cost forms of transportation. Although these services were unreliable they could be easily accommodated by stockpiling, a strategy common in the resource sector (e.g. power plants, grain elevators). An emerging freight market for rail mostly concerns intermodal services that require a much higher level of reliability, similar to what is expected in trucking. Commercial changes such as large volumes of retail import containerized cargo and just-in-time manufacturing require high reliability levels.

Pipelines

Pipelines represent a completely separate system of freight distribution where liquids (particularly oil) and gases can be pumped over long distances. Pipeline routes are practically unlimited as they can be laid on land or under water. Two main products dominate pipeline traffic: oil and gas, although locally pipelines are significant for the transport of water, and in some rare cases for the shipment of dry bulk commodities, such as coal in the form of slurry. The longest gas pipeline links Alberta to Sarnia (Canada), which is 2,911 km in length. The longest oil pipeline is the Trans-Siberian, extending over 9,344 km from the Russian arctic oilfields in eastern Siberia to Western Europe. Physical constraints are low and include the landscape and pergelisol in arctic or subarctic environments. Pipeline construction costs vary according to the diameter and increase proportionally with the distance and with the viscosity of fluids (from gas, low viscosity, to oil, high viscosity) and therefore the need for pumping stations. The Trans-Alaska pipeline, which is 1,300 km long, was built under difficult conditions and has to be above ground for most of its path. Pipeline terminals are very important since they correspond to refineries and harbors.

Pipelines are almost everywhere designed for a specific purpose only, to carry one commodity from a location to another. They are built largely with private capital and because the system has to be in place before any revenues are generated, represent a significant capital commitment. They are effective in transporting large quantities of products where no other feasible means of transport (usually water) is available. Pipeline routes tend to link isolated areas of production to major refining and manufacturing centers in the case of oil, or to major populated areas, as in the case of natural gas. Pipelines can even be used to carry small quantities of freight, such as in pneumatic tubes, but this use remains marginal and for short distances.

The routing of pipelines is largely indifferent to terrain, although environmental concerns frequently delay approval for construction. In sensitive areas, particularly in arctic/subarctic areas where the pipes cannot be buried because of permafrost, the impacts on migratory wildlife may be severe, and be sufficient to deny approval, as was the case of the proposed McKenzie Valley pipeline in Canada in the 1970s. Geopolitical factors play a very important role in the routing of pipelines that cross international boundaries. Pipelines from the Middle East to the Mediterranean have been routed to avoid Israel, and new pipelines linking Central Asia with the Mediterranean are being routed in response to the ethnic and religious mosaic of the republics in the Caucasus.

Pipeline construction costs vary; operating costs are very low, however, and as mentioned above, pipelines represent a very important mode for the transport of liquid and gaseous products. One major disadvantage of pipelines is the inherent inflexibility of the mode. Once built (usually at great expense), expansion of demand is not easily accommodated. There are specific limits to the carrying capacity. Conversely a lessening of supply or demand will produce a lowering of revenues that may affect the viability of the system. A further limit arises out of geographical shifts in production or consumption, in which a pipeline having been built from a location to another may not be able to easily adjust to changes.

Maritime transport

Maritime transportation, similar to land and air modes, operates on its own space, which is at the same time geographical by its physical attributes, strategic by its control and commercial by its usage. While geographical considerations tend to be constant in time, strategic and especially commercial considerations are much more dynamic. The physiography of maritime transportation is composed of two major elements, which are rivers and oceans. Although they are connected, each represents a specific domain of maritime circulation (Figure 3.5). The notion of maritime transportation rests on the existence of regular itineraries, better known as maritime routes.

Even if maritime transportation has experienced remarkable improvements in its safety and reliability, maritime routes are still hindered by dominant winds, currents and general weather patterns. The North Atlantic and the North Pacific (50 to 60 degrees north) are subject to heavy wave activity during the winter that sometimes impairs navigation, and may cause ships to follow routes at lower latitudes, thereby increasing route lengths. During the summer monsoon season (April to October), navigation may become more hazardous on the Indian Ocean and the South China Sea.

About 50 countries have an inland navigation network of more than 1,000 kilometers. Rivers may not be useful for commercial navigation if their orientation does not correspond to the directions of transport demand. Thus, many of the major rivers of Russia flow north–south, while the main trade and passenger flows are east–west. Shallow draught and extensive obstacles, such as rapids, may also limit navigation. However, many rivers, such as the Rhine or the Chang Jiang, are significant arteries for water transport because they provide access from the oceans to inland markets.

Maritime transportation is the most effective mode to move large quantities of cargo over long distances because the physical properties of water confer buoyancy and limited friction. Main maritime routes are composed of oceans, coasts, seas, lakes, rivers and channels. However, due to the location of economic activities maritime circulation takes place on specific parts of the maritime space, particularly over the North Atlantic and the North Pacific. The construction of channels, locks and dredging are attempts to facilitate maritime circulation by reducing discontinuity. Comprehensive inland

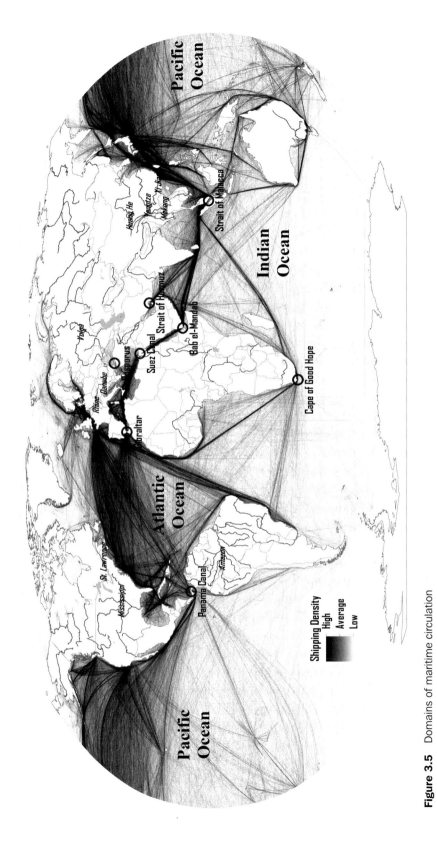

Figure 3.5 Domains of maritime circulation

Source: Shipping density data adapted from National Center for Ecological Analysis and Synthesis, A Global Map of Human Impacts to Marine Ecosystems.

waterway systems include Western Europe, the Volga/Don system, St. Lawrence/Great Lakes system, the Mississippi and its tributaries, the Amazon, the Panama/Paraguay and the interior of China. Maritime transportation has high terminal costs, since port infrastructures are among the most expensive to build, maintain and improve. High inventory costs also characterize maritime transportation, mainly because of the capital cost of ships, which is increasing with economies of scale.

Maritime traffic is predominantly focused on freight. Before the era of intercontinental air transportation, transcontinental passenger services were assumed by liner passenger ships, mainly over the North Atlantic. Long distance passenger movements are now a marginal leisure function solely serviced by cruise shipping. Several oceanic ferry services are also in operation over short distances, namely in Western Europe (Channel; Baltic Sea), Japan and Southeast Asia (Indonesia). The systematic growth of maritime freight traffic has been fueled by:

> **Increase in energy and mineral cargoes** derived from a growing demand from industrialized economies of North America, Europe, China and Japan. For instance, coal is mainly used for energy generation and steel-making.
>
> **Globalization** that went on par with an international division of the production and trade liberalization.
>
> **Technical improvements** in ship and maritime terminals have facilitated the flows of freight.
>
> **Economies of scale** permitted maritime transportation to remain a low cost mode, a trend which has been strengthened by containerization.

Maritime traffic is commonly measured in deadweight tons, which refers to the amount of cargo that can be loaded on an "empty" ship, without exceeding its operational design limits. This limit is often identified as a loadline, which is the maximal draft of the ship. Maritime freight is conventionally considered in two categories:

> **Bulk cargo** refers to freight, both in dry or liquid form, which is not packaged such as minerals (oil, coal, iron ore, bauxite) and grains. It often requires the use of specialized ships such as oil tankers as well as specialized transshipment and storage facilities. Conventionally, this cargo has a single origin, destination and client and is prone to economies of scale. Services tend to be irregular, except for energy trades, and part of vertically integrated production processes.
>
> **Break-bulk cargo refers** to general cargo that has been packaged in some way with the use of bags, boxes or drums. This cargo tends to have numerous origins, destinations and clients. Before containerization, economies of scale were difficult to achieve with break-bulk cargo as the loading and unloading process was very labor-intensive and time-consuming.

More than any other mode, maritime transportation is linked to heavy industries, such as steel and petrochemical facilities adjacent to port sites. Through the application of the principle of economies of scale maritime shipping has developed specialized ships to carry break bulk, dry bulk, liquids, vehicles (RORO) and even liquefied natural gas. Container shipping has also become a dominant maritime modal option supporting commercial transactions with multiple origins, destinations and cargo owners. The standard ISO containers of 20 and 40 feet are the main unit sizes, and they have been adapted to carry refrigerated goods (reefers) and even liquids (tank containers). Still, the dry maritime container is the most dominant container cargo unit. The role of maritime transportation to move passengers has substantially declined but remains important for ferry services. The cruise ship is not used as a form of transportation but as a touristic option between a network of ports of call.

The global maritime shipping industry is serviced by about 79,000 commercial vessels of more than 100 tons falling into four broad types:

Passenger vessels can be further divided into two categories: passenger ferries, where people are carried across relatively short bodies of water in a shuttle-type service, and cruise ships, where passengers are taken on vacation trips of various durations, usually over several days. The former tend to be smaller and faster vessels, the latter are usually very large capacity ships having a full range of amenities. In 2011, about 19 million passengers were serviced by cruise ships, underlining an industry with much growth potential since it services several seasonal markets where the fleet is redeployed to during the year.

Bulk carriers are ships designed to carry specific commodities, and are differentiated into liquid bulk and dry bulk vessels. They include the largest vessels afloat. The largest tankers, the Ultra Large Crude Carriers (ULCC) are up to 500,000 deadweight tons (dwt), with the more typical size being between 250,000 and 350,000 dwt; the largest dry bulk carriers are around 400,000 dwt, while the more typical size is between 100,000 and 150,000 dwt. The emergence of liquefied natural gas technology enabled the maritime trade of natural gas with specialized ships.

General cargo ships are vessels designed to carry non-bulk cargoes. The traditional ships were less than 10,000 dwt, because of extremely slow loading and offloading. Since the 1960s these vessels have been replaced by container ships because they can be loaded more rapidly and efficiently, permitting a better application of economies of scale. Like any other ship class, larger container ships require larger drafts with the current largest ships requiring a draft of 15.5 meters.

Roll-on-roll-off (RORO) vessels, which are designed to allow cars, trucks and trains to be loaded directly on board. Originally appearing as ferries, these vessels are used on deep-sea trades and are much larger than the typical ferry. The largest are the car carriers that transport vehicles from assembly plants to the main markets.

The distinctions in vessel types are further differentiated by the kind of services on which they are deployed. Bulk ships tend to operate on both a regular schedule between two ports or a voyage basis. In the latter case the ship may haul cargoes between different ports based on demand. General cargo vessels operate on liner services, in which the vessels are employed on a regular scheduled service between fixed ports of call, or as tramp ships, where the vessels have no schedule and move between ports based on cargo availability.

A global maritime freight transport system has been established since the late nineteenth century and expanded with containerization. It includes east–west and north–south routes and location enabling interconnectivity between these systems of circulation. Figure 3.6 illustrates the structure this system is likely to take in the coming years. The main components are:

Circum-equatorial route. With the expansion of the Panama Canal expected to come online in 2014, a relative parity will exist for the first time between the Panama

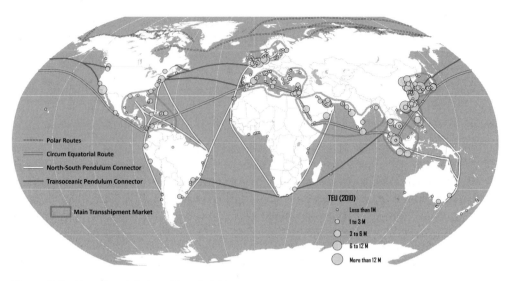

Figure 3.6 Emerging global maritime freight transport system

and Suez canals. In such a setting, maritime shipping companies may elect to establish circum-equatorial routes in both directions with the usage of high capacity (8,000 to 12,000 TEU) container ships. This high frequency "conveyor belt" could support a significant share of global east–west freight movements in a cost-effective way. This does not imply a homogeneous service as several different configurations of ports of call are possible along this route, particularly if a 300 nautical miles deviation is considered. This enables different circum-equatorial network configurations.

North–south pendulum connectors. These connectors reflect existing commercial relations, namely for raw materials (oil, minerals, agricultural goods), such as South America/North America, Africa/Europe or Australia/Asia. For container shipping, they are mostly based on the rationale that there is not enough volume to support transoceanic services, so cargo is collected/delivered along a latitudinal sequence of ports. This conventional network will be expanded with transshipment opportunities with the circum-equatorial route.

Transoceanic pendulum connectors. Connect through pendulum services selected ports of the facades of large oceanic masses. The three main transoceanic connectors are transpacific, Asia–Europe (through the Indian Ocean) and transatlantic. The industrialization of Asia (China in particular) has made the Asia–Europe and the transpacific connectors particularly important. Growth within the "BRIC" countries (Brazil, India and China) favors the emergence of a new connector in the southern hemisphere between the east coast of South America, the Cape of Good Hope and to Southeast Asia.

Polar routes. Consider the usage of circum-polar routes as shortcuts to link East Asia, Western Europe and North America (both east and west coasts). Even if the distance advantages of these polar routes appear significant, they are subject to the uncertainties of climate change.

Transshipment markets. They connect regional port systems to transoceanic and circum-equatorial routes, mainly through hub-and-spoke services. The relay function between long distance shipping services performed by those markets is also significant. The most important are Southeast Asia, the Mediterranean and the Caribbean.

They are referred to as markets because the transshipment function can be substituted to another port. Therefore a group of ports in a transshipment market are "bidding" for port calls as this type of traffic is difficult to anchor. The development of circum-equatorial routes is thus likely to expand the opportunities of transshipment, including interlining between these routes.

Air transport[1]

Air transport constraints are multidimensional and include the site (a commercial plane needs about 3,300 meters of runway for landing and takeoff), the climate, fog and aerial currents. Air routes are practically unlimited, but they are denser over the North Atlantic, inside North America and Europe and over the North Pacific. Speed is the major advantage of air transport compared to other modes. This feature has served to offset many of its limitations, among which operating costs, fuel consumption and limited carrying capacities are the most significant. Technology has worked to overcome some of the constraints, most notably the growth of capacity, with aircraft capable of transporting 500 passengers or 100 tons of freight. Technology has also significantly extended the **range** of aircraft, so that while in the 1960s aircraft were just beginning to be capable of crossing the Atlantic without stopping at intermediate places such as Newfoundland, they are now capable of making trips of up to 18 hours in duration. The speed of commercial aircraft has not progressed since the 1960s, when the prospect of supersonic speed was being anticipated with the development of the Anglo-French Concorde. This plane was removed from service in 2003.

Air transport services usually come as scheduled services offered by various competing air carriers, each within their respective networks. Based upon scheduled services posted several months in advance, a traveler (or someone acting on his/her behalf) will be able to book an itinerary that may include several flight segments. Charter air services as usually offered under specific circumstances such as seasonal flights towards resort areas or private jets servicing the mobility needs of a corporation or an individual. Air activities are linked to the tertiary and quaternary sectors, notably finance and tourism, which lean on the long distance mobility of people. Three major categories of passenger jet planes may be recognized:

Short-range aircraft. Bombardier's CRJ series and Embraer's ERJs are examples of planes with relatively small capacities (30–100 passengers) that travel over relatively short distances. They are usually referred to as regional jets that serve smaller markets and feed hub airports. They can also provide point-to-point services between large city-pairs.

Medium-range aircraft. The airbus A320, with a range of 3,700 km, and its Boeing equivalent, the B-737, are designed to service destinations within a continent. From New York, most of North America can be reached. This range can be applied to the European continent, South America, East Asia and Africa. This type of aircraft is also used for high demand regional services where many low cost air carriers compete.

Long-range aircraft. There are a variety of aircraft capable of crossing the oceans and linking together the continents. Early variants such as the B-707 have evolved into planes offering high capacity, such as the B-747 series, or long-range abilities, such as the B-777 series or the A350 series which have ranges of up to 17,400 km (Figure 3.7).

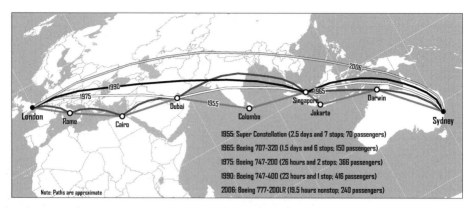

Figure 3.7   Shortest air route between London and Sydney, 1955–2006

Since the inauguration of the first commercial regular long distance air services in the 1950s, the relative distance between London and Sydney has been substantially reduced. When the route was serviced by a propeller plane (Super Constellation), the segment took two and a half days and seven stops to be serviced. The introduction of the 747 in the 1970s reduced this route to 26 hours and two stops. A more fuel-efficient and longer range 747–400 improved the route by three hours because only one stop was then required. In 2006, for the first time, a direct flight became a possibility with a new generation of long-range aircraft such as the 777–200LR. However, this yet to be serviced route is at the extreme limit of serviceability as dominant winds would only make possible an east–west non-stop full load leg. Thus a "direct" flight between London and Sydney still involves a technical refueling stop in Singapore or Dubai. Considering technical limitations linked with the ratio speed/fuel consumption, it is unlikely that the travel time between London and Sydney will become lower than the current 19.5 hours direct flight.

More recently, air transportation has been accommodating growing quantities of high value freight and is playing a growing role in global logistics. Air cargo is generally carried in unit load devices, either on dedicated freight planes (freighters) or in the bellyhold of scheduled passenger flights. Heavy loads, such as vehicles, require specialized cargo planes and are commonly used by the military. The air freight market is serviced by five types of operations:

Dedicated cargo operators maintaining a fleet of cargo-only aircraft and offering regular scheduled services between the airports they service. They also offer charter operations to cater to specific needs.

Combination services where an airline company will maintain a fleet of both specialized and passenger aircraft able to carry freight in their bellyhold. Most of the cargo operations involve long-haul services.

Passenger operators that will offer the freight capacity in the bellyhold of their aircraft. For these operators, freight services are rather secondary and represent a source of additional income. However, low cost airlines usually do not offer air cargo services.

Air freight integrators commonly operating hub-and-spoke freight services that reconcile short and long-haul flights. They offer comprehensive services that are usually door-to-door and can support the logistics requirements of their customers.

Photo 3.1 Boarding of a Ryanair flight

Ryanair is a salient example of a low cost airline business strategy relying on the usage of one type of aircraft (B737–800) and of secondary airports for hubs (e.g. Charleroi, Dublin) for lower gate fees. The booking is done online with the pricing system based on yield management, implying that price varies constantly based on the availability of seats. Carry-on luggage is limited to one piece weighing up to 9 kilos and any additional luggage is on a fee basis. Baggage policy is strictly enforced to ensure revenue generation as the pricing system is much higher if booking options are changed at the airport (e.g. boarding pass, additional luggage). Ryanair does not use sky-bridges, accelerating the boarding sequence which is done by using both the front and the back doors of the plane. A 737 can thus be boarded in less than 15 minutes, whereas it can take twice as long if a sky-bridge is used. This ensures fast turn-around times and a higher level of asset utilization. Passengers seat themselves with no pre-assigned seat, underlining the importance of priority boarding which customers wanting better seats can pay for, thus generating a source of revenue that comes at no cost for the airline. The 737s have a high density seat configuration with no reclining and seat pockets (no documentation or magazines). All inboard services are fee based with merchandizing common (e.g. duty free, phone cards, advertising on the overhead bins).

Specialized operators fulfilling niche services that cater to specific cargo requirements (e.g. heavy loads) that do not fit the capabilities of standard cargo aircraft.

Air transport makes use of air space that theoretically gives it great freedom of route choice. While the mode is less restricted than land transport to specific rights of way, it is nevertheless much more constrained than might be supposed. In part this is due to physical conditions, in which aircraft seek to exploit (or avoid) upper atmospheric winds, in particular the jet stream, to enhance speed and reduce fuel consumption. In addition, specific corridors have been established in order to facilitate navigation and safety. Strategic and political factors have also influenced route choice. For example, the flights of South African Airways were not allowed to over-fly many African nations during the apartheid period, and Cubana Airlines has been routinely prohibited from over-flying the United States. Even more significant was the opening up of Siberian airspace to Western airlines after the Cold War (Figure 3.8).

The advantage of the North Pole as a shortcut is evident but technical and geopolitical changes prevented the full usage of polar routes for commercial transportation. With the introduction of long-range aircraft such as the Boeing 747 and 777 and the Airbus A340, which have ranges exceeding 13,000 km, nonstop transpolar flights became a possibility. It is, however, the collapse of the Soviet Union and the end of the Cold War that marked the geopolitical possibility to use the Arctic as an intercontinental air route. The older, more circuitous routings between Asia and North America, typified by R222, came at the cost of either a reduced payload or a refueling stop en route, commonly Anchorage. New polar routes, like Polar 2, were opened in 2001 and permitted fuller payloads, nonstop flights and shorter trips. The comparative difference between R222 and Polar 2 is about 2,000 km in favor of Polar 2, resulting in a reduced flight time of more than two hours, excluding the refueling stop.

From its infancy, air transport was then seen as a public service and as an industry that should be regulated and protected. In many parts of the world, government intervention in the industry took the form of state-owned airlines. As recently as the early 1970s, Air Canada, Air France, British Airways, Japan Airlines, Qantas and most other flag carriers throughout the world were fully state-owned. In the United States, the government did not own any airlines but it did strongly affect the industry's development via regulation of fares, in-flight service, routes and mergers. Beginning in the 1970s, the relationship between the airline industry and the state changed, although the timing of liberalization (a term which refers to both deregulation and privatization) and its extent have varied among the world's main markets. Across the globe, dozens of airlines have been at least partially privatized, and many airline markets have been deregulated. In the United States, the Air Deregulation Act of 1978 opened the industry to competition.

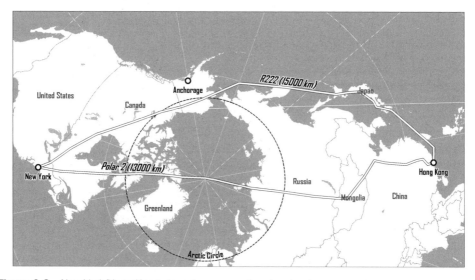

Figure 3.8 New York/Hong Kong air routes: conventional and polar

Geographically, a key outcome of airline deregulation has been the emergence of hub-and-spoke networks centered on a major airport where a single carrier is often dominant. Such networks existed before deregulation to some degree, but the Civil Aeronautics Board hampered the expansion of airlines and the rationalization of networks. United Airlines, for instance, was allowed to add only one city to its network between 1961 and 1978. Hub-and-spoke systems rely on the usage of an intermediate airport hub. They can either connect a domestic (or regional) air system if the market is large enough (e.g. United States, China, European Union) or international systems through longitudinal (e.g. Dubai, Reykjavik; see Figure 3.9) or latitudinal (Panama City; see Figure 3.10) intermediacy. An important aspect of an intermediate hub concerns maintaining schedule integrity since the majority of passengers are using connecting flights. Airports that are prone to delays due to congestion are not effective hubs since they compromise the schedule integrity.

Icelandair is a small airline company that carried around 1.75 million passengers in 2011. Because of the convenient intermediary location of Reykjavik along the transatlantic great circle route, the airline has been successful at establishing a pure hub-and-spoke system servicing mostly Northern Europe and nine North American cities. Scandinavian airports using Reykjavik to reach North America have a particularly low deviation. The same applies to western North American airports (Seattle, Denver and Minneapolis) having a low deviation to Western European airports. Additionally, Scandinavian countries may not generate enough traffic to justify regular direct services to American airports. Therefore, the consolidation of traffic at Reykjavik becomes an effective business proposition. Icelandair's fleet of 20 aircraft is exclusively composed of 757s whose range of 7,700 km is sufficient to support its service pairs. Since the intermediacy offered by Icelandair connects the eastern and western parts of the Atlantic, it is labeled as longitudinal. Iceland is the only effective location for such a pan-Atlantic

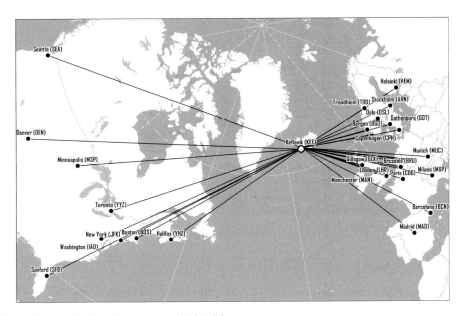

Figure 3.9   Longitudinal intermediacy: Icelandair

Source: Network from Icelandair website.

Figure 3.10 Latitudinal intermediacy: COPA Airlines

Source: Network from COPA Airlines website.

network as there are no other significant airports in the North Atlantic. It is expected with the growth of economic activity along the Arctic Circle that the role of Iceland as an intermediary hub will be reinforced. An indication of this trend is a direct service between Reykjavik and Anchorage that was inaugurated in 2012.

COPA Airlines is a medium-sized Panamanian company operating from its major hub in Panama City (Tocumen International Airport) (see Figure 3.10). The airline operates what can be labeled as a latitudinal intermediacy system since it offers connectivity mainly between airports along a north/south range. Panama City is strategically located as the intermediate location of the Americas; the main reason why the airline labels the hub as "Hub of the Americas in Panama". Its fleet is composed of Boeing 737–700/800s and Embraer E190s, which are medium-range aircraft having a capacity between 100 and 175 passengers and therefore well suited to the network configuration of the services.

Modal competition

Each transportation mode has key operational and commercial advantages and properties. However, contemporary demand is influenced by integrated transportation systems that require maximum flexibility. As a result, modal competition exists at various degrees and takes several dimensions. Different transportation modes have different

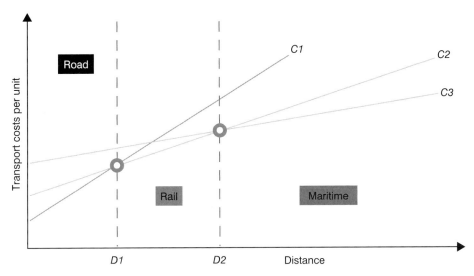

Figure 3.11 Distance, modal choice and transport costs

cost functions according to the serviced distance (Figure 3.11). Road, rail and maritime transport have respectively a C1, C2 and C3 cost function. While road has a lower cost function for short distances, its cost function climbs faster than rail and maritime cost functions. At a distance D1, it becomes more profitable to use rail transport than road transport while from a distance D2, maritime transport becomes more advantageous. Point D1 is generally located between 500 and 750 km of the point of departure while D2 is near 1,500 km. Although the above relation is rather straightforward, it does not fit reality well, mainly for the following reasons:

It assumes that modal options are **interchangeable**. For many origins and destinations, modal options such as rail or maritime may not be present.
Since rail and maritime transportation are discrete networks **only accessible through a terminal**, most locations will involve a road transportation segment, which changes the cost structure.

Modes can compete or complement one another in terms of cost, speed, accessibility, frequency, safety, comfort, etc. There are three main conditions that ensure that some modes are complementing one another:

Different geographical markets. It is clear that if different markets are involved, modes will permit a continuity within the transport system, particularly if different scales are concerned, such as between national and international transportation. This requires an interconnection, commonly known as a gateway, where it is possible to transfer from one mode to the other. Intermodal transportation has been particularly relevant to improve the complementarity of different geographical markets.
Different transport markets. The nature of what is being transported, such as passengers or freight, often indicates a level of complementarity. Even if the same market area is serviced, it may not be equally accessible depending on the mode used. Thus, in some markets rail and road transportation can be complementary as one may be focusing on passengers and the other on freight.

Different levels of service. For a similar market and accessibility, two modes that offer a different level of service will tend to complement the other. The most prevailing complementarity concerns costs versus time.

Thus, there is modal competition when there is an overlap in geography, transport and level of service. Cost is one of the most important considerations in modal choice. Because each mode has its own price/performance profile, the actual competition between the modes depends primarily upon the distance traveled, the quantities that have to be shipped and the value of the goods. While maritime transport might offer the lowest variable costs, over short distances and for small bundles of goods, road transport tends to be most competitive. A critical factor is the terminal cost structure for each mode, where the costs (and delays) of loading and unloading the unit impose fixed costs that are incurred independent of the distance traveled.

With increasing income levels the propensity for people to travel rises. At the same time, international trade in manufactured goods and parts has increased. These trends in travel demand act differentially upon the modes. Those that offer the faster and more reliable services gain over modes that might offer a lower cost, but slower, alternative. For passenger services, rail has difficulty in meeting the competition of road transport over short distances and aircraft for longer trips. For freight, rail and shipping have suffered from competition from road and air modes for high value shipments. While shipping, pipelines and rail still perform well for bulkier shipments, intense competition over the last thirty years has seen road and air modes capture an important market share of the high revenue-generating goods. Road transport clearly dominates.

Although intermodal transportation has opened many opportunities for a complementarity between modes, there is intense competition as companies are now competing over many modes in the transport chain. A growing paradigm thus involves supply chain competition with the modal competition component occurring over three dimensions:

Modal usage. Competition that involves the comparative advantage of using a specific or a combination of modes. Distance remains one of the basic determinants of modal usage for passenger transportation. However, for a similar distance, costs, speed and comfort can be significant factors behind the choice of a mode.
Infrastructure usage. Competition resulting from the presence of freight and passenger traffic on the same itineraries linking the same nodes.
Market area. Competition being experienced between transport terminals for using new space (terminal relocation or expansion) or capturing new markets (hinterland).

It is generally advocated that a form of modal equality (or modal neutrality) should be part of public policy where each mode would compete based upon its inherent characteristics. Since different transport modes are under different jurisdiction and funding mechanisms, modal equality is conceptually impossible as some modes will always be more advantageous than others. Modal competition is influenced by public policy where one mode could be advantaged over the others. This particularly takes place over government funding of infrastructure and regulation issues. For instance, in the United States the federal government financed 80 percent of the costs of a highway project, leaving the state government to supply the remaining 20 percent. For public transit, this share is 50 percent, while for passenger rail the federal government will not provide any funding. Under such circumstances, public policy shapes modal preferences.

The technological evolution in the transport industry aims at adapting the transport infrastructures to growing needs and requirements. When a transport mode becomes more advantageous than another over the same route or market, a modal shift is likely to take place. A modal shift involves the growth in the demand of a transport mode at the expense of another, although a modal shift can involve an absolute growth in both of the concerned modes. The comparative advantages behind a modal shift can be in terms of costs, convenience, speed or reliability. For passengers, this involves a transition in modal preferences as incomes go up, such as from collective to individual modes of transportation. For freight, this has implied a shift to faster and more flexible modes when possible and cost-effective, namely trucking and air freight.

China represents a salient example of modal shift brought by a fast process of economic growth, both for its passengers and freight sectors (Figure 3.12). For passengers, the road and air modes have experienced the most significant gain in market share. Considering the extensive fluvial and canal networks and the prominence of passengers for rail transportation, it is the road and water transport modes that have gained the most for freight transportation.

There are important geographical variations in modal competition. The availability of transport infrastructures and networks varies enormously. Some regions possess many different modes that in combination provide a range of transport services that ensure an efficient commercial environment. Thus, in contrast to the situation in the European Union, rail freight transport occupies a more important market share in North America but passenger rail has a negligible share. In many parts of

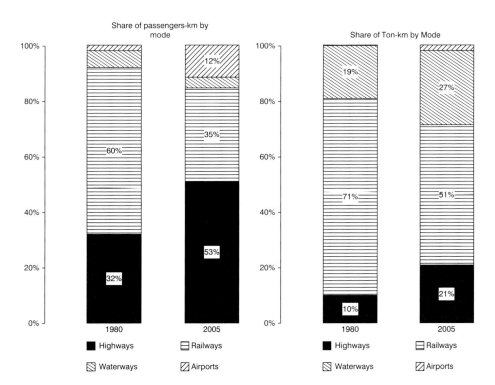

Figure 3.12 Modal shift in China, 1980–2005

Source: World Bank (2007) An Overview of China's Transportation Sector.

the world, however, there are only limited services, and some important modes such as rail may be absent altogether. This limits the choices for people and shippers, and acts to limit accessibility. People and freight are forced to use the only available modes that may not be the most effective to support the mobility needs of passengers or freight.

For these reasons, transport provision is seen as a major factor in economic development. Areas with limited modal choices tend to be among the least developed. The developed world, on the other hand, possesses a wide range of modes that can provide services to meet the needs of society and the economy. Since 2000 the price of fuel has increased significantly as well as its volatility. All modes are affected, from the individual car owner to the corporation operating a fleet of hundreds of aircraft or ships. The higher costs are being passed on to the customer, either directly, as is the case of shipping where freight rates are climbing, or indirectly as is the case of airlines, where passengers are being charged additional fuel surcharges. These cost increases are likely to have significant impacts on mobility and trade, as well as on the modal split:

> **Higher transport costs** increase the friction of distance and constrain mobility. As a major consumer of petroleum the transport industry has to increase rates. Across the board increases cause people to rethink their patterns of movement and companies to adjust their supply and distribution chains.
>
> Because the impact of higher fuel costs hits the modes differentially, a **modal shift is anticipated**. Road and air transport are more fuel-intensive than the other modes, and so fuel price increases are likely to impact upon them more severely than other modes. This could lead to a shift towards water and rail transport in particular.
>
> A further impact of fuel price increases is **greater fuel economy** across the modes. One of the best ways for all modes to reduce consumption is to lower speeds. A future of high energy prices is likely to have a major impact on just-in-time deliveries, and lead to a restructuring of supply chains.

Concept 2 – Intermodal transportation

The nature of intermodalism

History as well as competition between modes has tended to produce a transport system that is segmented and un-integrated. Each mode, particularly the carriers that operate them, has sought to exploit its own advantages in terms of cost, service, reliability and safety. Carriers try to retain business and increase revenue by maximizing the line-haul under their control. All the modes saw the other modes as competitors, and were viewed with a level of suspicion and mistrust. The lack of integration between the modes was also accentuated by public policy that has frequently barred companies from owning firms in other modes (as in the United States before deregulation), or has placed a mode under direct state monopoly control (as in Europe). Modalism was also favored because of the difficulties of transferring goods from one mode to another, thereby incurring additional terminal costs and delays.

Since the 1960s major efforts have been made to integrate separate transport systems through intermodalism, which took place in several stages. What initially began as improving the productivity of shipping evolved into an integrated supply chain management system across modes. This involves the use of at least two different modes in a trip from an origin to a destination through an intermodal transport chain (Figure 3.13).

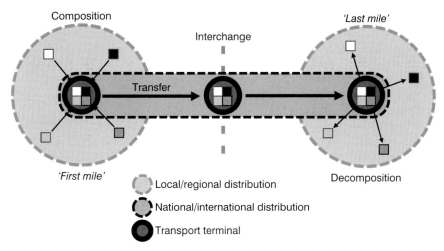

Figure 3.13 Intermodal transport chain

Four major functions define an intermodal transport chain:

Composition. The process of assembling and consolidating freight at a terminal that offers an intermodal interface between a local/regional distribution system and a national/international distribution system. It is commonly referred as the "first mile". Ideally, loads of freight coming from different suppliers are assembled at distribution centers so they can be forwarded to high capacity modes such as rail and maritime shipping. The dominant mode for such a process tends to be trucking as it offers flexibility and door-to-door services. Activities such as packaging and warehousing are also included in the composition process, which is closely linked with the function of production.

Connection (transfer). Involves a consolidated modal flow, such as a freight train or a container ship (or even fleets of trucks), between at least two terminals, which takes place over national or international freight distribution systems. The efficiency of a connection is mainly derived from economies of scale, such as double-stacking or post-Panamax container ships, coupled with an adequate frequency of service.

Interchange. The major intermodal function takes place at terminals whose purpose is to provide an efficient continuity within a transport chain. Those terminals are predominantly within the realm of national or international freight distribution systems, with ports (transshipment hubs) being the most notable example.

Decomposition. Once a load of freight has reached a terminal close to its destination, it has to be fragmented and transferred to the local/regional freight distribution system. Commonly referred as the "last mile", it often represents one of the most difficult segments of distribution. This function, which is linked with the function of consumption, predominantly occurs within metropolitan areas and involves unique distribution problems also known as urban logistics.

Intermodality enhances the economic performance of a transport chain by using modes in the most productive manner. Thus, the line-haul economies of rail may be exploited for long distances, with the efficiencies of trucks providing flexible local pickup and delivery.

The key is that the entire trip is seen as a whole, rather than as a series of legs, each marked by an individual operation with separate sets of documentation and rates. From a functional and operational perspective, two components are involved in intermodalism:

> **Intermodal transportation**. The movements of passengers or freight from one mode of transport to another, commonly taking place at a terminal specifically designed for such a purpose.
>
> **Transmodal transportation**. The movements of passengers or freight within the same mode of transport. Although "pure" transmodal transportation rarely exists and an intermodal operation is often required (e.g. ship to dockside to ship), the purpose is to ensure continuity within the same modal network.

The emergence of intermodalism has been brought about in part by technology and requires management units for freight such as containers, swap bodies, pallets or semi-trailers. In the past, pallets were a common management unit, but their relatively small size and lack of protective frame made their intermodal handling labor-intensive and prone to damage or theft. Better techniques and management units for transferring freight from one mode to another have facilitated intermodal transfers.

Figure 3.14 illustrates two alternatives to freight distribution. The first is a conventional point-to-point multimodal network where origins (A, B and C) are independently linked to destinations (D, E and F). In this case, two modes (road and rail) are used. The second alternative involves the development of an integrated intermodal transport network with common load units (containers). Traffic converges at two transshipment points, rail terminals, where loads are consolidated. This can result in higher load factors and/or higher transport frequency, especially between terminals. Under such circumstances, the efficiency of such a network mainly resides in the transshipment capabilities of transport terminals.

While handling technology has influenced the development of intermodalism, another important factor has been changes in public policy. Deregulation in the United States in the early 1980s liberated firms from government control. Companies were no longer prohibited from owning across modes, which developed a strong impetus towards intermodal cooperation. Shipping lines in particular began to offer integrated rail and

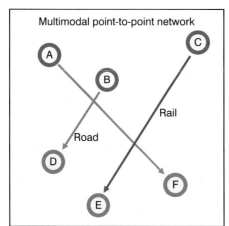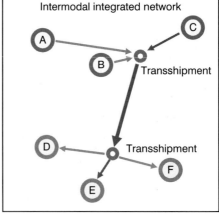

Figure 3.14 Intermodal transportation as an integrative force

road services to customers. The advantages of each mode could be exploited in a seamless system, which created multiplying effects. Customers could purchase the service to ship their products from door to door, without having to concern themselves with modal barriers. With one bill of lading clients can obtain one through-rate, despite the transfer of goods from one mode to another.

The most important feature of intermodalism is the provision of a service with one ticket (for passengers) or one bill of lading (for freight). This has necessitated a revolution in organization and information control. At the heart of modern intermodalism are data handling, processing and distribution systems that are essential to ensure the safe, reliable and cost-effective control of freight and passenger movements being transported by several modes. Electronic Data Interchange (EDI) is an evolving technology that is helping companies and government agencies (customs documentation) cope with an increasingly complex global transport system. Intermodal transport is transforming a growing share of the medium- and long-haul freight flows across the globe where large integrated transport carriers provide door-to-door services. The limits of intermodality are imposed by factors of space, time, form, pattern of the network, the number of nodes and linkages, and the type and characteristic of the vehicles and terminals.

Containerization

The driver of intermodal transportation has undoubtedly been the container, which permits easy handling between modal systems.

> **Container**. A large standard size metal box into which cargo is packed for shipment aboard specially configured transport modes. It is designed to be moved with common handling equipment enabling high speed intermodal transfers in economically large units between ships, railcars, truck chassis and barges using a minimum of labor. The container, therefore, serves as the load unit rather than the cargo contained therein, making it the foremost expression of intermodal transportation. The usage of containers shows the complementarity between freight transportation modes by offering a higher fluidity to movements and a standardization of loads. Thus, the relevance of containers is not what they are – simple boxes – but what they enable – intermodalism. The reference size is the 20-foot box, 20 feet long, 8'6" high and 8 feet wide, or 1 Twenty-foot Equivalent Unit (TEU). Since the great majority of containers are now 40-foot long, the term Forty-foot Equivalent Unit (FEU) is also used, but less commonly.

Intermodalism originated in maritime transportation, with the development of the container in the late 1960s and has since spread to integrate other modes. It is not surprising that the maritime sector should have been the first mode to pursue containerization. It was the mode most constrained by the time taken to load and unload vessels. A conventional break-bulk cargo ship could spend as much time in a port as it did at sea. Containerization permits the mechanized handling of cargoes of diverse types and dimensions that are placed into boxes of standard sizes. In this way goods that might have taken days to be loaded or unloaded from a ship can now be handled in a matter of minutes.

As of 2009, the global container fleet was estimated to be at 26.37 million TEUs, or about 17.25 million units. Containers are either made of steel (the most common for maritime containers) or aluminum (particularly for domestic routes) and their structure confers flexibility and hardiness. The development of intermodal transportation and containerization are mutually inclusive, self-strengthening and rely on a set of driving forces linked with technology, infrastructures and management. One of the initial issues concerned the different sizes and dimensions of containers used by shipping lines, which were a source of much confusion in compiling container shipping statistics. A lift could involve different volumes since different box sizes were involved. As a result, the term TEU (Twenty-foot Equivalent Unit) was first used by Richard F. Gibney in 1969, who worked for the Shipbuilding and Shipping Record, as a measure of comparison. Since then, the TEU remains the standard measure for containerized traffic.

Another factor behind the diffusion of the container is that an agreement about its base dimensions and latching system was reached through the International Organization for Standardization (ISO) within 10 years of its introduction. From this standard, a wide variety of container sizes and specifications have been put in use. The most prevalent container size is, however, the 40-foot box with 2,400 cubic feet that can carry up to 22 tons of cargo (Figure 3.15). International containers are either owned by shipping lines that tend to use them as a tool to help fill up their ships or by leasing companies using containerized assets for revenue generation. In the United States, a large amount of domestic containers of 53 feet are also used. Double-stacking of containers on railways (COFC: Containers On Flat Cars) has doubled the capacity of trains to haul freight with minimal cost increases, thereby improving the competitive position of the railways with regards to trucking for long-haul shipments.

The initial container sizes were the "20 footer" and the "40 footer", dimensions that were agreed upon in the 1960s and became an ISO standard. Initially, the "20 footer" was the most widely used container. However, as containerization became widely adopted in the 1990s, shippers switched to larger container sizes, notably the "40 footer". Larger sizes confer economies of scale in loading, handling and unloading, which are preferred for long distance shipping as well as by customers shipping large batches of

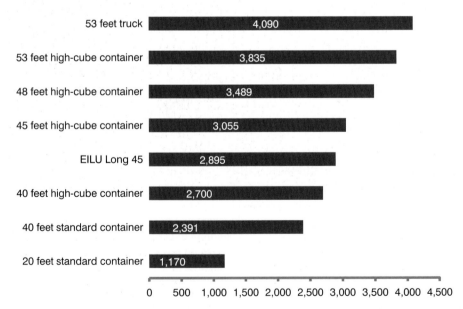

Figure 3.15 Carrying capacity of containers (in cubic feet)

Source: Adapted from Robert C. Leachman (2005) Port and Modal Elasticity Study, Dept. of Industrial Engineering and Operations Research, University of California at Berkeley.

containerized commodities. The same ship capacity would take in theory twice as much time to load or unload if 20 footers were used instead of 40 footers. There is thus an evident rationale to use the largest container size possible. "High-cube" containers have also been put in use, notably since they do not require different handling equipment or road clearance. They are one foot higher (9′6″) than the standard 8′6″ height and a 40-footer high-cube container provides about 12 percent more carrying capacity than its standard counterpart. Most North American double-stack rail corridors can handle two stacked high-cube containers, creating an additional multiplying effect in terms of total capacity per rail car (Photo 3.2). The 53-foot high-cube container, which is the maximum length permitted on the American Interstate highway system, is a load unit that enables even more low weight cargo to be carried (42 percent more volume than a 40-foot high-cube container). However, it is not commonly used since it can only be stacked in the upper section of container ships and does not fit into their bellyhold designed to accommodate 40-foot containers.

Photo 3.2   Forty-foot containers double-stacked on a rail car

Double-stacking rail services were first introduced in North America in 1984, multiplying the productivity of inland container transportation. The advantages are obvious since two 40-foot containers (or four 20-foot) can be stacked on a rail car, essentially doubling the capacity of a unit train. The drawback is a more complex loading or unloading procedure and higher clearance requirement so that double-stacked cars can fit under bridges and tunnels – as shown here, in the port of Vancouver, Canada. Until recently, North America remained the only region of the world where double stacking services were possible. In April 2004 double-stack container train services using 40-foot boxes between Shanghai and Beijing were launched. Both cities can be serviced in about 38 hours. There is also the possibility to use double-stacked trains on the Betuweroute, a 160-km rail segment between Rotterdam and the German border, but no such trains are currently using the corridor.

The European Union is trying to implement a new container labeled the European Intermodal Load Unit (EILU), which would have a length of 45 feet and a width of 8.5 feet. The rationale behind this initiative is that it would allow two of the standard European pallets to be loaded in containers side by side as existing containers are based on North American pallet dimensions. While the new dimensions would still meet clearances for road and rail transport in Europe as well as abroad, the EILU is being strongly opposed by maritime shipping lines, because they have huge investments in current equipment and new ships under construction are optimized for existing ISO container sizes. Because containers have a useful life of about 12 to 15 years, intermodal carriers are reluctant to adopt any new standard because of prior commitments in capital investment in modal and intermodal infrastructures.

While it is true that the maritime container has become the workhorse of international trade, other types of containers are found in certain modes, most notably in the airline industry. High labor costs and the slowness of loading planes, that require a very rapid turnaround, made the industry very receptive to the concept of a loading unit of standard dimensions designed to fit the specific shape of the bellyhold. The maritime container was too heavy and did not fit the rounded configuration of a plane's fuselage, and thus a box specific to the needs of the airlines was required. The major breakthrough came with the introduction of wide-bodied aircraft in the late 1970s. Lightweight aluminum boxes, called unit load devices, could be filled with passenger's baggage or parcels and freight, and loaded into the holds of the planes using tracking that requires little human assistance.

Containerized traffic has surged since the 1990s, underlining its adoption as a privileged means to ship products on international and national markets, particularly for non-bulk commodities where the container accounts for about 90 percent of all movements. Containerization leans on growth factors mainly related to globalization, substitution from break bulk and more recently the setting of intermediate transshipment hubs. The diffusion and adaptation of transport modes to containerization is an ongoing process which will eventually reach a level of saturation. Containers have thus become the most important component for rail and maritime intermodal transportation. The challenge remains about the choice of modes in an intermodal transport chain as well as minimizing the costs and delays related to moving containers between modes.

The advantages of containerization

Among the numerous advantages related to the success of containers in international and hinterland transport, it is possible to note the following:

Standard transport product. A container can be manipulated anywhere in the world as its dimensions are an ISO standard. Indeed, transfer infrastructures allow all elements (vehicles) of a transport chain to handle it with relative ease. Standardization is a prevalent benefit of containerization as it conveys a ubiquity to access the distribution system and reduces the risks of capital investment in modes and terminals. The rapid diffusion of containerization was facilitated by the fact that its initiator, Malcolm McLean, purposely did not patent his invention. Consequently all segments of the industry, competitors alike, had access to the standard. It necessitated the construction of specialized ships and of lifting equipment, but in several instances existing transport modes can be converted to container transportation.

Flexibility of usage. It can transport a wide variety of goods ranging from raw materials (coal, wheat), manufactured goods and cars to frozen products. There are specialized containers for transporting liquids (oil and chemical products) and perishable food items in refrigerated containers (called "reefers", which now account for 50 percent of all refrigerated cargo being transported). About 1.6 million TEUs of reefers were being used by 2009. Discarded containers are often used as storage, housing, office and retail structures.

Management. The container, as an indivisible unit, carries a unique identification number and a size type code enabling transport management not in terms of loads, but in terms of unit. This identification number is also used to ensure that it is carried by an authorized agent of the cargo owner and is verified at terminal gates. Computerized management enables waiting times to be reduced considerably and the location of containers (or batches of containers) known at any time. It enables containers to be assigned according to the priority, the destination and the available transport capacities. Transport companies book slots in maritime or railway convoys that they use to distribute containers under their responsibility. As such, the container has become a production, transport and distribution unit.

Economies of scale. Relatively to bulk, container transportation reduces transport costs considerably, about 20 times less. While before containerization maritime transport costs could account for between 5 and 10 percent of the retail price, this share has been reduced to about 1.5 percent, depending on the goods being transported. The main factors behind cost reductions reside in the speed and flexibility incurred by containerization. Similar to other transportation modes, container shipping is benefiting from economies of scale with the usage of larger container ships (the 6,000 TEU landmark was surpassed in 1996 with the *Regina Maersk* and in 2006 the *Emma Maersk* surpassed the 14,000 TEU landmark). A 5,000 TEU container ship has operating costs per container 50 percent lower than a 2,500 TEU vessel. Moving from 4,000 TEU to 12,000 TEU reduces operating costs per container by 20 percent, which is very significant considering the additional volume involved. System-wide the outcome has been cost reductions of about 35 percent by the use of containerization.

Speed. Transshipment operations are minimal and rapid, which increase the utilization level of the modal assets. A modern container ship has a monthly capacity of three to six times more than a conventional cargo ship. This is notably attributable to gains in transshipment time as a crane can handle roughly 30 movements (loading or unloading) per hour. Port turnaround times have thus been reduced from three weeks to less than 24 hours since it is uncommon for a ship to be fully loaded or unloaded along pendulum routes. It takes on average between 10 and 20 hours to unload 1,000 TEUs compared to between 70 and 100 hours for a similar quantity of bulk freight. With larger container ships, more cranes can be allocated to transshipment. Five or six cranes can service a 5,000 TEU container ship implying that larger ship sizes do not have much difference in loading or unloading time. A regular freighter can spend between half and two-third of its useful life in ports. With less time in ports, container ships can spend more time at sea, and thus be more profitable to operators. Further, container ships are on average 35 percent faster than regular freighter ships (19 knots versus 14 knots). Put all together, it is estimated that containerization has reduced travel time for freight by 80 percent.

Warehousing. The container limits damage risks for the goods it carries because it is resistant to shocks and weather conditions. The packaging of goods it contains is therefore simpler, less expensive and can occupy less volume. Besides, containers fit together permitting stacking on ships, trains (double-stacking) and on the ground. It is possible to superimpose three loaded and six empty containers on the ground. The container is consequently its own warehouse.

Security. The contents of the container are anonymous to outsiders as it can only be opened at the origin, at customs and at the destination. Thefts, especially those of valuable commodities, are therefore considerably reduced. This was a serious issue at ports before containerization as longshoremen had ready access to cargo.

The challenges of containerization

In spite of numerous advantages in the usage of containers, some challenges are also evident:

Site constraints. Containerization involves a large consumption of terminal space. A container ship of 5,000 TEU requires a minimum of 12 hectares of unloading space, while unloading its containers entirely would require the equivalent of about seven double-stack trains of 400 containers each. Conventional port areas are often not adequate for the location of container transshipment infrastructures, particularly because of draft issues as well as required space for terminal operations. Many container vessels require a draft of at least 14 meters (45 feet). A similar challenge applies to container rail terminals, and many have been relocated at the periphery of metropolitan areas. Consequently, major container handling facilities have modified the local geography of containers by forcing relocation to new sites at the periphery.

Infrastructure costs. Container handling infrastructures, such as gantry cranes, yard equipment, road and rail access, represent important investments for port authorities and load centers. For instance, the costs of a modern container crane (portainer) are in the range of 4 to 10 million US dollars depending on the size. Several developing countries cannot afford these infrastructures with local capital and so have difficulties in participating effectively in international trade as efficient load centers unless concession agreements are reached with terminal operators.

Stacking. The arrangement of containers, both at terminals and on modes (container ships and double-stack trains) is a complex problem. At the time of loading, it becomes imperative to make sure that containers that must be taken out first are not below the pile. Further, container ships must be loaded in a way to avoid any restacking along its numerous port calls where containers are loaded and unloaded (Photo 3.3).

Thefts and losses. While many theft issues have been addressed because of the freight anonymity a container confers, it remains an issue for movements outside terminals where the contents of the container can be assessed based upon its final destination. It is estimated that about 10,000 containers per year (27 per day) are lost at sea when they fall overboard from container ships. Rough weather is the major cause, but improper container stacking also plays a role (distribution of heavy containers). Yet, the loss rate remains very low since 5 to 6 million containers are being transported at any given time.

Empty travel. Maritime shippers need containers to maintain their operations along the port networks they service. The same number of containers brought into a market must thus eventually be relocated, regardless if they are full or empty. On average containers will spend about 56 percent of their 10–15-year lifespan idle or being repositioned empty, which is not generating any income but conveys a cost that must be assumed in one way or the other. Either full or empty, a container takes the same amount of space on the ship or in a storage yard and takes the same amount of time to be transshipped. Due to a divergence between production and consumption, it is uncommon to see equilibrium in the distribution of containers. About 2.5 million

Photo 3.3 Stacked container ship

Stacking containers on a ship is a complex operation involving a close interaction between the container yard, drayage operations between the ship and the yard and crane operations. This photo depicts a typical stacking configuration of different container sizes and types taking place on a Panamax container ship (13 container rows) at the port of Veracruz in Mexico (ICAVE terminal operated by HPH). Each row (more likely several rows) generally has containers bound to a specific port. This is done to facilitate crane operations since cranes are designed to quickly move along a row but require the entire crane to be displaced laterally to access another row. Since a ship is usually serviced by several cranes, they can all operate at once on their ship section without impeding one another. This also enables a distribution of the weight across several sections of the ship. The empty rows are likely to be loaded at another port of call. Twenty-foot and 40-foot containers are usually loaded in separate rows to facilitate stacking and loading/unloading efficiency. Container loads bound for a single large customer are usually bundled together, as evidenced by the 20-foot tanker containers. Specialized and oversized containers are stacked on top of the pile since their structure forbids the stacking of containers on top of them. In this case top-loaded containers are used to carry oversized cargo, with some of them covered with a tarpaulin.

TEUs of empty containers are stored in yards and depots around the world, underlining the issue of the movement and accumulation of empty containers. They represent about 20 percent of the global container port throughput and of the volume carried by maritime shipping lines. Most container trade is imbalanced, and thus containers "accumulate" in some places and must be shipped back to locations where there are deficits (mostly locations having a strong export function). This is particularly the case for American container shipping. As a result, shipping lines waste substantial amounts of time and money in repositioning empty containers.

Illicit trade. By its confidential character, the container is a common instrument used in the illicit trade of drug and weapons, as well as for illegal immigrants. Concerns have also been raised about containers being used for terrorism. These fears have given rise to an increasing number of regulations aimed at counteracting illegal use of containers. In 2003, following US inspection requirements the International Maritime Organization (IMO) introduced regulations regarding the security of port sites and the vetting of workers in the shipping industry. The United States established a 24-hour rule, requiring all shipments destined for the USA to receive clearance from US authorities 24 hours prior to the departure of the vessel. In 2008, the US Congress passed a regulation requiring all US-bound containers to be electronically scanned at the foreign port of loading, prior to departure. Needless to say, these measures incur additional costs and delays that many in the industry oppose.

Yet, the advantages of containerization have far outweighed the drawbacks, transforming the global freight transport system and along with it the global economy.

Concept 3 – Passengers and freight: complementarity and competition

Passengers vs. freight

There is a complementarity between passenger and freight transport systems. With some exceptions, such as buses and pipelines, most transport modes have developed to handle both freight and passenger traffic. In some cases both are carried in the same vehicle, as for instance in air transport where about 80 percent of the freight is transported in the cargo holds of passenger aircraft. In others, different types of vehicle have been developed for freight and passenger traffic, but they both share the same road infrastructure, as for example in rail and road traffic. In shipping, passengers and freight used to share the same vessels and often the same terminals. Since the 1950s specialization has occurred, and the two are now quite distinct, except for ferries and some RORO services.

The sharing by freight and passengers of a mode is not without difficulties, and indeed some of the major problems confronting transportation occur where the two compete for the use of scarce transport infrastructure. For example, trucks in urban areas are seen as a nuisance and a cause of congestion by passenger transport users. Daytime deliveries and double-parked trucks are a particular nuisance. The poor performance of some modes, such as rail, is seen as the outcome of freight and passengers having to share routes. There is also a growing interest in using segments of transit systems to move freight, particularly in central areas. This raises the question as to what extent and under which circumstances freight and passengers are compatible. The main advantages of joint-operations are:

High capital costs can be justified and amortized more easily with a diverse revenue stream (rail, airlines, ferries).
Maintenance costs can be spread over a wider base (rail, airlines).
The same **modes or traction sources** can be used for both freight and passengers, particularly for rail.

The main disadvantages of joint operations are:

- **Locations of demand rarely match** since the origins and destinations of freight flows are usually quite distinct spatially from passenger traffic.

- **Frequency of demand** is different, as for passengers the need is for high frequency service; for freight it tends to be somewhat less critical.
- **Timing of service**. Demand for passenger services has specific peaks during the day; for freight it tends to be more evenly spread throughout the day.
- **Traffic balance**. On a daily basis passenger flows tend to be in equilibrium, irrespective of the distance involved (e.g. commuting or air transportation). For freight, market imbalances produce empty flows that require the repositioning of assets.
- **Reliability**. Although freight traffic increasingly demands quality service, for passengers delays (diversion from posted schedules) are unacceptable. Sharing routes favors passenger traffic with passenger trains often given priority or trucks excluded from specific areas at certain times of the day.
- **Different operational speeds** where passengers demand faster service but specific cargo, such as parcels, face similar requirements.
- **Security screening** measures for passengers and freight require totally different procedures.

The ongoing separation of passengers and freight on specific gateways and corridors is consequently a likely outcome, involving a growing divergence of flows, modes and terminals.

A growing divergence

Passengers and freight are increasingly divergent activities as they reflect different transportation markets. In several modes and across many regions passenger and freight transport is being unbundled. **Maritime** passenger services have become separated from freight operations. The exception being ferry services where the use of RORO ships on high frequency services adapt to the needs of both market segments. Deep sea passenger travel is now dominated by cruise shipping which has no freight-handling capabilities, and bulk and general cargo ships rarely have an interest or the ability to transport passengers.

Most **rail systems** improved passenger and freight services. Where both segments are maintained the railways give priority to passengers, since rail persists as the dominant mode for intercity transport in India, China and much of the developing world. In Europe the national rail systems and various levels of government have prioritized passenger service as a means of checking the growth of the automobile, with its resultant problems of congestion and environmental degradation. Significant investments have occurred in improving the comfort of trains and in passenger rail stations, but most notable have been the upgrading of track and equipment in order to achieve higher operational speeds. Freight transport has tended to lose out because of the emphasis on passengers. Because of their lower operational speeds, freight trains are frequently excluded from day-time slots, when passenger trains are most in demand. Overnight journeys may not meet the needs of freight customers. This incompatibility is a factor in the loss of freight business by most rail systems still trying to operate both freight and passenger operations. It is in North America where the separation between freight and passenger rail business is the most extensive. The private railway companies could not compete against the automobile and airline industry for passenger traffic, and consequently withdrew from the passenger business in the 1970s. They were left to operate a freight only system, which has generally been successful, especially with the introduction of intermodality.

Freight and passenger vehicles still share the **roads**. The growth of freight traffic is increasing road congestion and in many cities concerns are being raised about the presence of trucks. Already, restrictions are in place on truck dimensions and weights in certain parts of cities, and there are growing pressures to limit truck access to non-daylight hours. Certain highways exclude truck traffic – the parkways in the USA for example. These are examples of what is likely to become a growing trend: the need to separate truck from passenger vehicle traffic. Facing chronic congestion around the access points to the port of Rotterdam and at the freight terminals at Schiphol airport, Dutch engineers have worked on feasibility studies of developing separate underground road networks for freight vehicles.

Air transport is the mode where freight and passengers are most integrated. Yet even here a divergence is being noted. The growth of all-freight airlines and the freight-only planes operated by some of the major carriers, such as Singapore Airlines, are heralding a trend. The interests of the shippers, including the timing of the shipments and the destinations, are sometimes better served than in passenger aircraft. The divergence between passengers and freight is also being accentuated by the growing importance of charter and "low-cost" carriers. Their interest in freight is very limited, especially when their business is oriented towards tourism, since tourist destinations tend to be lean freight-generating locations.

CASE STUDY Information technologies and mobility

Information technologies and transportation

The emergence and diffusion of information and telecommunication technologies (ICT) have several economic and social impacts, notably on functions related to information processing and diffusion. One of the ongoing tenets is that ICT can offer a form of substitution for physical mobility. Particularly, it can be expected that work-related movements can be mitigated through telecommuting.

> **Telecommuting**. Using information and telecommunication technologies to perform work at a location away from the traditional office location and environment. Commuting is thus substituted and it takes place remotely.

There are obviously various degrees of telecommuting ranging from a partial substitution where a worker may spend one day per week performing work at another location, to a complete substitution where the work is performed elsewhere, such as in an offshore call center. The latter is much less likely as the great majority of work tasks tend to be collaborative and require face-to-face meetings. Yet, with the emergence of an information society, the transactional structures of the economy have changed drastically towards a networked organizational form to which ICT supports improved and more intensive interactions. These interactions involve three major spheres:

> **Personal**. ICT enables individuals to maintain contact through additional mediums (e.g. email). This may lead to more interactions. The diffusion of mobile personal computing devices (e.g. laptops, smart phones and tablet computers) have also enabled individuals to enrich their mobility by enabling them to perform various

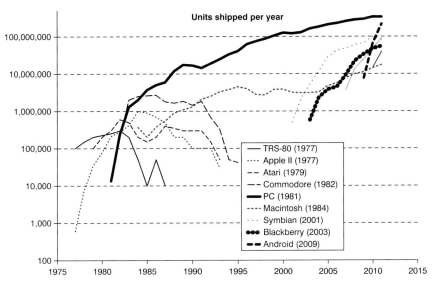

Units shipped per year

TRS-80 (1977)
Apple II (1977)
Atari (1979)
Commodore (1982)
PC (1981)
Macintosh (1984)
Symbian (2001)
Blackberry (2003)
Android (2009)

Figure 3.16 Diffusion of personal computing devices, 1977–2011

Source: Data compiled by H. Dediu and J. Reiner.

tasks while in transit or outside a conventional office setting (Figure 3.16). Several applications, such as global positioning systems, enable individuals to better manage their mobility.

Customer/Retailer. Online retailing has opened a whole new array of commercial opportunities as a complement or a substitution to conventional shopping. It does not necessarily imply that there would be more consumption, but that a growing share of retailing transactions takes place online, resulting in parcels to be delivered.

Business-to-business. The increasing scale and intensity of business transactions is commonly linked with new efficiencies, particularly through supply chain management.

Personal computing devices, enabling their users to execute customizable programs, started to become available in the late 1970s and became mass market products by the mid-1980s. By 2011, more than 335 million PC platform devices alone were being sold around the world. The diffusion of personal computing undertook three distinct phases:

The setting of standards. By the late 1970s several different platforms using different standards and operating systems were being introduced. Interoperability was close to nonexistent, which meant that each platform required its own hardware and software. The main contenders were Apple (Apple II), Commodore (64 and Amiga), Atari (400/800) and Tandy (TRS-80).The introduction of the IBM PC in 1981 marked the downfall of competing standards with the adoption of the MSDOS operating system from which the PC platform would evolve. In 1984 MacOS, the first graphical user interface available for a home computer, carved a niche on which the Macintosh platform would evolve. By the late 1980s, non-PC and MacOS devices were disappearing from the market.

Performance, interface and interconnectivity. From 1985, the two prevailing platforms, with the PC dominating (selling eight to ten times more platforms than

Macintosh), undertook a massive diffusion in the consumer and corporate markets. Performance in terms of processing, memory and storage capacity was growing exponentially (Moore's law) while costs were declining. Graphical user interfaces were standard (e.g. Windows), enabling users to operate complex applications in a relatively easy fashion. By the late 1990s the development of the Internet opened an entirely new range of services to personal computing devices, such as telecommunications (e.g. email), e-commerce, information access and entertainment. Portable computing devices (laptops) also became widely available.

Mobile computing. By 2005, cellular phones, which emerged as mass market products in the mid-1990s, saw their integration with features that were previously only available to desktop and laptop computers or to specialized devices such as digital cameras or global positioning system receivers. The cellular phone evolved from being solely a telecommunication device (with basic features such as an address book, a clock and a calendar) to a true mobile personal computing device offering a wide range of customizable features (apps). The most salient are Symbian (Nokia), Blackberry, iPhone and Android platforms. Their massive diffusion was helped by the ubiquity of wireless networks in developed and developing countries alike. With the introduction of the iPad in 2010, new forms of personal portable computing devices became available which created new niches (e.g. ebooks) that are more complementing than competing with conventional personal computing devices such as the PC.

Telecommuting and mobility

Computers, networking and related information and communication technologies continue to push the centrifugal forces that have shaped the physical structure of metropolitan areas around the world outward. Reducing automobile use is one of the primary expected benefits of telecommuting, as it is assumed that corresponding home-to-work trips will not be made. Telecommuters usually eliminate two trips per day when they work at home. If this commuting was done driving, the reduction of trips has obvious environmental benefits, including reduced emissions of pollutants and reduced fossil fuel use.

Telecommunications, like the automobile, has become a force shaping land use and transportation in urban areas. Cheaper space in the suburbs is an important requirement for newer and smaller firms that are users of new telecommunications technologies. The growing capability of telecommunications allows businesses and other organizations to locate operations more flexibly, but this may be perceived as paradoxical as telecommuting may be supporting an energy inefficient spatial structure.

This all said, telecommuting has often failed to meet expectations and its share remains relatively unchanged. There are many reasons, ranging from activities that cannot be easily substituted to a loss of direct control from management. One major factor behind this continuing low usage level is that if a job has the potential to be complemented by telecommuting it is also a target to be relocated in a low cost location either through outsourcing or offshoring. Thus a large amount of telecommuting takes place as offshoring instead.

Telecommuting and office space

Location and larger building sizes have dominated retailing and offices since the Second World War. Indeed, newer and larger stores overtook smaller rivals and established new distribution structures based on mass retailing. The standard 2,000-square foot market of the 1950s, became the 20,000-square foot supermarket in the 1960s and evolved into the 50,000-square foot superstore of the 1990s. Following a similar trend, the small office of a company has become several floors in a skyscraper located in downtown areas and in time the amount of space devoted to administrative functions has increased significantly.

Competition and technological changes are forcing corporations to recognize real estate as an undermanaged asset as occupancy cost is the second largest corporate expense after wages. Telecommunications advances may lead to a change in the space needs of some corporations, as more services can be performed with less office and branch space. Telecommuting workers also tend to be more productive as they systematically spend more time working, mainly because they are not spending that time commuting but also ICT makes them more available to perform tasks.

Most corporations see telecommuting as a way to reduce costs, not necessarily to promote the welfare of their office employees, even if they are doing both at the same time. The costs of providing office space to employees are very high, far more than just the cost of leasing or building the space and maintaining it. It also concerns, for instance, parking which tends to be more expensive in high density areas often related to office activities. In some cases, it can run as high as 20 to 30 percent of disbursed salaries per employee. In 2007, nearly 20 million workers in the United States (about 3 percent of the total workforce) were telecommuting at least once a month.

The expected impacts of telecommuting on office space and on the general urban environment can be summarized by the following:

- The rapid proliferation of electronic communications, such as cellular phones, intranets and teleconferencing, **promotes telecommuting**.
- Telecommuting is part of a paradigm where **organizational structures are being transformed** from a hierarchy to a network of collaborators.
- Telecommuting is **changing parts of the retailing sector** by rendering some location structures obsolete and by minimizing inventory costs.
- Telecommuting helps **lower office space requirements** by decentralizing a set of tasks to a low cost environment, such as the suburbs or at home, or by permitting their complete offshoring to low cost locations. They are a corporate strategy to improve the productivity of labor.
- Telecommuting reduces the **usage of the urban transport system** because fewer trips are generated and distances are shorter and may reduce congestion. This impact has so far been marginal.
- Telecommuting improves the **flexibility of office activities** by offering a wider array of locational choices.

Note

1 John Bowen, Department of Geography, Central Washington University, contributed to this section.

Bibliography

Agusdinata, B. and W. de Klein (2002) "The dynamics of airline alliances", *Journal of Air Transport Management*, 8: 201–11.

Air Transport Association (2010) *The Airline Handbook*, Washington, DC: Air Transport Association of America.

Allaz, C. (2005) *History of Air Cargo and Airmail from the 18th Century*, London: Christopher Foyle Publishing.

Bilstein, R.E. (1983) *Flight Patterns: Trends of Aeronautical Development in the United States, 1918–29*, Athens: University of Georgia Press.

Bohlman, M.T. (2001) "ISO's container standards are nothing but good news", *ISO Bulletin*, Geneva: International Organization for Standardization, pp. 12–15.

Bowen, J. (2010) *The Economic Geography of Air Transportation: Space, Time, and the Freedom of the Sky*, London: Routledge.

Brooks, M. (2000) *Sea Change in Liner Shipping*, New York: Pergamon.

Brueckner, K. (2003) "Airline traffic and urban economic development", *Urban Studies*, 40(8): 1455–69.

BTS [Bureau of Transportation Statistics] (2006) *America on the Go: Long Distance Transportation Patterns: Mode Choice*, Washington, DC: US DOT.

Cullinane, K. and M. Khanna (2000) "Economies of scale in large containerships", *Journal of Transport Economics and Policy*, 33: 185–207.

Dick, R. and D. Patterson (2003) *Aviation Century: The Early Years*, Erin, Ontario: Boston Mills Press.

Fremont, A. (2007) *Le monde en boîtes. Conteneurisation et mondialisation*, Paris: Les collections de l'Inrets.

Givoni, M. (2006) "Development and impact of the modern high-speed train: a review", *Transport Reviews*, 26(5): 593–611.

Graham, B. (1995) *Geography and Air Transport*, Chichester: Wiley.

Hayuth, Y. (1987) *Intermodality: Concept and Practice*, London: Lloyds of London Press.

Kaluza, P., A. Kölzsch, M.T. Gastner and B. Blasius (2010) "The complex network of global cargo ship movements", *Journal of the Royal Society Interface*, 7(48): 1093–103.

Lee, J.J., S.P. Lukachko, I.A. Waitz and A. Schafer (2001) "Historical and future trends in aircraft performance, cost and emissions", *Annual Review Energy Environment*, 26: 167–200.

Levinson, M. (2006) *The Box: How the Shipping Container Made the World Smaller and the World Economy Bigger*, Princeton, NJ: Princeton University Press.

Mokhtarian, P.L. (2009) "If telecommunication is such a good substitute for travel, why does congestion continue to get worse?", *Transportation Letters*, 1(1): 1–17.

Notteboom, T. (2004) "Container shipping and ports: an overview", *Review of Network Economics*, 3(2): 86–106.

—— (2012) "Container shipping", in W. Talley (ed.) *The Blackwell Companion to Maritime Economics*, New York: Wiley-Blackwell, pp. 230–62.

Notteboom, T. and J-P. Rodrigue (2009) "The future of containerization: perspectives from maritime and inland freight distribution", *Geojournal*, 74(1): 7–22.

Slack, B. (1998) "Intermodal transportation", in B.S. Hoyle and R. Knowles (eds) *Modern Transport Geography*, 2nd edition, Chichester: Wiley, pp. 263–90.

Smith, R.A. (2003) "The Japanese Shinkansen", *Journal of Transport History*, 24(2): 222–37.

Stopford, M. (2009) *Maritime Economics*, 3rd edition, London: Routledge.

Taaffe, E., H.L. Gauthier and M.E. O'Kelly (1998) *Geography of Transportation*, 2nd edition, Upper Saddle River, NJ: Prentice Hall.

van Klink, A. and G.C. van den Berg (1998) "Gateways and intermodalism", *Journal of Transport Geography*, 6: 1–9.

4 Transportation terminals

All spatial flows, with the exception of personal vehicular and pedestrian trips, involve movements between terminals. Transport modes require assembly and distribution of their traffic, both passenger and freight. For example, passengers have to go to bus terminals and airports first in order to reach their final destinations, and freight has to be consolidated at a port or a rail yard before onward shipment. Terminals are, therefore, essential links in transportation chains with many representing substantial infrastructure and capital investments. The goal of this chapter is to examine the spatial and functional characteristics of transport terminals. They occupy specific locations and they exert a strong influence over their surroundings. At the same time they perform specific economic functions and serve as clusters of specialized activities.

Concept 1 – The function of transport terminals

Authors: Jean-Paul Rodrigue and Brian Slack

The nature of transport terminals

A terminal may be defined as any facility where passengers and freight are assembled or dispersed. Both cannot travel individually, but in batches. Passengers have to go to bus terminals and airports first, where they are "assembled" in busloads or planeloads to reach their final destinations where they are dispersed. Freight has to be consolidated at a port or a rail yard before onward shipment. Terminals may also be points of interchange involving the same mode of transport. Thus, a passenger wishing to travel by train from Paris to Rotterdam may have to change trains in Brussels, or an air passenger wishing to fly between Montreal and Los Angeles may have to change planes in Toronto. Terminals may also be points of interchange between different modes of transportation, so that goods being shipped from the American Midwest to the Ruhr in Germany may travel by rail from Cincinnati to the port of New York, be put on a ship to Rotterdam, and then placed on a barge for delivery to Duisburg. Transport terminals, therefore, are central and intermediate locations in the movements of passengers and freight.

> **Terminal**. Any location where freight and passengers either originate, terminate or are handled in the transportation process. Terminals are central and intermediate locations in the movements of passengers and freight. They often require specific facilities and equipment to accommodate the traffic they handle.

Terminals may be points of interchange within the same modal system and which ensure a continuity of the flows. This is particularly the case for modern air and port operations. Terminals, however, are also very important points of transfer between modes. Buses and cars deliver people to airports, trucks haul freight to rail terminals, and rail brings freight to docks for loading on ships. One of the main attributes of transport terminals, international and regional alike, is their convergence function. They are indeed obligatory points of passage having invested in their geographical location which is generally intermediate to commercial flows. Thus, transport terminals are either created by the centrality or the intermediacy of their respective locations. In some cases, large transport terminals, particularly ports, confer the status of gateway or hub to their location since they become obligatory points of transit between different segments of the transport system.

Three major attributes are linked with the importance and the performance of transport terminals:

Location. The major locational factor of a transport terminal is obviously to serve a large concentration of population and/or industrial activities, representing a terminal's market area. Specific terminals have specific locational constraints, such as port and airport sites. New transport terminals tend to be located outside central areas to avoid high land costs and congestion.

Accessibility. Accessibility to other terminals (at the local, regional and global scale) as well as how well the terminal is linked to the regional transport system is of importance. For instance, a maritime terminal has little relevance if it is efficiently handling maritime traffic but is poorly connected to its market areas through an inland transport system (rail, road or barge).

Infrastructure. The main function of a terminal is to handle and transship freight or passengers since modes and passengers or cargo are physically separated. They have a nominal capacity which is related to the amount of land they occupy and their level of technological, labor and managerial intensity. Infrastructure considerations are consequently important as they must accommodate current traffic and anticipate future trends and also technological and logistical changes. Modern terminal infrastructures consequently require massive investments and are among the largest structures ever built. A utilization rate of 75 to 80 percent is considered to be the optimal since above this level congestion starts to arise, undermining the reliability of the terminal facility.

A transport terminal is composed of a set of intermodal infrastructures taking advantage of a geographical location, conferring a higher level of accessibility to local, regional and global markets. Depending on the mode being considered, terminals are bound to various degrees to the site. For instance, maritime transportation terminals are particularly dependent on local conditions, especially for large port activities which can be accommodated in a limited number of locations. Airport terminals are more flexible in their locations, but still bound to specific locational constraints. Terminals fulfill three general functions within transport systems (Figure 4.1):

Connectivity. Transport terminals provide connectivity within a transport network as they are the only locations where a network can be entered or exited. For instance, subway stations are the connecting elements of a transit network.

Interface. Transport terminals provide an interface between transport modes enabling passengers and cargo to transit. A port or an airport are points of interface between maritime or air and land transport systems.

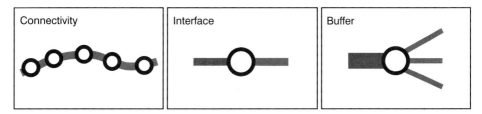

Figure 4.1 The functions of transport terminals

Buffer. Transport terminals provide a buffer between the different capacity and frequency of the transport modes they connect, such as a port does for the maritime and land transportation systems. A container ship may call at a port once every two days while trucks carrying containers may come in and out of the terminal every minute.

Passenger terminals

With one exception, passenger terminals require relatively little specific equipment. This is because individual mobility is the means by which passengers access buses, ferries or trains. Certainly, services such as information, shelter, food and security are required, but the layouts and activities taking place in passenger terminals tend to be simple and require relatively little equipment. They may appear congested at certain times of the day, but the flows of people can be managed successfully with good design of platforms and access points, and with appropriate scheduling of arrivals and departures. The amount of time passengers spend in such terminals tends to be brief. As a result bus termini and railway stations tend to be made up of simple components, from ticket offices and waiting areas to limited amounts of retailing.

Airports are of a completely different order. They are among the most complex of terminals. Moving people through an airport has become a very significant problem, not least because of security concerns. Passengers may spend several hours transiting, with check-in and security checks on departure, and baggage pickup and in many cases customs and immigration on arrival. Planes may be delayed for a multitude of reasons, which involves a complex management of gates and scheduling of flights. The result is that a wide range of services have to be provided for passengers not directly related to the transfer function, including restaurants, bars, stores, hotels, in addition to the activities directly related to operations such as check-in halls, passenger loading ramps and baggage handling facilities (Photo 4.1). At the same time airports have to provide the very specific needs of the aircraft, from runways to maintenance facilities, from fire protection to air traffic control.

Measurement of activities in passenger terminals is generally straightforward. The most common indicator is the number of passengers handled, sometimes differentiated according to arrivals and departures. Transfer passengers are counted in the airport totals even though they do not originate there, and so airports that serve as major transfer facilities inevitably record high passenger totals. This is evident in airports such as Atlanta and Chicago where in-transit passengers account for over 50 percent of the total passenger movements. High transfer passenger activity has been enhanced by the actions of many of the leading airlines adopting hub-and-spoke networks. This results in

Photo 4.1 Modern airport terminal, Barajas, Madrid, Spain

The Barajas terminal in Madrid, opened in 2006, is typical of a modern airport terminal design. Very high ceilings convey an impression of spaciousness, which has a calming effect on passengers. Since the linear design conveys long distances between gates, mechanized walkways are a common feature. The growing size of planes has also necessitated the provision of larger waiting areas at the gate.

many passengers being forced to change planes at the hub airports. By selecting certain airports as hubs, the carriers are able to dominate activity at those airports, thereby controlling most landing and departure slots and the best gates, thus fending off rival airlines. In this way they are able to extract monopoly profits.

A further measure of airport activity is the number of aircraft movements, a figure that must be used with some caution because it pays no regard to the capacity of planes. A 50-seat regional jet and a 300-seat wide-body aircraft both count as one movement. High numbers of aircraft movements thus may not be highly correlated with passenger traffic totals.

Freight terminals

Freight handling requires specific loading and unloading equipment. In addition to the facilities required to accommodate ships, trucks and trains (berths, loading bays and freight yards, respectively) a very wide range of handling gear is required that is determined by the kinds of cargoes handled. Freight transport terminals have a set of characteristics linked with core and ancillary activities. The result is that terminals are differentiated functionally both by the mode involved and the commodities transferred. A basic distinction is that between bulk, general cargo and containers:

Bulk refers to goods that are handled in large quantities that are unpackaged and are available in uniform dimensions. Liquid bulk goods include crude oil and refined products that can be handled using pumps to move the product along hoses and pipes. Relatively limited handling equipment is needed, but significant storage facilities may be required. Dry bulk includes a wide range of products, such as ores, coal and cereals. More equipment for dry bulk handling is required, because the material may have to utilize specialized grabs and cranes and conveyer-belt systems.

General cargo refers to goods that are of many shapes, dimensions and weights such as machinery, processed materials and parts. Because the goods are so uneven and irregular, handling is difficult to mechanize. General cargo handling usually requires a lot of labor.

Containers are standard units that have been designed for simplicity and functionality. Container terminals have minimal labor requirements and perform a wide variety of intermodal functions. They, however, require a significant amount of storage space, that is simple paved areas where containers can be stacked and retrieved with intermodal equipment (cranes, straddlers and holsters). Depending on the intermodal function of the container terminal, specialized cranes are required, such as portainers (container cranes; Photo 4.2). Intermodal terminals and their related activities are increasingly seen as agents of added value within supply chains.

Photo 4.2   Portainer, APM Terminal, Port Newark, New York

Container cranes, also known as portainers, require a high level of capital equipment, such as this post-Panamax crane at the APM terminal (Maersk) at the port of New York (Newark facilities). It was manufactured by ZPMC (Shanghai Zhenhua Port Machinery Company), which is the world's largest gantry crane supplier and accounts for about 60 percent of the world's manufactured cranes.

A feature of most freight activity is the need for storage. Assembling the individual bundles of goods may be time-consuming and thus some storage may be required. This produces the need for terminals to be equipped with specialized infrastructures such as grain silos, storage tanks and refrigerated warehouses, or simply space to stockpile, such as for containers. Containerization, because of its large volumes, has forced a significant modal and temporal separation at terminals and thus the need for a buffer in the form of storage areas. In addition, a variety of transloading activities can take place in the vicinity of terminals, particularly if long distance inland transportation is involved.

Measurement of freight traffic through terminals is more complicated than for passengers. Because freight is so diverse, standard measures of weight and value are difficult to compare and combine. Because bulk cargoes are inevitably weighty, terminals specialized in such cargoes will record higher throughputs measured in tons than others more specialized in general cargoes. This is evident for the world's two leading ports, Singapore and Rotterdam, which are dominated by petroleum. The reverse may be true if value of commodities handled is the measure employed. The problem of measurement involving weight or volumes becomes very difficult when many types of freight are handled, because one is adding together goods that are inherently unequal. Care must be taken in interpreting the significance of freight traffic totals, therefore. For container terminals a common measure of productivity concerns the number of lifts per container gantry crane-hour, which are usually 25–40 moves per hour for quay cranes and 40–60 for rail cranes.

The difficulty of comparing traffic totals of different commodities has led to attempts to "weight" cargoes based upon some indication of the value added they contribute to the terminal. The most famous is the so-called "Bremen rule". It was developed in 1982 by the port of Bremen and was based on a survey of the labor cost incurred in the handling of one ton of different cargoes. The results found that handling one ton of general cargo equals three tons of dry bulk and 12 tons of liquid bulk. Although this is the most widely used method, other 'rules' have been developed by individual ports, such as the Antwerp and Rotterdam rules. The "Antwerp rule" indicates that the highest value added is the handling of fruit. Using this as a benchmark, forest products handling requires 3.0 tons to provide the same value added as fruit, cars 1.5 tons, containers 7 tons, cereals 12 tons and crude oil 47 tons. The "Rotterdam rules" are more recent (2009) and relate to common practices to ensure the transport of freight "door-to-door" in which a sea transport leg is concerned.

Terminal costs

Terminals jointly perform transfer and consolidation functions, and are important economically because of the costs incurred in carrying out these activities. The traffic they handle is a source of employment and benefit to regional economic activities, notably by providing accessibility to suppliers and customers. Terminal costs represent an important component of total transport costs. They are fixed costs that are incurred regardless of the length of the eventual trip, and vary significantly between modes. They can be considered as:

- **Infrastructure costs**. Include construction and maintenance costs of structures such as piers, runways, cranes and facilities (warehouses, offices, etc.).
- **Transshipment costs**. The costs of loading and unloading passengers or freight.
- **Administration costs**. Many terminals are managed by institutions such as port or airport authorities or by private companies (e.g. terminal operators). In both cases administration costs are incurred.

Figure 4.2 represents a simplified assumption concerning transport costs for three modes: road, rail and maritime. The cost functions all begin at some point up the cost axis, which represents terminal costs. Because of terminal costs, maritime shipping (T3) and rail (T2) are at a significant disadvantage compared to road (T1) over short distances. Because ships have the largest carrying capacities, they incur the largest terminal costs, since it may take many days to load or unload a vessel. Conversely, a truck or a passenger bus can be loaded much more quickly, and hence the terminal costs for road transport are the lowest. Terminal costs play an important role in determining the competitive position between the modes. Because of their high freight terminal costs, ships and rail are generally unsuitable for short-haul trips.

Competition between the modes is frequently measured by cost comparisons. Efforts to reduce transport costs can be achieved by using more fuel-efficient vehicles, increasing the size of ships, and reducing the labor employed on trains. However, unless terminal costs are reduced as well, the benefits would not be realized. For example, in water transportation, potential economies of scale realized by ever larger and more fuel-efficient vessels would be negated if it took longer to load and offload the jumbo ships.

Over the last decades, very significant steps to reduce terminal costs have been made. These have included introducing information management systems such as EDI (electronic data interchange) that have greatly speeded up the processing of information and removed delays typical of paper transactions. The most significant development has been the mechanization of loading and unloading activities. Mechanization has been facilitated by the use of units of standard dimensions such as the pallet and, most importantly, the container. The container, in particular, has revolutionized terminal operations. For the mode most affected by high terminal costs, ocean transport, ships used to spend as much as three weeks in a port undergoing loading and loading. The much larger ships of today spend less than a couple of days in port. A modern container ship requires approximately 750 man/hours to be loaded and unloaded. Prior to containerization it would have required 24,000 man/hours to handle the same volume of cargo.

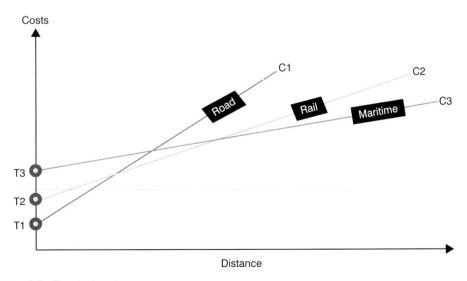

Figure 4.2 Terminal costs

The rail industry too has benefited from the container, which permits trains to be assembled in freight yards in a matter of hours instead of days.

Reduced terminal costs have had a major impact on transportation and international trade. Not only have they reduced overall freight rates, thereby reshaping competition between the modes, but they have had a profound effect on transport systems. Ships spend far less time in port, enabling ships to make many more revenue-generating trips per year. Efficiency in the airports, rail facilities and ports greatly improves the effectiveness of transportation as a whole.

Activities in transport terminals represent not just exchanges of goods and people, but constitute an important economic activity. Employment of people in various terminal operations represents an advantage to the local economy. Dockers, baggage handlers, crane operators and air traffic controllers are examples of jobs generated directly by terminals. In addition there are a wide range of activities that are linked to transportation activity at the terminals. These include the actual carriers (airlines, shipping lines, etc.) and intermediate agents (customs brokers, freight forwarders) required to carry out transport operations at the terminal. It is no accident that centers that perform major airport, port and rail functions are also important economic poles.

Concept 2 – The location of terminals

Authors: Brian Slack, Jean-Paul Rodrigue and Theo Notteboom

The relative location of terminals

Geographers have long recognized situation, or relative location, as an important component of location. It refers to the position of places with regards to other places. Accessibility is relative, because the situation of places changes over time. For example, ports in the Mediterranean used to be in the heart of the Western world during the Greek and Roman eras, and Genoa and Venice prospered during the Middle Ages. The exploitation of the Americas changed the location of these places, since the Mediterranean became a backwater. The opening of the Suez Canal in the nineteenth century refocused the relative location of the Mediterranean again. So the importance of locations changes with the fluctuations in trade and growth opportunities.

Although the term "terminal" implies an end, a final destination, because they are transfer points terminals in fact are typically intermediate locations in the global flows of passengers and freight. Differences in the nature, composition and timing of transfer activities give rise to significant differentiations in the form and function between terminals. A basic distinction is between passenger and freight transfers, because in order to carry out the transfer and bundling of each type, specific equipment and infrastructures are required. Consequently passenger terminals and freight terminals refer to substantially different entities and often have different locational attributes.

Spatial relationships between terminals are a vital element in competition, particularly for ports and rail terminals, and geographers have developed a number of concepts to explore these locational features. One of particular interest concerns the function of centrality and intermediacy performed by transport terminals:

> **Centrality**. Focus on the terminal as a point of origin and destination of traffic. Thus, centrality is linked with the generation and attraction of movements, which are related to the nature and the level of economic activities within the vicinity of the

terminal. The function of centrality also involves a significant amount of intermodal activities.

Intermediacy. Focus on the terminal as an intermediate point in the flows of passengers or freight. This term is applied to the frequent occurrence of places gaining advantage because they are between other places. The ability to exploit transshipment has been an important feature of many terminals.

One of the most enduring concepts in urban geography is the central place theory, with its emphasis on centrality as a feature of the urban hierarchy. Cities more centrally located to markets are larger with a wider range of functions. Transport accessibility is equated with size, and thus many large terminals arise out of centrality. Examples include Heathrow Airport, London, whose traffic preeminence is related to the city's location in the heart of the most developed part of Britain, as one of the world's most important financial centers, as well as Britain's functional centrality to its former empire to a lesser degree. The port of New York owes its preeminence in part to the fact that it is at the heart of the largest market area in the United States; the Boston–Washington corridor. A similar observation applies to the port of Shanghai serving a large market and industrial base.

The concept of centrality is straightforward as the vicinity of the terminal is either the origin or the destination of the movement, which in turn is linked with the level of economic activity. The extent of this vicinity remains to be fully assessed, but a range of about 100 km appears suitable as it corresponds to a commuting or drayage range. Intermediacy, however, is a multifaceted concept that relates to a number of issues (Figure 4.3):

Range. This is more a conventional aspect of intermediacy that tends to be of lesser importance in the contemporary situation. Due to technical limitations of the modes, such as the range of an aircraft or the need to refuel a coal-powered ship, intermediate locations were used as stages to overcome the range gap as two locations could

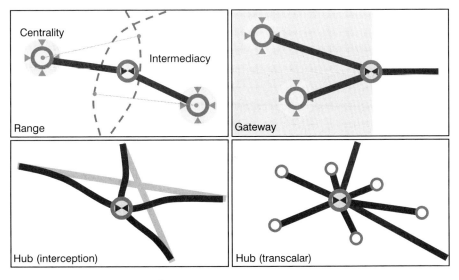

Figure 4.3   Centrality and intermediacy

not be reached in a single trip. Intermediate airports such as Anchorage, Alaska (transpacific flights) or Gander, Newfoundland (transatlantic flights), are good examples of locations that were used to overcome the technical limitations of aircraft before the introduction of long-range couriers. Although range plays a more limited role for air passenger transportation, particularly with long-range aircraft, it is still prevalent for air freight transportation as the range of freight aircraft is more limited due to heavier loads. Therefore, transpacific air cargo routes commonly involve a stop at Anchorage while Asia–Europe routes will have a stop at a Middle Eastern airport such as Dubai. To a lesser extent, intermediate locations are used for long distance trucking since rest periods are required for drivers, but the required facilities are very basic.

Gateway. Connects two systems of circulation and thus represents an intermediate location imposed by geographical constraints. Thus, in order to reach its final destination, a movement must use an intermediate location often implying a transfer from one mode to the other. Many gateways have also a significant centrality component as they represent industrial zones and large urban agglomerations.

Hub (Interception). A location near, or at, the convergence of several long distance routes can develop an intermediacy by "intercepting" some of the traffic. This is notably the case for intermediate hub terminals located along major long distance maritime corridors, such as Algeciras (Spain) or Singapore.

Hub (Transcalar). A location is specifically used to serve as a connection between different scales of a transport system. Air transportation is a notable example with the emergence of hub-and-spoke network structures where the hub is an intermediate location between regional and international flights. Freight distribution, particularly cross-docking distribution centers, also rely on the usage of an intermediate location to service specific market segments often supplied by distant sources.

Hinterlands and forelands

One of the most enduring concepts in transport geography, especially applied to ports, is the hinterland. Transport terminals are within a transport system which includes the notions of foreland and hinterland binding import and export activities:

The **hinterland** is a land space over which a transport terminal, such as a port, sells its services and interacts with its clients. It accounts for the regional market share that a terminal has relative to a set of other terminals servicing a region. It regroups all the customers directly bound to the terminal and the land areas from which it draws and distributes traffic. The terminal, depending on its nature, serves as a place of convergence for the traffic coming by roads, railways or by sea/fluvial feeders.

The concept of foreland is a mirror image of the hinterland:

The term **foreland** is the ocean-ward mirror of hinterland, referring to the ports and overseas markets linked by shipping services from the port. It is above all a maritime space with which a port performs commercial relationships, namely its overseas customers. With the emergence of feeder services and hub ports, the concept of foreland has been expanded as a port can service a hinterland through a maritime link.

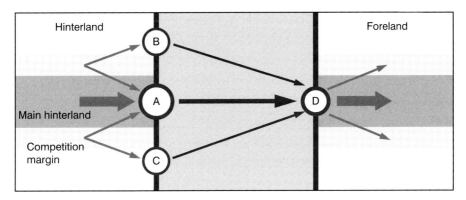

Figure 4.4 Port foreland and hinterland

Two types of hinterlands are often noted (Figure 4.4). The term main or **fundamental hinterland** refers to the market area for which a terminal is the closest. It is assumed that traffic will normally pass through the terminal, because of proximity and the lack of competitive alternatives. The **competitive hinterland** (or competitive margin) is used to describe the market areas over which the terminal has to compete with others for business.

The main nature of a hinterland is commercial and its importance is linked with the level of economic activity as well as the level of competition from other modes not linked to the terminal. Hinterlands vary significantly for the same location if the flows concern passengers or freight. For airports, like most passenger terminals, the hinterland is well delimited and corresponds to a commuting range where customers can access the terminal within a couple of hours. The level of activity is proportional to the population density, the level of income and the prominence of tertiary activities. For ports, like most freight terminals, the level of activity corresponds to the dynamics of the land they are connected to, which is subject to changes in the nature of its activities and in the level of accessibility. Any change implies either new opportunities to generate additional port traffic, a decline, or a change in the nature and composition of the traffic. Inbound hinterland traffic tends to be consumption based, except in the case when commodities and parts are involved in the fabrication of a product, while outbound hinterland traffic is an outcome of extraction or production. Hinterlands can further be discriminated by the type of commodity as each is part of a specific supply chain with its own spatial relationships:

Bulk products (minerals, chemicals, raw materials, wood, grain, etc.). In this case distance is one of the most important factors shaping hinterlands. Due to the nature of the products and the high transport costs involved, hinterlands tend to be small and serviced by high capacity corridors to the direct location of extraction or production.

Parts and manufactured goods. Mostly concerns containerized traffic. Improvements in intermodal transportation and globalization have considerably expanded the hinterland for this type of traffic. In many cases, the hinterland can encompass large economic regions, particularly if transport corridors are involved.

The hinterlands of three major economic regions can be synthetically represented in terms of their intensity and the importance of gateways and corridors that service them (Figure 4.5):

In **North America**, there is a high level of concentration of economic activities along the coastal areas (East and West coasts) with significant resource and manufacturing hinterlands. From coastal gateways long distance rail corridors, often taking the form of a landbridge, are servicing a continental hinterland. This hinterland is articulated by major transportation and industrial hubs such as Chicago.

In **Western Europe**, the hinterland is the most intense in the interior, notably along the Rhine river system. This hinterland is accessed from coastal gateways, such as Rotterdam, Antwerp, Hamburg and Le Havre, through medium distance corridors involving a variety of combinations of road, barge and rail services. Almost all the major European capitals are interior cities located along rivers.

In **East and Southeast Asia**, a significant share of the economic activity takes place along the coast, with a few high population density interior hinterlands, such as in China. Hinterland access is commonly problematic, linked to the fact that a large share of the accumulation of new economic activities has taken place in the vicinity of major gateways. There is thus a strong contrast between coastal gateways equipped with modern (container) terminals and hinterlands usually poorly serviced by rail freight services.

In recent years the validity of the hinterland concept has been questioned, especially in the context of contemporary containerization. The mobility provided by the container has greatly facilitated market penetration, so that many ports compete over the same market areas for business. Therefore, hinterlands may be overlapping. The notion of discrete hinterlands with well-defined boundaries is questionable since many hinterlands have become discontinuous, a process facilitated by the development of corridors and inland terminals. Also, the extension and strengthening of hinterlands follows a vertical or horizontal integration process depending on whether the port establishes

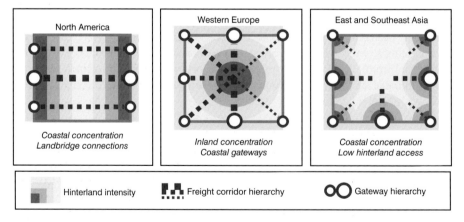

Figure 4.5 Hinterland setting and major economic regions

Source: Adapted from Lee, S.W., D.W. Song and C. Ducruet (2008) "A tale of Asia's world ports: the spatial evolution in global hub port cities", *Geoforum*, 39: 373–85.

more effective functional linkages with inland intermodal terminals (vertical) or other maritime terminals (horizontal). Nevertheless, the concept of hinterland is still widely employed, and port authorities continue to emphasize their port's centrality to hinterland areas in their promotional literature.

The provision of services to a wide range of markets around the world is considered to be an advantage. With the growth in maritime traffic and congestion in proximity of port terminal facilities, several port authorities have become more involved in the development of strategies aimed at better servicing their hinterland. In academic studies there have been far fewer assessments of foreland than hinterland, yet in port publicity documents the foreland is usually one of the elements stressed. Geographers have long criticized the distinction, arguing that foreland and hinterland should be seen as a continuum, rather than separate and distinct elements. This point has achieved greater weight recently, with the emergence of door-to-door services and networks, where the port is seen as one link in through-transport chains. In such a context, the port becomes one element of the maritime/land interface which ensures the continuity of global freight circulation.

Port terminals

Ports are points of convergence between two domains of freight circulation (sometimes passengers): the land and maritime domains. The term port comes from the Latin *portus*, which means gate or gateway. Ports are bound by the need to serve ships, and so access to navigable water has been historically the most important site consideration. Before the industrial revolution, ships were the most efficient means of transporting goods, and thus port sites were frequently chosen at the head of water navigation, the most upstream site. Many major cities owed their early preeminence to this fact, such as London on the Thames, Montreal on the St. Lawrence River or Guangzhou on the Pearl River. Ship draft was small, so many sites were suitable. Sites on tidal waterways created a particular problem for shipping because of the twice-daily rise and fall of water levels at the berths, and by the eighteenth century the technology of enclosed docks with lock gates was developed to mitigate this problem. Because ship transfers were slow, and vessels typically spent weeks in ports, a large number of berths were required. This frequently gave rise to the construction of piers and jetties, often called finger piers, to increase the number of berths per given length of shoreline.

The gradual shift from conventional break-bulk terminals to container terminals from the early 1960s brought about a fundamental change in layout of terminals as well as site selection. Containerized transportation has substantially changed port dynamics to favor the emergence of specialized container ports. Compared to conventional break-bulk cargo ships, container ships do not have onboard cranes, and so container terminal facilities had to provide capital-intensive cranes as well as ample storage space to stack containers dockside. Finger piers were no longer adequate and berths were redesigned to accommodate quick ship turnaround and more effective dockside operations between the crane and the container storage areas. Containerization has consequently become a fundamental function of global port operations and has changed the structure and configuration of port terminals that tend to occupy more space. While inland port sites (such as at the end of a bay or along a river) generally have the advantage of being closer to the final market they entail longer deviations from maritime shipping routes. Therefore the most successful inland port sites are those that act as gateways (e.g. Antwerp, Montreal, Constanza).

As terminals, ports handle the largest amounts of freight, more than any other types of terminals combined. To handle this freight, port infrastructures jointly have to accommodate transshipment activities both on ships and inland and thus facilitate convergence between land transport and maritime systems. In many parts of the world, ports are the points of convergence from which inland transport systems, particularly rail, were laid. Considering the operational characteristics of maritime transportation, the location of ports is constrained to a limited array of sites, mostly defined by geography. Most ports, especially those that have ancient origins, owe their initial emergence to their site; the great majority of harbors are taking advantage of a natural coastline or a natural site along a river. Many port sites are constrained by:

Maritime access, which refers to the physical capacity of the site to accommodate ship operations. It includes the tidal range, which is the difference between the high and low tide, as normal ship operations cannot handle variations of more than three meters. Channel and berth depths are also very important to accommodate modern cargo ships. A standard Panamax ship of 65,000 deadweight tons requires more than 12 meters (40 feet) of depth. However, about 70 percent of world ports have depths of less than 10 meters and are unable to accommodate ships of more than 200 meters in length. In view of the construction of larger ships, such as tankers and container ships, many port sites found themselves unable to provide maritime access to modern cargo operations. Since container terminals were constructed much more recently, they have a better nautical profile as depth and available space were fundamental factors in site selection. There is thus a pressure to increase channel depth where possible, but this is a costly and environmentally controversial endeavor. Many ports are also impacted by sedimentation, particularly ports in river deltas. This requires continuous dredging, which adds to the costs of port operations.

Maritime interface. Indicates the amount of space that is available to support maritime access, namely the amount of shoreline that has good maritime access. This attribute is very important since ports are linear entities. Even if a port site has an excellent maritime access, namely deep water waterways, there may not be enough land available to guarantee its future development and expansion. Containerization has expanded the land consumption requirements of many ports. It is therefore not surprising to see that modern port expansion projects involve significant capital investments to create artificial port facilities.

Infrastructures and equipment. The site, to be efficiently used, must have infrastructures such as piers, basins, stacking or storage areas, warehouses and equipment such as cranes, all of which involve high levels of capital investment. In turn, these infrastructures consume land which must be available to ensure port expansion. Keeping up with the investment requirements of modern port operations has become a challenge for many ports, particularly in light of containerization which requires substantial amounts of terminal space to operate.

Land access. Access from the port to industrial complexes and markets ensure its growth and importance. This requires efficient inland distribution systems, such as fluvial, rail (mainly for containers) and road transportation. The land access to ports located in densely populated areas is facing increasing congestion. For instance, the ports of Los Angeles and Long Beach invested massively to develop the Alameda rail corridor in an attempt to promote inland access and reduce truck congestion. A similar trend has taken place in Europe where ports such as Rotterdam and Antwerp have been involved in the setting of inland barge and rail shuttle services.

There is also an array of problems related to port infrastructures. Ports along rivers are continuously facing dredging problems and the width of the river can strongly limit capacity if it constrains navigation. Rarely does a port along a river have the capacity to handle the new generation of giant ships, namely post-Panamax container ships, which have put additional pressures on port infrastructures to accommodate the transshipment generated by these ships. Ports next to the sea are commonly facing a lateral spread of their infrastructures. Several ports have growth problems that force them to spread their infrastructures far from the original port sites. Since ports are generally old, and in several cases were responsible for urban growth, they are located near central areas. This is creating congestion problems where the transport network has the least capacity to be improved.

The city and the port are often competing for the same land, which can create priority problems (port versus waterfront development for instance). Ports thus have a complex set of relationships, sometimes conflicting, with the cities they service, often a function of the port and city size. The pressure of many ports on their sites is even more demanding than those of airports because they have to be adjacent to deep water. Such sites are very limited, and may give rise to conflicts with the city at large that sees waterfront land as potential park space, or as environmentally sensitive. Many ports are now constrained by urban and environmental pressures, which did not exist when the existing facilities were developed.

The main function of a port is to supply services to freight (warehousing, transshipment, etc.) and ships (piers, refueling, repairs, etc.). Consequently, it is misleading to consider a port strictly as a maritime terminal since it acts concomitantly as a land terminal where inland traffic originates or ends. Ports are becoming increasingly regional in their dynamics, which represents a new development from their traditional local function, namely as industrial complexes. For instance, the port of Hong Kong owes its wealth to its natural site and its geographical position of a transit harbor for southern China. A similar function is assumed by Shanghai for central China with the Yangtze river system. Singapore, for its part, is favored by its location at the outlet of the strategic Strait of Malacca and is therefore a point of convergence of Southeast Asian transportation. More than 90 percent of the traffic it handles is strictly transshipments. New York has traditionally acted as the gateway of the North American Midwest through the Hudson/Erie Canal system, a function which Western European ports such as Rotterdam or Antwerp perform with their access to the Rhine system.

About 4,600 commercial ports are in operation worldwide, but fewer than one hundred ports have a global importance (Figure 4.6). There are about 500 container ports with 180 handling a traffic of more than half a million TEU. Maritime traffic thus has a high level of concentration in a limited number of large ports, a process mainly attributed to constraints related to maritime access and infrastructure development. Major ports have established themselves as gateways of continental distribution systems and have access to high capacity inland freight distribution corridors, notably rail. Such a position is very difficult to challenge unless a port is facing acute congestion forcing maritime shipping companies to seek alternatives.

Container ports are reflective of the world's commercial geography particularly since they predominantly handle finished goods and parts. The world's most important ports were North American (e.g. New York) and Western European (e.g. Rotterdam). Containerization completely changed the world's commercial geography with the emergence of port locations reflecting changes in the global geography of production and consumption. This geography indicates a high level of traffic concentration around

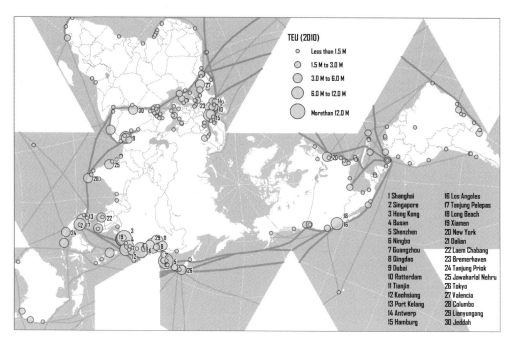

TEU (2010)

○ Less than 1.5 M
◎ 1.5 M to 3.0 M
◉ 3.0 M to 6.0 M
⬤ 6.0 M to 12.0 M
⬤ More than 12.0 M

1 Shanghai	16 Los Angeles
2 Singapore	17 Tanjung Pelepas
3 Hong Kong	18 Long Beach
4 Busan	19 Xiamen
5 Shenzhen	20 New York
6 Ningbo	21 Dalian
7 Guangzhou	22 Laem Chabang
8 Qingdao	23 Bremerhaven
9 Dubai	24 Tanjung Priok
10 Rotterdam	25 Jawaharlal Nehru
11 Tianjin	26 Tokyo
12 Kaohsiung	27 Valencia
13 Port Kelang	28 Columbo
14 Antwerp	29 Lianyungang
15 Hamburg	30 Jeddah

Figure 4.6 World's major container ports, 2010

large port facilities, notably Pacific Asian ports along the Tokyo–Singapore corridor. As export-oriented economic development strategies took shape, containers handled in Pacific Asian ports, notably Chinese ports, surged. The comparative size of ports requires caution as several ports can be considered more statistical agglomerations than functional entities. For instance, the port of Shenzhen in the Pearl River Delta is composed of several large port facilities (e.g. Yantian, Chiwan, Shekou) that act as distinct entities within their operations and are even servicing different hinterlands. The same observation applies to Guangzhou and Shanghai that are multiport (terminal) entities.

The world container port system is characterized by a high level of traffic concentration with the 20 largest container ports handling more than 49 percent of global traffic in 2010. There is also an emerging geography of container ports where there is a specialization between container ports acting as gateways and container ports acting as intermediate hubs. Gateway ports command the access of large manufacturing or market regions. Hong Kong, Los Angeles and Rotterdam are notable examples of ports that command access to a vast hinterland. Intermediate hub ports (or offshore hubs) act as intermediary locations where containers are transshipped between different segments of the global maritime transport system in a manner similar to hubs in air transportation. Singapore and Dubai are among the most prominent transshipment hubs, each servicing a specific transshipment market.

A port throughput is linked to a variety of local and regional industrial activities as the largest ports in the world are all gateways to massive industrial regions. However, comparing ports on a tonnage basis requires caution as it does not indicate the nature and the value of the cargo. For instance, a mineral port (e.g. iron ore), an energy port (e.g. coal or oil) and a commercial port (containers) could handle a similar tonnage but significantly different value levels.

Rail terminals

The use of the transport capacity offered by rail transportation requires purposely designed terminals where passengers can embark and disembark and where freight can be transferred. Rail terminals, while not quite as space-extensive as airports and ports, suffer less from site constraints. This involves two major issues:

Location. An important distinction concerns passengers and freight rail terminals, which commonly involve very different locations. Many rail terminals were established in the nineteenth century during the heyday of rail development. While the sites may have been on the edge of urban areas at the time, they now find themselves surrounded by urban development, leaving limited opportunities for expansion. Passenger terminals tend to occupy central locations and are commonly the defining element of urban centrality while freight terminals have seen a growing separation from central locations, with new facilities often built in an exurban location, particularly for high speed train stations.

Setting. Because of the linear characteristic of the mode they serve, rail terminals are predominantly rectangular-shaped facilities. Their capacity is a function of the number of track spurs available, which is a characteristic difficult to change once the terminal has been built. Individually rail terminals may not be as extensive as airports or ports, but cumulatively the area of all the rail sites in a city may exceed those of the other modes. For example, in Chicago the combined area of rail freight yards exceeds that of the airports.

Rail terminals have a unique characteristic related to shunting (or switching), which requires separate yard facilities often adjacent to the terminal. The wagons composing a train often need to be assembled or broken down in shunting (switching) yards. This is particularly the case for freight trains that need to be assembled at their origin, switched at intermediary locations (if long distance hauling is concerned) and broken down at their destination. While this is less of an issue for passenger rail as trains tend to remain assembled the way they are, shunting remains fundamental to rail operations.

Rail terminals have significant agglomeration effects that have had an impact in urban land markets since their introduction. This includes related activities such as retail, restaurants and hotels for passenger terminals or distribution centers for freight terminals. This is in part due to the accessibility they provide and in part because of the traffic they generate. Before the prominence of the automobile and trucking, economic activities tended to cluster around their respective rail terminals. Whole urban districts emerged around rail terminals. However, as the trucking industry matured and highway infrastructure was expanded and improved, rail terminals lost a great deal of their primacy. Even if rail transportation is generally more fuel-efficient than other modes, the mobility of passengers and freight quickly responded to the availability of the ubiquitous highway infrastructure.

In the second half of the twentieth century, as rail passenger traffic declined, the need for many of these stations diminished, particularly in North America. A rationalization has resulted in the conversion of many stations to other uses, sometimes with striking effects, such as the Musée D'Orsay in Paris and Windsor Station in Montreal. Rail yard conversion has been less spectacular, partly because the sites are less interesting from an architectural standpoint, but nonetheless important. Many former downtown freight facilities have been completely redeveloped in residential developments or commercially. In other cases, yards can be converted to related activities such as warehouses or even urban logistics centers. The current setting underlines an almost complete separation between rail passenger and freight terminals. Although they can share access to the same rail network they service completely different mobility requirements. Any proximity tends to be coincidental.

Passenger rail terminals tend to be functionally simple facilities and in their most basic form, they include a quay for passengers to embark or disembark and a common area for ticket purchase, waiting, and for activities servicing large volumes of passengers (e.g. retail and restoration). While some are along a line that requires a stop of a few minutes so that passengers can embark or disembark, others are terminal locations at the head of an intercity corridor. Like any other terminal facility, rail terminals have a size and complexity directly related to the amount of passengers they service. There is a hierarchy of the importance of passenger rail terminals which is illustrated in the rail network structure. It ranges from simple stops with only a platform available for passengers to embark or disembark to central rail stations composed of enclosed facilities with multiple piers and amenities.

Central railway stations are typically in the heart of downtown cores and primary elements of national or regional passenger rail systems. At one time their sites may have been on the edge of the pre-industrial city, as is the case for London and Paris, but today they are very much part of the central business district. The stations are typically imposing buildings reflecting the power and importance represented by the railway in the nineteenth and early twentieth centuries. For many cities, railway stations are the key elements of urban centrality and activity and represent an impressive architectural achievement unmatched in any other type of transportation terminal and occupying a large amount of real estate. Notable examples include the Grand Central Station in New York, St. Pancras station in London, the Gare de Lyon in Paris, or the Shinjuku train station in Tokyo (the world's busiest with more than 3.5 million people per day). Many central rail stations handle large numbers of commuters, partly because they also tend to be the nexus of public transit systems as subway stations are directly connected to the terminal facility. Even if in several cases, particularly in North America, the long distance function has subsided, the imprint of passenger rail terminals on the structure of urban transit systems has endured.

Still, the development of high speed rail systems has offered a new opportunity for rail terminals with the renovation of existing facilities, many of which are central railway stations, or the construction of new facilities in suburban areas. The centrality of rail stations became a positive factor in the development of high speed rail systems as it confers a direct accessibility to core business activities. In many cases the high speed rail station has become a new nexus of activity with co-located real estate development such as office buildings, retail stores, hotels and parking facilities. An additional level of integration concerns the design of airport terminals with high speed train stations, such as the case of Charles de Gaulle (Paris) and Schiphol (Amsterdam). This enables long distance air travel to connect with regional accessibility. Over specific corridors in France, Spain and Germany, high speed rail stations are effectively competing with airports.

Unlike passenger terminals, rail freight yards did not have to be quite so centrally located, and because they required a great deal of space for multiple tracks for marshaling they were more likely located on entirely greenfield sites than passenger terminals. However, rail yards tended to attract manufacturing activities able to use the distribution capabilities of rail, and thus became important industrial zones. When dealing with bulk commodities, rail terminals will locate in proximity of the source. They also vary in complexity because of the different freight markets they service (e.g. grain, coal, cars, containers) which require specialized loading/unloading facilities and equipment. Rail freight terminals perform four major functions:

Bulk. These rail terminals are linked with extractive industries such as agriculture, mining and wood products. For instance, grain elevators are bulk terminals commonly used to store, mix and load grain into railcars. Another important characteristic of bulk rail terminals is their unidirectional flows, implying that they are designed specifically to either load or unload bulk. Rail terminals doing both are uncommon. This is reflective of the nature of bulk trades.

Roll-on-roll-off. Used to transport vehicles such as cars, trucks or construction equipment where the vehicles are rolled in to a railcar using a ramp. Such terminals commonly require a large amount of parking space to store vehicles, particularly if they concern cars bound for retail outlets. Such terminals also serve as storage facilities supplying regional markets.

Intermodal. The function of loading and unloading unitized freight from railcars. Containerization has greatly expanded the intermodal productivity of rail terminals since it permits quick loading and unloading sequences, but at the expense of more trackside space being available. Depending on the type of operation, specific intermodal equipment will be used. An intermodal terminal can be part of a port facility (on-dock or near-dock facilities) or be a standalone inland terminal.

Shunting. The function of assembling, sorting and breaking of freight trains. Since trains can be composed of up to about 100 railcars (even more in North America), often of various natures, origins and destinations, shunting can be a complex task performed on several occasions. Comparatively, unit trains which carry the same commodity, such as coal, cars or containers, require less shunting. Bailey Yard in North Platte, Nebraska, operated by Union Pacific, is the largest classification yard in the world and handles 10,000 railcars per day.

The first forms of intermodal application to rail appeared in the late nineteenth century with practices dubbed "circus trains" because lorries were rolled in on flatcars using a ramp, a practice that was pioneered by circuses (Barnum in 1872). This simple ramp-based technique enabled many rail terminals to become "intermodal" by offering "piggy back" services. By the end of the twentieth century many of the industries around rail freight yards had relocated or disappeared, and in many cities these former industrial parks were targets of urban revitalization. At the same time, new intermodal practices emerged, notably lifting trailers or containers directly onto a flatcar. However, this required capital investments in intermodal equipment as well as paved terminal surfaces for storage. Only terminals with sufficient volume could be profitable. This has been accompanied by closure of some of the rail yards, either because they were too small for contemporary operating activities, or because of shrinkage of the local traffic base. In spite of a growth of intermodal traffic, the number of intermodal terminals declined.

In North America and Europe many older rail freight yards have been converted into intermodal facilities because of the burgeoning traffic involving containers and road trailers, a process which started in the 1960s. The ideal configuration for these terminals is different from the typical general freight facility with their need for multiple spurs to permit the assembling of wagons to form train blocks. The loading and unloading of wagons tended to be a manual process, often taking days, tying up terminal rail capacity.

Retrofitting conventional rail yards for contemporary intermodal operations proved problematic. Intermodal trains tend to serve a more limited number of cities and are more likely to be dedicated to one destination. They offer the notable advantage of being able to be quickly loaded or unloaded, thus tying up less terminal rail capacity. The need here is for long but fewer rail spurs. The configuration typically requires a site over three kilometers in length and over 100 hectares in area. In addition, good access to the highway system is a requisite as well as a degree of automation to handle the transshipment demands of modern intermodal rail operations.

One of the important growth factors of rail transportation has been its closer integration with maritime shipping. This is particularly the case at port terminals with new on-dock container rail facilities. The term "on-dock" can itself be misleading since a direct ship-to-rail transshipment actually does not take place (with the exception of the port of Montreal where containers can be directly unloaded from a ship to a railcar). A dray carries the container from alongside the ship to alongside the rail track (and vice versa), but frequently the containers are brought back and forth from a stack. Transloading, the practice of transferring loads between truck and rail transportation, has also experienced a remarkable growth in recent years. As long distance trucking is becoming increasingly expensive due to growing energy costs and congestion, many shippers see the advantages of using rail transportation to a location in the vicinity of their markets. At this location, freight loads are broken down into a Less than Truckload (LTL) and then shipped by short distance trucks to their final destinations.

Airports[1]

The rapid expansion of air passengers and air freight flows fostered by globalization has made everything about the world's great airports bigger. They are bigger in the volumes of traffic they handle, their sizes and the distances that separate them from the cities they serve, their costs and economic impacts, their environmental consequences, and the political controversies they engender. The very importance of airports globally has exacerbated the local conflicts they provoke. Indeed, a fundamental feature of airports is the degree to which they are embedded at several scales:

> **Regional/National/Global.** Airports are the articulation points of the circulatory system of the global economy. They mediate currents of people and goods. The importance of an airport in this regard is a function of its centrality and its intermediacy. The term centrality refers to a node's role as an origin and destination gateway to a surrounding region, and intermediacy refers to the degree to which a node serves as an interchange between different regions. The most important airports enjoy either centrality within one of the world's foremost city-regions, intermediacy among key markets, or both. The stretching of production linkages across the world has increased the importance of intermediacy on a global scale. For example, one factor propelling the growth of Dubai as an air transport hub is the fact that with ultra-long-range aircraft like the A340–500 any two locations on earth can be linked via a stop in Dubai.
>
> **Local.** Airports, especially large ones, are defining features of the communities in which they are set. A large airport generates thousands of jobs directly and thousands more via forward and backward linkages. Airports are not just features of a community's economic geography, however. An airport the size of Schiphol is a critically important source of noise pollution and other local environmental effects, a large

consumer of land, and a signature piece of the built environment. Indeed, the newest airport terminals feature extremely long roofs and are impressive architectural achievements.

There are three major concentrations of airports around which the world's air traffic is articulated: Eastern North America, Western Europe and Japan (Figure 4.7). The key airports of these platforms, or rather the main airport cities since they count more than one airport, are New York, London and Tokyo. They correspond to the world's most prominent cities and the most important financial centers. Yet, this supremacy is being challenged by new hubs of activity such as Beijing and Dubai. There is thus a direct relationship between the level of air passenger traffic and the primacy of a city in the world urban system. Large airport terminals also attract a substantial concentration of related activities such as distribution centers, just-in-time manufacturers, office parks, hotels, restaurants and convention centers.

The level of freight activity at airports tends to be different from that of passenger traffic, especially in the United States (Figure 4.8). The Midwest being the demographic and economic center of the United States, many air freight forwarders have located their hubs at airports such as Memphis (Federal Express) and Louisville (UPS) that generate little passenger traffic. The importance of Pacific Asian airports is linked with the specific role of the region in the global economy, especially for electronics. Since these products tend to have a high value-to-weight ratio, air transport is particularly suitable for their shipping to North American and Western European markets. Because long distance cargo planes have less range than passenger planes, two airports play a notable intermediate role, Anchorage (Pacific Asia–North American traffic) and Dubai (Pacific Asia–Western Europe traffic).

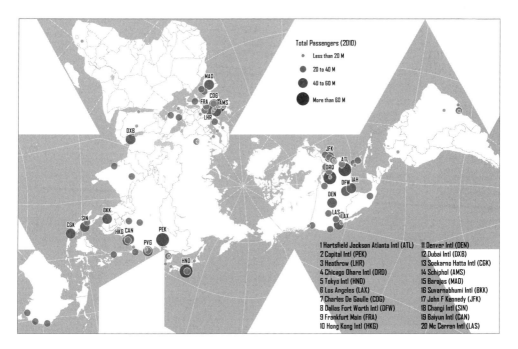

Figure 4.7   Passenger traffic at the world's largest airports, 2010

Source: Airport Council International.

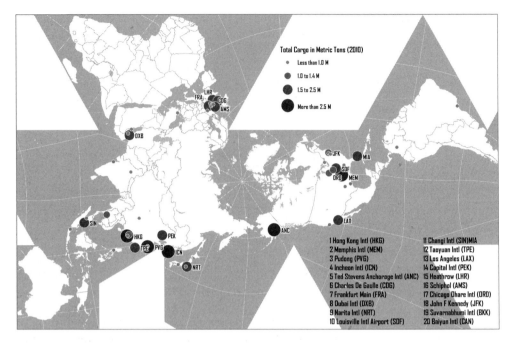

Figure 4.8 Freight traffic at the world's largest airports, 2010

Source: Airport Council International.

Airports require very large sites; they need space for runways, for terminal buildings, maintenance hangars and for parking. The runway remains one of the most vital elements of air transportation as it dictates the capacity of the system. While there are considerable variations in the scale of different airports, minimum sizes in excess of 500 hectares represent enormous commitments of urban land. Thus, airports are sited at the periphery of urban areas, because it is only there that sufficient quantities of land are available. Many airports built in the 1940s and 1950s on the periphery now find themselves surrounded by subsequent metropolitan development. New site development today, in North America and Europe at least, is becoming very difficult because available sites are frequently so far from the urban core that even if planning permission could be obtained, it would lead to very significant diseconomies because of the distance from business and demographic cores. It is significant that there have been few new large-scale airport developments in North America over the last 30 years. The result has been that most airports have to adjust to their existing sites, by reconfiguring runways and renovating existing terminal facilities.

Suburbanization in general is one big reason it has become more difficult to place major airports with each passing decade. Local site requirements are extremely important for air terminals for the two major components, the airfields and the terminals. The increasing physical size of airports and the difficulty of fitting in with neighboring land uses have encouraged the development of airports at increasingly remote locations. Indeed, the more recently an airport was constructed, the more likely it is to be located far from the center of the metropolitan area it services. In the most extreme cases, land has been reclaimed from the sea to make space for airports.

Asia is, in fact, home to several of the most extreme examples of airport "terra-forming". More distinctive than Kansai's roof, for instance, is its location on a man-made island in Japan's Inland Sea. The island, which was a prime contributor to the stratospheric cost of Kansai, is an extreme example of the lengths to which airport-builders have had to go to meet the spatial requirements of key hub airports. Overall, the four most expensive new airports in the world (Chek Lap Kok, Osaka–Kansai, Nagoya–Central Japan, and Seoul–Incheon) share three characteristics: their location in fast-growing Asia, proximity to densely populated metropolitan areas and their construction atop land reclaimed from the sea.

Concept 3 – Transport terminal governance

Authors: Jean-Paul Rodrigue, Brian Slack and Theo Notteboom

The nature of governance in transportation

Governance is associated with an effective usage of existing resources as well as a better allocation of new resources. Like all sectors of activity, transportation has a unique set of characteristics about its governance as both the public and private sectors are actively involved.

> **Governance** is the exercise of authority and institutional resources to manage activities in society and the economy. It concerns the public as well as the private sectors, but tends to apply differently depending if public or private interests are at stake. In both cases a significant concern is performance, which is how effectively available resources are used.

For transport infrastructure, such as port terminals, airports, inland ports or logistics zones, many different forms of governance are in place which shape modes of financing, operations, functioning and external relationships. This is particularly important as large terminal infrastructure involved in global flows of passengers and freight are complex, capital-intensive and of strategic importance to the economic welfare of whole regions. There are two main components of terminal governance: ownership and operations. Ownership involves who is the owner of the terminal site and facilities (including equipment):

> **Public ownership** is common because of the economic and strategic importance of many types of terminals. In several countries passenger railroads are owned by the national government, and the passenger stations are thus under the control of the state-owned railway company, such as is the case in China, Europe and North America. Public ownership of airports is also prevalent, although in the United States this takes place at the state or municipal levels of government. Under public ownership, investment in infrastructure and planning future expansion is carried out by the public authority using public monies or public guarantees for capital subscribed on private markets. The private sector is then offered leasing opportunities with terms and duration negotiated.
>
> **Private ownership** is less evident in transport terminals. There are numerous exceptions for certain modes, such as road freight (distribution centers), rail freight transport in North America (terminals and rights of way), and where privatization has taken place, as for example in ports and airports in the United Kingdom and New Zealand. Here, private capital is used to provide infrastructure.

Operations involve the day-to-day management and carrying out of terminal activities:

Public control of operations is typical in many ports, such as Singapore and Hampton Roads, in many state-controlled railroads such as China, and at publicly owned airports such as in the United States. Here the public authority provides the handling equipment, contracts with the labor force, and operates the rail, airport and port terminals.

Private companies manage and carry out operations in privately owned terminals. They are also active as operators in many publicly owned facilities under a concession agreement. The latter is a growing trend in ports and airports, where facilities are leased to terminal operators for fixed terms. The types of concession vary considerably, in terms of duration and conditions. Some are short term, a few years or so; more typically they are long-term concessions of 15 to 30 years. In some the owner provides some equipment, such as gantry cranes in ports, in others the concession holders are expected to invest in equipment. In some they are required to use public employees, while in others they may use their own workers. In Canada a halfway private/public system of governance of major airports and ports is in place. The airports and ports are leased to locally managed non-profit corporations that have to operate the facilities commercially, without access to public funds. Surpluses have to be reinvested.

Public ownership and operations have been important in many modes because of the strategic importance of transport and the long-term investments required that the private sector may be incapable or unwilling to make. In this way the terminals can be owned and operated as public goods, and can be integrated with public regional and national economic policies. On the other hand, public facilities are seen by some as slow to respond to market conditions, with a propensity to over-invest in non-economic developments, and with high costs to the users.

There is a growing tendency towards privatization in transport as a whole, particularly with deregulation. Transport terminals increasingly became an attractive form of investment for private equity firms seeking valuable assets and a return on their investments. This is manifested in the sale of ports and airports in some countries such as the UK, and in the break-up of state rail monopolies, as in the EU. Privatization is most evident, however, in the awarding of operational concessions to private companies. The trend towards concessions is warranted in part by the belief that the private sector is more efficient than the public in operating terminals, and that this form of governance keeps the ownership still under public control. It is also seen as a means of reducing public expenditures at a time when states are becoming less willing (or able) to make large investments. Thus, the setting of public–private partnerships is seen as a dominant trend in the governance of transport terminals.

Port devolution and global terminal operators

Even as late as the 1980s, ports around the world were the types of terminal most dominated by public ownership and operation. While the forms of port governance differed greatly, from the municipally owned ports in Northern Europe and the USA, to the state-owned ports in France, Italy and much of the developing world, public ownership was dominant and publicly managed port operations were prevalent. This contrasted with the shipping industry, where private ownership was almost universal. The development of containerization particularly underlined how operationally deficient public port authorities were to the growing time and performance requirements intermodalism imposed on transport chains.

These developments helped create what has become a global snowball of port reform, commonly known as port devolution since the public sector was relinquishing its role from a function it formerly assumed. It made governments around the world more open to considering reforming port governance and offering better conditions to ensure privatization. The growing demands for public and private investment in ports, precipitated by the growth in world trade, and the limited abilities of governments to meet these needs because of competing investment priorities, were key factors. Thus, while few were willing to go as far as the UK in the total privatization of ports, many countries were willing to consider awarding concessions as an intermediate form of privatization, leading to various forms of public–private partnerships. The result has been an almost global trend towards the award of port operational concessions, especially for container terminals.

If the opportunities to award operational concessions can be seen as an increase in demand, growth has also been greatly affected by an increase in the supply of companies seeking concessions. In Northern Europe and the USA many ports had already operated through concessions, awarded to local terminal handling companies. Because they were relatively small and locally based, with only few exceptions, they did not participate in the global growth of opportunities for concession awards. The exceptions were Stevedore Services of America (SSA), which was already active in several US West Coast ports and which obtained concessions to operate facilities in Panama and several other smaller ports in Central America, and Eurogate, a joint company formed by terminal handling companies from Bremen and Hamburg, which obtained concessions in Italy and Morocco.

Over a short period a few companies were able to become major global terminal operators controlling a multinational portfolio of terminal assets. They mostly came from Asia with four large companies dominating; three coming from a stevedore background and one from a shipping line (Figure 4.9):

- Hong Kong-based firm, **Hutchison Port Holdings** (HPH), part of a major conglomerate Hutchison Whampoa.
- **Port of Singapore Authority** (PSA), the government-owned operator of the port of Singapore.
- **Dubai Ports World** (DPW), mainly part of a sovereign wealth fund created to invest the wealth derived from oil trade.
- **AP Moller Terminals** (APM), a parent company of the world's largest shipping line, Maersk.

HPH, which originated as a terminal operator in Hong Kong, first purchased Felixestowe, the largest UK container port, and today has a portfolio of 51 terminals around the world, including in Rotterdam and Shanghai. PSA has been active securing concessions in China and Europe, including Antwerp. These two terminal operators take their origin from globally oriented ports offering limited local terminal expansion opportunities. The local operators were thus incentivized to manage the constrained assets efficiently and to look abroad for expansion opportunities. DPW has grown through purchases, such as P&O Ports and CSX World Terminals, and by securing concessions elsewhere.

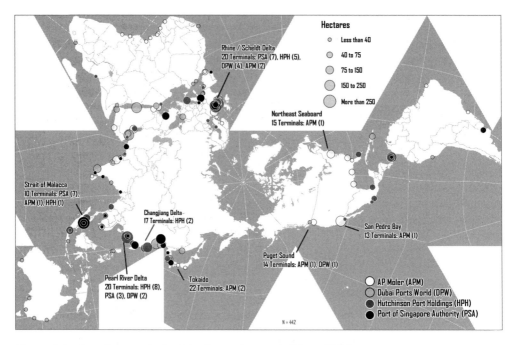

Figure 4.9 Container terminals of the four major port holdings, 2010

Shipping lines have also participated in terminal concessions, but to a lesser extent. The most important is the in-house terminal operating company of Maersk, APM Terminals. In addition, Evergreen, COSCO, MSC, NYK, and CMA-CGM hold port terminal leases. Between the dedicated terminal operating companies and the shipping lines, a global pattern of concessions is evident.

CASE STUDY Inland ports

Authors: Jean-Paul Rodrigue and Theo Notteboom

A new role for inland terminals

In many places around the world bimodal and trimodal inland terminals have become an intrinsic part of the transport system, particularly in gateway regions having a high reliance on trade. Transport development is gradually shifting inland after a phase that focused on the development of port terminals and maritime shipping networks. The complexity of modern freight distribution, the increased focus on intermodal and co-modal transport solutions and capacity issues appear to be the main drivers behind a renewed focus on hinterland logistics. While trucking tends to be sufficient in the initial phase of the development of inland freight distribution systems, at some level of activity, diminishing returns such as congestion, energy consumption and empty movements become strong incentives to consider the setting of inland terminals as the next step in regional freight planning. Also the massification of flows in networks, through a concentration of cargo on a limited set of ports of call and associated trunk lines to the hinterland, have created the right conditions for nodes to appear along and at the end of these trunk lines.

The evolution of inland freight distribution can be seen as a cycle in the ongoing developments of containerization and intermodal transportation. The geographical characteristics linked with modal availability, capacity and reliability of regional inland access have an important role to play in shaping this development. As maritime shipping networks and port terminal activities become better integrated, particularly through the symbiotic relationship between maritime shipping and port operations, the focus shifted on inland transportation and the inland terminal as a fundamental component of this strategy. Thus, after a phase that relied on the development of port terminals and maritime shipping networks, the integration of maritime and inland freight distribution systems has favored the setting of inland ports.

> **Inland port**. A rail or a barge terminal that is linked to a maritime terminal with regular inland transport services. An inland port has a level of integration with the maritime terminal and supports a more efficient access to the inland market both for inbound and outbound traffic. This implies an array of related logistical activities linked with the terminal, such as distribution centers, depots for containers and chassis, warehouses and logistical service providers.

Since the inland terminal is essentially an extension of some port activities inland, the term "dry port" has gained acceptance. However, using this term to define an inland terminal is subject to debate since many inland terminals are in fact "wet" given their direct access to inland waterway systems. Moreover, the inland location can effectively be a port if a barge service is concerned, but fundamentally cannot be considered a port if it involves a rail terminal or more simply truck depots. Thus, there seems to be no consensus on the terminology resulting in a wide range of terms including dry ports, inland terminals, inland ports, inland hubs, inland logistics centers, inland freight villages, etc. The reason for this lies in the multiple shapes, functions and network positions these nodes can have. A similar issue applies with the inclusion of airport terminals, mainly the freight component, as an element of an inland port. A whole array of transport terminal infrastructures is therefore often presented as a dry port. Regardless of the terminology used, three fundamental characteristics are related to an inland node:

- An **intermodal terminal**, either rail or barge that has been built or expanded.
- A **connection with a port terminal** through rail, barge or truck services, often through a high capacity corridor.
- An array of **logistical activities** that support and organize the freight transited, often co-located with the intermodal terminal.

The functional specialization of inland terminals has been linked with cluster formation of logistical activities. Inland terminals in many cases have witnessed a clustering of logistics sites in the vicinity, leading to a process of logistics polarization and the creation of logistics zones. They have become excellent locations for consolidating a range of ancillary activities and logistics companies. In recent years, the dynamics in logistics networks have created the right conditions for a large-scale development of such logistics zones.

Driving forces

Each inland port remains the outcome of the considerations of a transport geography pertaining to modal availability and efficiency, market function and intensity as well as the regulatory framework and governance. Their emergence underlines some deficiency in conventional inland freight distribution that needed to be mitigated. This mitigation includes:

Land value. Many deep sea terminal facilities have limited land available for expansion. This favors the intensification of activities at the main terminal and the search for lower value locations supporting less intensive freight activities.

Capacity and congestion. Capacity issues appear to be the main driver of inland port development since a system of inland terminals increases the intermodal capacity of inland freight distribution. While trucking tends to be sufficient in the initial phase of the development of inland freight distribution systems, at some level of activity, diminishing returns such as congestion, energy and empty movements become strong incentives to consider the setting of inland terminals as the next step in regional freight planning.

Hinterland access. Inland locations tend to be less serviced by intermodal transportation than coastal regions. Through long distance transport corridors, inland ports confer a higher level of accessibility because of lower distribution costs and improved capacity. These high capacity inland transport corridors allow ports to penetrate the local hinterland of competing ports and thus to extend their cargo base. In such a setting, the inland port becomes a commercial and trade development tool that jointly increase imports, exports and intermodal terminal use.

Supply chain management. In addition to standard capacity and accessibility issues in the hinterland, an inland port is a location actively integrated within supply chain management practices, particularly in view of containerization. This takes many forms such as the agglomeration of freight distribution centers, custom clearance, container depots and logistical capabilities. The inland terminal can also become a buffer in supply chains, acting as a temporary warehousing facility often closely connected to the warehouse planning systems of nearby distribution centers. Purchasers can even be advantaged by such a strategy since they are not paying for their orders until the container leaves the terminal, delaying settlement even if the inventory is nearby and available.

The growth of hinterland traffic, if only supported by drayage (truck) operations, leads to increasing diseconomies such as congestion. The level of disorder in the transport system involves higher transport costs and unreliability in freight distribution. The formation of an inland load center network therefore aims at coping with these diseconomies through a massification of several inland flows (Figure 4.10). It involves a series of inland terminals (IT) linked to the port facilities by high capacity rail or barge corridors. Also, a supporting land use structure needs to be established, mostly concerning the clustering of logistics activities (e.g. distribution centers) and often in co-location with the terminal facilities.

For instance, **port-centric logistics zones** support freight distribution activities related to maritime shipping and have a dominant international trade orientation. Port authorities tend to be proactive in this type of development since it supports and provides added value to port activities. An **inland port** is an intermodal terminal (commonly rail) built or updated concomitantly with the development of adjacent (co-located) logistical and service activities. An **intermodal industrial park** is a similar structure although in proximity (not co-located) to the terminal facility.

The geographical characteristics linked with modal availability and the capacity of regional inland access have an important role to play in shaping the emergence and

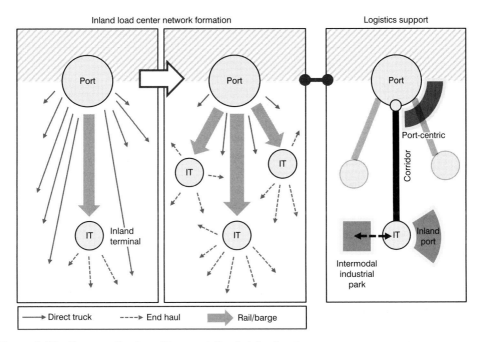

Figure 4.10 The massification of transportation in inland systems

Source: Rodrigue, J-P. and T. Notteboom (2011) "Port regionalization: improving port competitiveness by reaching beyond the port perimeter", *Port Technology International*, 52: 11–17.

development of inland ports. Each inland market has its own potential requiring different transport services. Thus, there is no single strategy for an inland port in terms of modal preferences as the regional effect remains fundamental. In developed countries, namely North America and Europe, which tended to be at the receiving end of many container-ized supply chains, a number of inland ports have been developed with a focus on inbound logistics (Photo 4.3).

The setting of global supply chains and the strategy of Pacific Asian countries around the export-oriented paradigm have been powerful forces shaping contemporary freight distribution. Indirectly, this has forced players in the freight transport industry (shipping companies, terminal operators, logistics providers) to examine supply chains as a whole and to identify legs where capacity and reliability were an issue. Once maritime ship-ping networks and port terminal activities had been better integrated, particularly through the symbiotic relationship between maritime shipping and port operations, inland transportation became the obvious focus and the inland terminal a fundamental component of this strategy. This initially took place in developed countries, namely North America and Europe, which tended to be at the receiving end of many container-ized supply chains. The focus has also shifted to considering inland terminals for the early stages of global supply chains (outbound logistics), namely in countries having a marked export-oriented function.

Inland terminals have evolved from simple intermodal locations to their incorpora-tion within logistics zones. Inland terminals (particularly rail) have always been present since they are locations from which specific market coverage is achieved. Containeriza-tion has impacted this coverage through the selection of terminals that were servicing a

Photo 4.3   Terminal Intermodal Logística de Hidalgo

TILH is an inland port facility owned and operated by HPH (80 percent stake), which opened in 2012. It is located in the southern part of the state of Hidalgo about 50 km north of Mexico City, one of the world's largest metropolitan areas with a population of more than 21 million people.

wider market area. This spatial change also came with a functional change as intermodal terminals began to experience a specialization of roles based on their geographical location but also based on their "location" within supply chains.

Note

1 John Bowen, Department of Geography, Central Washington University, contributed to this section.

Bibliography

Bird, J.H. (1963) *The Major Seaports of the United Kingdom*, London: Hutchison.
—— (1971) *Seaports and Seaport Terminals*, London: Hutchison.
Bowen, J. (2010) *The Economic Geography of Air Transportation: Space, Time, and the Freedom of the Sky*, London: Routledge.
Caves, R.E. and G.D. Gosling (1999) *Strategic Airport Planning*, Oxford: Pergamon.
Charlier, J. (1992) "The regeneration of old port areas for new port uses", in B.S. Hoyle and D. Hilling (eds) *Seaport Systems and Spatial Change*, Chichester: Wiley, pp. 137–54.
De Langen, P.W. (2004) "Analysing seaport cluster performance", in D. Pinder and B. Slack (eds) *Shipping and Ports in the Twenty-first Century*, London: Routledge, pp. 82–98.
Dempsey, P.S., A.R. Goetz and J.S. Szyliowicz (1997) *Denver International Airport: Lessons Learned*, New York: McGraw-Hill.

Ducruet, C. and S.W. Lee (2006) "Frontline soldiers of globalization: port-city evolution and regional competition", *GeoJournal*, 67(2): 107–22.

Federal Aviation Administration (2007) *Capacity Needs in the National Airspace System, 2007–25*, Washington, DC: FAA.

Fleming, D.K. and Y. Hayuth (1994) "Spatial characteristics of transportation hubs: centrality and intermediacy", *Journal of Transport Geography*, 2(1): 3–18.

Fremont, A. (2007) "Global maritime networks: the case of Maersk", *Journal of Transport Geography*, 15(6): 431–42.

Fuller, G. and R. Harley (2004) *Aviopolis: A Book about Airports*, London: Black Dog Publishing.

Goetz, A.R. and J-P. Rodrigue (1999) "Transport terminals: new perspectives", *Journal of Transport Geography*, 7: 237–40.

Graham, B. (1995) *Geography and Air Transport*, Chichester: Wiley.

Haezendonck, E. (2001) *Essays on Strategy Analysis for Seaports*, Leuven: Garant.

Hakfoort, J., T. Poot and P. Rietveld (2001) "The regional economic impact of an airport: The case of Amsterdam Schiphol Airport", *Regional Studies*, 35(7): 595–604.

Hayuth, Y. (1988) "Rationalization and deconcentration of the U.S. container port system", *Professional Geographer*, 40: 279–88.

Hoyle, B.S. (1967) "East African seaports: an application of the concept of Anyport", *Transactions of the Institute of British Geographers*, pp. 163–83.

—— (1988) "Development dynamics at the port-city interface", in B.S. Hoyle, D.A. Pinder and M.S. Husain (eds) *Revitalising the Waterfront*, Chichester: Wiley.

—— (1989) "The port-city interface: trends problems and examples", *Geoforum*, 20: 429–35.

Kasarda, J.D. and G. Lindsay (2011) *Aerotropolis: The Way We'll Live Next*, New York: Farrar, Straus & Giroux.

McCalla, R.J. (2004) "From 'Anyport' to 'Superterminal'", in D. Pinder and B. Slack (eds) *Shipping and Ports in the Twenty-first Century*, London: Routledge, pp. 123–42.

—— (2008) "Site and situation factors in transshipment ports: the case of the Caribbean Basin", *Tijdschrift voor Economische en Sociale Geografie*, 99(4): 440–53.

McCalla, R.J., B. Slack and C. Comtois (2001) "Intermodal freight terminals: locality and industrial linkages", *Canadian Geographer*, 45(3): 404–13.

Notteboom, T. and J-P. Rodrigue (2005) "Port regionalization: towards a new phase in port development", *Maritime Policy and Management*, 32(3): 297–313.

Olivier, D. and B. Slack (2006) "Rethinking the port", *Environment and Planning A*, 38: 1409–27.

Pearman, H. (2004) *Airports: A Century of Architecture*, New York: Harry N. Abrams.

Pinder, D. and B. Slack (eds) (2004) *Shipping and Ports in the Twenty-first Century: Globalisation, Technological Change and the Environment*, London: Routledge.

Robinson, R. (2002) "Ports as elements in value-driven chain systems: the new paradigm", *Maritime Policy and Management*, 29(3): 241–55.

Rodrigue, J-P. (2008) "The thruport concept and transmodal rail freight distribution in North America", *Journal of Transport Geography*, 16: 233–46.

Roso, V. and K. Lumsden (2010) "A review of dry ports", *Maritime Economics and Logistics*, 12(2): 196–213.

Slack, B. and A. Fremont (2005) "Transformation of port terminal operations: from the local to the global", *Transport Reviews*, 25(1): 117–30.

Slack, B. (1994) "Pawns in the game: ports in a global transport system", *Growth and Change*, 24: 597–8.

—— (1999) "Satellite terminals: a local solution to hub congestion?", *Journal of Transport Geography*, 7: 241–6.

5 International trade and freight distribution

Globalization, trade and freight transportation are interrelated and concern a mobility scale that spans nations and often continents. This transnational mobility is subject to many geopolitical considerations, such as who controls trade routes and what forms of competition and cooperation have emerged with expanded trade relations. Processes related to economic integration, the fragmentation of production systems due to outsourcing and offshoring are interdependent and have favored to setting of global commodity chains, from the extraction of raw materials, manufacturing, to final consumption. This requires an understanding of logistics and the growing level of integration between production, distribution and consumption.

Concept 1 – Transportation, globalization and international trade

The flows of globalization

In a global economy, no nation is self-sufficient. Each is involved at different levels in trade to sell what it produces, to acquire what it lacks and also to produce more efficiently in some economic sectors than its trade partners. As supported by conventional economic theory, trade promotes economic efficiency by providing a wider variety of goods, often at lower costs, notably because of specialization, economies of scale and the related comparative advantages. International trade is also subject to much contention since it can at times be a disruptive economic and social force as it changes how wealth is distributed within a national economy, particularly due to changes in prices and wages.

The globalization of production is concomitant with the globalization of trade as one cannot function without the other. Even though international trade took place centuries before the modern era, as ancient trade routes such as the Silk Road can testify, trade occurred at an ever increasing scale over the last 600 years to play an even more active part in the economic life of nations and regions. This process has been facilitated by significant technical changes in the transport sector. The scale, volume and efficiency of international trade have all continued to increase since the 1970s. As such, space/time convergence was an ongoing process implying a more extensive market coverage that could be accessed in less time. It has become increasingly possible to trade between parts of the world that previously had limited access to international transportation systems. Further, the division and the fragmentation of production that went along with these processes also expanded trade. Trade thus contributes to lower manufacturing costs (Figure 5.1).

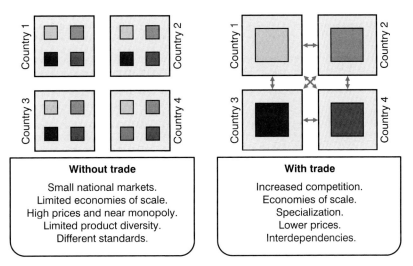

Figure 5.1 Economic rationale of trade

Without international trade, few nations could maintain an adequate standard of living. With only domestic resources being available, each country could only produce a limited number of products and shortages would be prevalent. Global trade allows for an enormous variety of resources – from Persian Gulf oil, Brazilian coffee to Chinese labor – to be made more widely accessible. It also facilitates the distribution of a wide range of manufactured goods that are produced in different parts of the world to what can be labeled as the global market. Wealth becomes increasingly derived through the regional specialization of economic activities. This way, production costs are lowered, productivity rises and surpluses are generated, which can be transferred or traded for commodities that would be too expensive to produce domestically or would simply not be available. As a result, international trade decreases the overall costs of production worldwide. Consumers can buy more goods from the wages they earn, and standards of living should, in theory, increase.

International trade consequently demonstrates the extent of globalization with increased spatial interdependencies between elements of the global economy and their level of integration. These interdependencies include numerous relationships where flows of capital, goods, raw materials and services are established between regions of the world. At the beginning of the twenty-first century, the flows of globalization have been shaped by four salient trends (Figure 5.2):

- An ongoing **growth of international trade**, both in absolute terms and in relation to global national income. From 1970 to 2010 the value of exports grew by a factor of 48 times if measured in current dollars, while GDP increased 22 times and population increased 1.8 times.
- A substantial level of **containerization** of commercial flows. Containerization tends to grow at a rate faster than trade and GDP growth.
- A higher relative **growth of trade in Pacific Asia** as many economies developed an export-oriented development strategy that has been associated with imbalances in commercial relations.

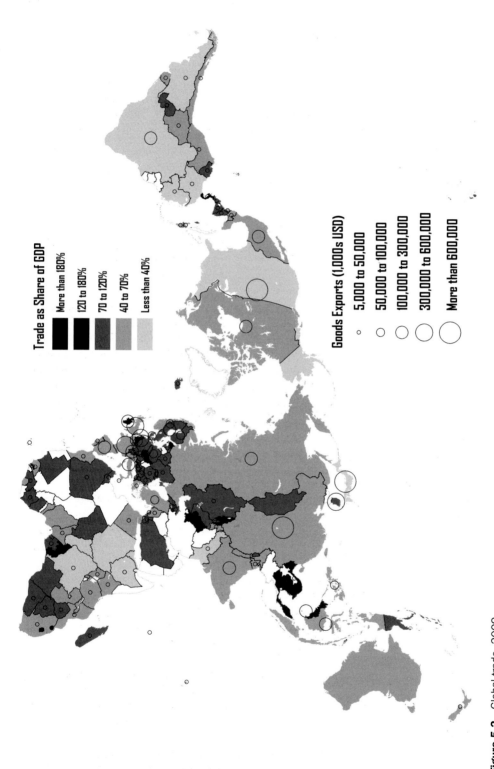

Figure 5.2 Global trade, 2009

Source: World Trade Organization.

- The growing role of **multinational corporations** as vectors for international trade, particularly in terms of the share of international trade taking place within corporations and the high level of concentration of their head offices.

Geographically, global trade takes place around three major poles: North America, Europe and Pacific Asia (the Japan, Korea and China triad). Trade volumes are not necessarily correlated with trade as a share of GDP. Some countries, like the United States, have significant trade volumes, but this volume accounts for a small share of the national economic activity (about 25 percent). Inversely, exporting countries, such as the Netherlands, China and Canada, have a higher dependence on international trade. Countries having a very high share of trade in relation to GPD (above 100 percent) are usually focusing on natural resources exports (for example the Middle East, Africa).

Trade facilitation

The volume of exchanged goods and services between nations is taking a growing share of the generation of wealth, mainly by offering economic growth opportunities in new regions and by reducing the costs of a wide array of manufacturing goods. By 2007, international trade surpassed for the first time 50 percent of global GDP, a twofold increase in its share since 1950. The facilitation of trade involves how the procedures regulating the international movements of goods can be improved. It depends on the reduction of the general costs of trade, which considers transaction, tariff, transport and time costs, often labeled as the "Four Ts" of international trade. United Nations estimates have underlined that for developing countries a 10 percent reduction in transportation cost could be accompanied with a growth of about 20 percent in international and domestic trade. Thus, the ability to compete in a global economy is dependent on the transport system as well as a trade facilitation framework with activities, including:

- **Distribution-based**. A multimodal and intermodal freight transport system composed of modes, infrastructures and terminals that spans across the globe. It ensures a physical capacity to support trade and its underlying supply chains.
- **Regulation-based**. Customs procedures, tariffs, regulations and handling of documentation. They ensure that trade flows abide by the rules and regulations of the jurisdictions they cross. Cross-border clearance, particularly in developing countries, can be a notable trade impediment with border delays, bottlenecks and long customer clearance times.
- **Transaction-based**. Banking, finance, legal and insurance activities where accounts can be settled and risk mitigated. They ensure that the sellers of goods and services are receiving an agreed-upon compensation and that the purchasers have a legal recourse if the outcome of the transaction is judged unsatisfactory or are insured if a partial or full loss incurs.

The quality, cost and efficiency of these services influence the trading environment as well as the overall costs linked with the international trade of goods. Many factors have been conducive to trade facilitation in recent decades, including integration processes, standardization, production systems, transport efficiency and transactional efficiency:

Integration processes, such as the emergence of economic blocks and the decrease of tariffs at a global scale through agreements, promoted trade as regulatory regimes were harmonized. One straightforward measure of integration relates to custom delays, which can be a significant trade impediment since they add uncertainty in supply chain management. The higher the level of economic integration, the more likely the concerned elements are to trade. International trade has consequently been facilitated by a set of factors linked with growing levels of economic integration, the outcome of processes such as the European Union or the North American Free Trade Agreement. The transactional capacity is consequently facilitated with the development of transportation networks and the adjustment of trade flows that follow increased integration. Integration processes have also taken place at the local scale with the creation of free trade zones where an area is given a different governance structure in order to promote trade, particularly export-oriented activities. In this case, the integration process is not uniform as only a portion of a territory is involved. China is a salient example of the far-reaching impacts of the setting of special economic zones operating under a different regulatory regime.

Standardization concerns the setting of a common and ubiquitous frame of reference over information and physical flows. Standards facilitate trade since those abiding by them benefit from reliable, interoperable and compatible goods and services which often results in lower production, distribution and maintenance costs. Measurement units were among the first globally accepted standards (metric system) and the development of information technologies eventually led to common operating and telecommunication systems. It is, however, the container that is considered to be the most significant international standard for trade facilitation. By offering a load unit that can be handled by any mode and terminal with the proper equipment, access to international trade is improved.

Production systems are more flexible and embedded (see concept 3). It is effectively productive to maintain a network of geographically diversified inputs, which favors exchanges of commodities, parts and services. Information technologies have played a role by facilitating transactions and the management of complex business operations. Foreign direct investments are commonly linked with the globalization of production as corporations invest abroad in search of lower production costs and new markets. China is a leading example of such a process, which went on a par with a growing availability of goods and services that can be traded on the global market.

Transport efficiency has increased significantly because of innovations and improvements in the modes and infrastructures in terms of their capacity and throughput. Ports are particularly important in such a context since they are gateways to international trade through maritime shipping networks. As a result, the transferability of commodities, parts and finished goods has improved. Decreasing transport costs does more than increasing trade; it can also help change the location of economic activities. Yet, transborder transportation issues remain to be better addressed in terms of capacity, efficiency and security.

Transactional efficiency. The financial sector also played a significant role in integrating global trade, namely by providing investment capital and credit for international commercial transactions. For instance, a letter of credit may be issued based upon an export contract. An exporter can thus receive a payment guarantee from a bank until its customer finalizes the transaction upon delivery. This is particularly important since the delivery of international trade transactions can take several weeks due to the long distances involved. During the transfer, it is also common that the cargo is insured in the event of damage, theft or delays, a function supported by insurance companies. Also, global financial systems enable currency conversion according to exchange rates that are commonly set by market forces, although some currencies, such as the Chinese Yuan, are set by policy. Monetary policy can thus be a tool, albeit contentious, used to influence trade.

Global trade flows

Global trade flows have recently shifted with many developing countries having a growing participation in international trade. The nature of what can be considered international trade has also changed, particularly with the emergence of global value chains and the trade of intermediary goods they involve. This trend obviously reflects the strategies of multinational corporations positioning their manufacturing assets in order to lower costs and maximize new market opportunities, while maintaining the cohesion of their supply chains and the freight distribution systems supporting them. In addition, another emerging trade flow concerns the imports of resources to developing countries, namely energy, commodities and agricultural products, which diverge from the conventional role of developing countries as exporters of resources.

The dominant factor behind the growth in international trade has been an increasing share of manufacturing activities taking place in developing countries with manufacturers seeking low cost locations for many stages of the supply chain. The evolution of international trade thus has a concordance with the evolution of production. There are, however, significant fluctuations in international trade that are linked with economic cycles of growth and recession, fluctuations in the price of raw materials, as well as disruptive geopolitical and financial events. The international division of production has been accompanied by growing flows of manufactured goods, which take a growing share of international trade. There are relatively fewer bulk liquids (such as oil) and more dry bulk goods and general cargo being traded.

The geography of international trade still reveals the dominance of a small number of countries, mainly in North America, Europe and Asia, which are commonly referred to as the triad. Alone, the United States, Germany and Japan account for about a quarter of all global trade, with this supremacy being seriously challenged by emerging economies. Further, G7 countries account for half of the global trade, a dominance which has endured for over 100 years. A growing share is being accounted for by the developing countries of Asia, with China accounting for the most significant growth both in absolute and relative terms. Those geographical and economic changes are also reflected over transoceanic trade with transpacific trade growing faster than transatlantic trade.

Neo-mercantilism is reflective of global trade flows as several countries have been actively pursuing export-oriented economic development policies using infrastructure development, subsidies and exchange rates as tools. This strategy has been followed by developing economies and resulted in growing physical and capital flow imbalances in international trade. This is particularly reflective in the American container trade structure, which is highly imbalanced and having acute differences in the composition of imports and exports. Still, these imbalances must be looked at with caution as products are composed of parts manufactured in several countries with assembly often taking place in a low cost location. In international trade statistics, this location assumes the full value of finished goods imported elsewhere while it may have only contributed to a small share of the total added value. Electronic devices are illustrative of this issue.

Regionalization has been one of the dominant features of global trade as the bulk of international trade has a regional connotation, promoted by proximity and the establishment of economic blocs such as the North American Free Trade Agreement (NAFTA) and the European Union (EU). The closer economic entities are, the more likely they are to trade, which explains that the most intense trade relations are within Western Europe and North America. A similar but more recent trend has also emerged in Asia, particularly between Japan, China, Korea and Taiwan.

International transportation

The growth of the amount of freight being traded as well as a great variety of origins and destinations promotes the importance of international transportation as a fundamental element supporting the global economy. Economic development in Pacific Asia and in China in particular has been the dominant factor behind the growth of international transportation in recent years. Since the trading distances involved are often considerable, this has resulted in increasing demands on the maritime shipping industry and on port activities. As its industrial and manufacturing activities develop, China is importing growing quantities of raw materials and energy and exporting growing quantities of manufactured goods. The ports in the Pearl River Delta in Guangdong province now handle almost as many containers as all the ports in the United States combined.

International transportation systems have been under increasing pressure to support additional demands in freight volume and the distance at which this freight is being carried. This could not have occurred without considerable technical improvements enabling larger quantities of passengers and freight to be transported, and this more quickly and more efficiently. Few other technical improvements than containerization have contributed to this environment of growing mobility of freight. Since containers and their intermodal transport systems improve the efficiency of global distribution, a growing share of general cargo moving globally is containerized.

Consequently, transportation is often referred to as an enabling factor that is not necessarily the cause of international trade, but without its capacity and efficiency globalization could not have occurred. A common development problem is the inability of international transportation infrastructures to support flows, undermining access to the global market and the benefits that can be derived from international trade. International trade also requires distribution infrastructures that can support trade between several partners. Three components of international transportation that facilitate trade:

- **Transportation infrastructure**. Concerns physical infrastructures such as terminals, vehicles and networks. Efficiencies or deficiencies in transport infrastructures will either promote or inhibit international trade.
- **Transportation services**. Concerns the complex set of services involved in the international circulation of passengers and freight. It includes activities such as distribution, logistics, finance, insurance and marketing.
- **Transactional environment**. Concerns the complex legal, political, financial and cultural setting in which international transport systems operate. It includes aspects such as exchange rates, regulations, quotas and tariffs, but also consumer preferences.

About half of all global trade takes place between locations of more than 3,000 km apart. Because of this geography, most international freight movements involve several modes since it is impossible to have a physical continuity in freight flows. Transport chains must thus be established to service these flows which reinforce the importance of intermodal transportation modes and terminals at strategic locations (Figure 5.3).

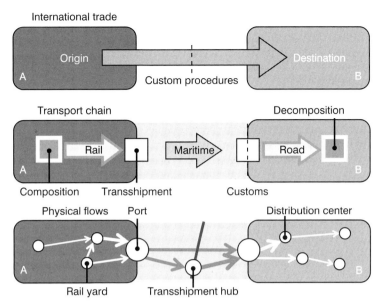

Figure 5.3 International trade, transportation chains and flows

Source: Rodrigue, J-P. (2012) "Supply chain management, logistics changes and the concept of friction", in P.V. Hall and M. Hesse (eds) *Cities, Regions and Flows*, London: Routledge.

The physical realization of international trade requires a transport chain. This is a series of logistical activities that organize modes and terminals, such as railway, maritime and road transportation systems, and thus continuity along the supply chain through a set of stages along an intermodal transport chain (see Figure 3.13). In the operational reality of modes and terminals, international trade is a series of physical flows that may not necessarily use the most direct path, but the path of least cost. The existence of inland corridors where economies of scale are more effective shapes the structure of freight flows as well as the selection of the port of exit. On the maritime side, transshipments hubs have become strategic intermediary locations helping consolidate maritime flows and connecting different maritime systems of circulation. In such a setting, the container has become the fundamental element facilitating transfers between modes and supporting international trade flows. Distribution centers play an important role in physical flows since they can act as a buffer, helping reconcile the temporal and spatial requirements of demand.

Among the numerous transport modes, two are specifically concerned with international trade:

Ports and maritime shipping. The importance of maritime transportation in global freight trade is unmistakable, particularly in terms of tonnage, as it handles about 90 percent of the global trade. Thus, globalization is the realm of maritime shipping, with containerized shipping at the forefront of the process. The global maritime transport system is composed of a series of major gateways granting access to major production and consumption regions. Between those gateways are major hubs acting as points of interconnection and transshipment between systems of maritime circulation.

Airports and air transport. Although in terms of tonnage air transportation carries an insignificant amount of freight (0.2 percent of total tonnage) compared with maritime transportation, its importance in terms of the total value is much more significant: 15 percent of the value of global trade. International air freight is about 70 times more valuable than its maritime counterpart and about 30 times more valuable than freight carried overland, which is linked with the types of goods it transports (e.g. electronics). The location of freight airports corresponds to high technology manufacturing clusters as well as intermediary locations where freight planes are refueled and/or cargo is transshipped.

Road and railway modes tend to occupy a more marginal portion of international transportation since they are above all modes for national or regional transport services. Their importance is focused on their role in the "first and last miles" of global distribution. Freight is mainly brought to port and airport terminals by trucking or rail. There are, however, notable exceptions in the role of overland transportation in international trade. A substantial share of the NAFTA trade between Canada, the United States and Mexico is supported by trucking, as well as a large share of the Western European trade. In spite of this, these exchanges are a priori regional by definition, although intermodal transportation confers a more complex setting in the interpretation of these flows.

Still, many challenges are impacting future developments in international trade and transportation, mostly in terms of demographic, energy and environmental issues. While the global population and its derived demand will continue to grow and reach around 9 billion by 2050, the aging of the population, particularly in developed countries, will transform consumption patterns as a growing share of the population shifts from wealth producing (working and saving) to wealth consuming (selling saved assets). The demographic dividend in terms of peak share of working age population that many countries benefited from, particularly China, will recede. As both maritime and air freight transportation depend on petroleum, the expected scarcity of this fossil fuel will impose a rationalization of international trade and its underlying supply chains. Environmental issues have also become more salient with the growing tendency of the public sector to regulate components of international transportation that are judged to have negative externalities. Also, international trade enables several countries to mask their energy consumption and pollutant emissions by importing goods that are produced elsewhere and where environmental externalities are generated. Thus international trade has permitted a shift in the international division of production, but also a division between the generation of environmental externalities and the consumption of the goods related to these externalities.

Concept 2 – Commodity chains and freight transportation

Contemporary production systems

Production and consumption are the two core components of economic systems and are both interrelated through the conventional supply/demand relationship. Basic economic theory underlines that what is being consumed has to be produced and what is being produced has to be consumed. Any disequilibrium between the quantity being produced and the quantity being consumed can be considered as a market failure. On one side, insufficient production involves shortages and price increases, while on the other, over-production and overcapacity involves waste, storage and price reductions. It is mainly through the corporation and its perception of market potential that a set of decisions are made about how to allocate scarce resources, reconciling production and consumption. The realization of production and consumption cannot occur without flows of freight within a complex system of distribution that includes modes and terminals, and also facilities managing freight activities, namely distribution centers.

Contemporary production systems are the outcome of significant changes in production factors, distribution and industrial linkages:

Production factors. In the past, the three dominant factors of production – land, labor and capital – could not be effectively used at the global level. For instance, a corporation located in one country had difficulties taking advantage of cheaper inputs (e.g. labor and land) in another country, notably because regulations would not permit full (and often dominant) ownership of a manufacturing facility by foreign interests. Such difficulties have been overcome by economic integration and trade agreements. The European Union established a structure that facilitates the mobility of production factors, which in turn enabled a better use of the comparative productivity of the European territory. Similar processes are occurring in North America (NAFTA), South America (Mercosur) and in Pacific Asia (ASEAN) with various degrees of success. Facing integration processes and massive movements of capital coordinated by global financial centers, factors of production have an extended mobility, which can be global in some instances. To reduce their production costs, especially labor costs, many firms have relocated segments (sometimes the entire process) of their manufacturing activities to new locations.

Distribution. In the past, the difficulties of overcoming distances were related to constraints in physical distribution as well as to telecommunications. Distribution systems had limited capabilities to ship merchandise between different parts of the world and it was difficult to manage fragmented production systems due to inefficient communication systems. In such a situation, freight alone could cross borders, while capital flows, especially investment capital, had more limited ranges. The tendency was to trade finished goods. Trade could be international, but production systems were predominantly regionally focused and mainly built through regional agglomeration economies with industrial complexes as an outcome. With improvements in transportation and logistics, the efficiency of distribution has reached a point where it is possible to manage large-scale production and consumption.

Industrial linkages. In the past, the majority of relationships between elements of the production system took place between autonomous entities, which tended to be smaller in size. As such, those linkages tended to be rather uncoordinated. The emergence of multinational corporations underlines a higher level of linkages within production systems, as many activities that previously took place over several entities now occur within the same corporate entity. While in the 1950s, the share of the global economic output attributable to multinational corporations was in the 2 to 4 percent range, by the early twenty-first century this share had surged to between 25 and 50 percent. About 30 percent of all global trade occurs within elements of the same corporation, with this share climbing to 50 percent for trade between advanced countries.

The development of global transportation and telecommunication networks, ubiquitous information technologies, the liberalization of trade and multinational corporations are all factors that have substantially impacted production systems. Products are getting

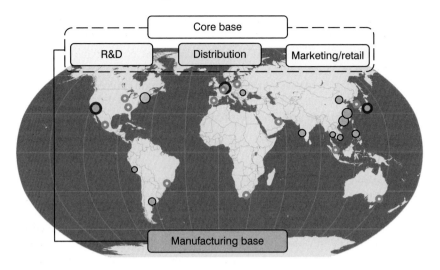

Figure 5.4 Disconnection of global production and distribution

increasingly sophisticated requiring a vast array of skills for their fabrication. One key issue is the array of expansion strategies available in a global economy, including horizontal and vertical integration, as well as outsourcing. In many cases, so-called "platform companies" have become new paradigms where the function of manufacturing has been removed from the core of corporative activities (Figure 5.4). Corporations following this strategy, particularly mass retailers, have been active in taking advantage of the "China effect" in a number of manufacturing activities.

A growing disconnection is being observed between the manufacturing base and what can be called the "core base", which mainly includes research and development (R&D), distribution and marketing/retail. The term platform corporation has been used to describe a variety of multinational corporations that have removed the manufacturing component from their core activities, or never had manufacturing in the first place. They have done so by focusing on activities that provide the most added value and outsourced or offshored the manufacturing of the products they design. Their core activities include R&D, finance, marketing, retail and distribution. Many of them own globally recognized brand names and are actively involved in the development of new products.

The net worth of a platform corporation is thus more a function of their brand names and capacity at innovation than from tangible assets (like factories), outside those heavily involved in mass retailing where commercial real estate assets can be very significant. Low margin work is usually outsourced, leading to a very flexible supplier base. This is reminiscent of the cottage production system that took place in the early phases of the industrial revolution when many labor-intensive activities (especially in garments) were subcontracted to households looking for additional income.

Commodity chains

Commodities are resources that can be consumed. They can be accumulated for a period of time (some are perishable while others can be virtually stored for centuries), exchanged as part of transactions or purchased on specific markets (such as the futures market). Some commodities are fixed, implying that they cannot be transferred, except for the title. This includes land, mining, logging and fishing rights. In this context, the value of a fixed commodity is derived from the utility and the potential rate of extraction. Bulk commodities are commodities that can be transferred, and includes for instance grains, metals, livestock, oil, cotton, coffee, sugar and cocoa. Their value is derived from utility, supply and demand, which is established through major commodity markets involving a constant price discovery mechanism.

The global economy and its production systems are highly integrated, interdependent and linked through commodity chains.

> **Commodity Chain (also known as value chain).** A functionally integrated network of production, trade and service activities that covers all the stages in a supply chain, from the transformation of raw materials, through intermediate manufacturing stages, to the delivery of a finished good to a market. The chain is conceptualized as a series of nodes, linked by various types of transactions, such as sales and intrafirm transfers. Each successive node within a commodity chain involves the acquisition or organization of inputs for the purpose of added value.

There are several stages through which a multinational corporation (or a group of corporations in partnership) can articulate its commodity chain (Figure 5.5). These stages are in large part conditioned by the location and availability of raw materials, production costs and the location of main consumption markets. Commodity chains are also integrated by a transport chain routing goods, parts and raw materials from extraction and transformation sites to markets. Obviously, the nature of what is being produced and the markets where it is consumed will correspond to a unique geography of flows. Three major stages can be considered within a commodity chain:

> **First Stage (raw materials).** The availability of raw materials often imposes sourcing at the international level, a process which has accelerated in recent years. It predominantly concerns the procurement of commodities. The flows occurring at this stage are mainly supported by international transportation systems relying on bulk shipping. Distribution tends to involve high volumes and low frequency. This system is also being impacted by the containerization of several commodity markets.

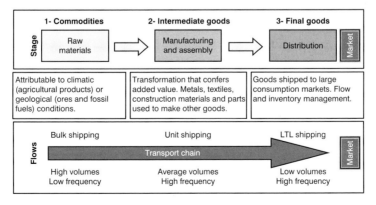

Figure 5.5 The commodity chain (or value chain)

Second Stage (manufacturing and assembly). Mainly concerns intermediate goods. Some capital-intensive manufacturing and assembly activities will take place inside of the national economy while labor-intensive activities are outsourced. Flows are either containerized or on pallets, with average volumes and rather high frequencies, notably for commodity chains relying on timely deliveries.

Third Stage (distribution). Distribution of final goods mainly takes place on the national market, although globally oriented distribution implies that a national market is serviced from a major gateway. Depending on the scale of the distribution (international, national or regional), flows can be coordinated by distribution centers each having their own market areas. Flows are actively managed, often in low volumes (less than truckload, LTL), but with a high frequency since they are related to retailing.

Commodity chains are thus a sequential process used by corporations within a production system to gather resources, transform them in to parts and products and, finally, distribute manufactured goods to markets. Each sequence is unique and dependent on product types, the nature of production systems, where added value activities are performed, market requirements as well as the current stage of the product life cycle (Figure 5.6).

The manufacturing function of many corporations has been hollowed out by the process of globalization, in which manufacturing accounts for one of the least added value activities, particularly if it takes place within an outsourcing and offshoring framework. The massive entry of low cost manufacturers led to a high level of competitiveness in fabrication, reducing profit margins as well as its overall level of contribution to added value. This led to a growing contribution and importance of the concept and logistics segments of the commodity chain. Consequently, in a global production and consumption market R&D, branding and design (creating a product) can be a significant component of the added value function of a commodity chain. On the logistics segment (making a specified product available on markets), distribution, marketing and sales/after sales services (such as customer support) are the activities generating the most added value.

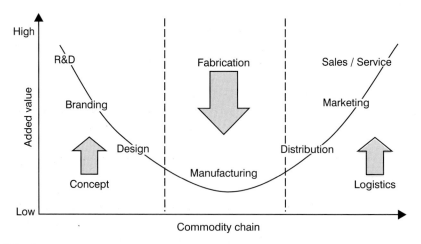

Figure 5.6 Commodity chains and added value

Source: Adapted from the Stan Shih "Smile Curve" concept.

Commodity chains enable a sequencing of inputs and outputs between a range of suppliers and customers, mainly from a producer and buyer-driven standpoint. They also offer adaptability to changing conditions, namely an adjustment of production to adapt to changes in price, quantity and even product specification. The flexibility of production and distribution becomes particularly important, with a reduction of production, transaction and distribution costs as the logical outcome. The three major types of commodity chains involve:

Raw materials. The origin of these goods is linked with environmental (agricultural products) or geological (ores and fossil fuels) conditions. The flows of raw materials (particularly ores and crude oil) are dominated by a pattern where developing countries export towards developed countries. Transport terminals in developing countries are specialized in loading while those of developed countries unload raw materials and often include transformation activities next to port sites. Industrialization in several developing countries has modified this standard pattern with new flows of energy and raw materials.

Semi-finished products. These goods already have had some transformation performed conferring them an added value. They involve metals, textiles, construction materials and parts used to make other goods. Depending on the labor-intensiveness and comparative advantages, segments of the manufacturing process have been offshored. The pattern of exchanges is varied in this domain. For ponderous parts, it is dominated by regional transport systems integrated to regional production systems. For lighter and high value parts, a global system of suppliers tends to prevail.

Manufactured goods. These include goods that are shipped towards large consumption markets and require a high level of organization of flows to fulfill the demand. The majority of these flows concern developed countries, but a significant share is related to developing countries, especially those specializing in export-oriented manufacturing. Containerization has been the dominant transport paradigm for manufactured goods with production systems organized around terminals and their distribution centers.

Integration in commodity chains

Transport chains are being integrated into production systems. As manufacturers are spreading their production facilities and assembly plants around the globe to take advantage of local factors of production, transportation becomes an ever more important issue. The integrated transport chain is itself being integrated into the production and distribution processes. Transport can no longer be considered as a separate service that is required only as a response to supply and demand conditions. It has to be built into the entire supply chain system, from multisource procurement, to processing, assembly and final distribution. Supply Chain Management (SCM) has become an important facet of international transportation. As such, the container has become a transport, production and distribution unit.

A significant trend has thus been a growing level of embeddedness between production, distribution and market demand. Since interdependencies have replaced relative autonomy and self-sufficiency as the foundation of the economic life of regions and firms, high levels of freight mobility have become a necessity. The presence of an efficient distribution system supporting global commodity chains (also known as global production networks) is sustained by:

Functional integration. Its purpose is to link the elements of the supply chain in a cohesive system of suppliers and customers. A functional complementarity is then achieved through a set of supply/demand relationships, implying flows of freight, capital and information. Functional integration relies on distribution over vast territories where "just-in-time" and "door-to-door" strategies are relevant examples of interdependencies created by new freight management strategies. Intermodal activities tend to create heavily used transshipment points and corridors between them, where logistical management is more efficient.

Geographical integration. Large resource consumption by the global economy underlines a reliance on supply sources that are often distant, as for example crude oil and mineral products. The need to overcome space is fundamental to economic development and the development of modern transport systems have increased the level of integration of geographically separated regions with a better geographical complementarity. With improvements in transportation, geographical separation has become less relevant, as comparative advantages are exploited in terms of the distribution capacity of networks and production costs. Production and consumption can be more spatially separated without diminishing economies of scale, even if agglomeration economies are less evident.

The level of customization of a product can also be indicative of how commodity chains are integrated. For products requiring a high level of customization (or differentiation) the preference is usually to locate added value components relatively close to the final market. For products that can be mass produced and that require limited customization, the preference leans on locating where input costs (e.g. labor) are the least.

Freight transport and commodity chains

As the range of production expanded, transport systems adapted to the new operational realities in local, regional and international freight distribution. Freight transportation offers a whole spectrum of services catering to cost, time and reliability priorities and has consequently taken an increasingly important role within commodity chains. Among the most important factors:

- Improvements in transport efficiency facilitated an **expanded territorial range of commodity chains**.
- A reduction of telecommunication costs and the development of information technologies enabled corporations to establish a **better level of control over their commodity chains**. Information technologies have a wide array of impacts on the management of freight distribution systems.
- Technical improvements, notably for intermodal transportation, enabled a more efficient **continuity** between different transport modes (especially land/maritime) and thus within commodity chains.

The results have been an improved velocity of freight, a decrease of the friction of distance and a spatial segregation of production. This process is strongly embedded with the capacity and efficiency of international and regional transportation systems, especially maritime and land routes. It is uncommon for the production stages of a good to occur at the same location. Consequently, the geography of commodity chains is integrated with the geography of transport systems.

Most commodity chains are linked to regional transport systems, but with globalization, international transportation accounts for a growing share of flows within production systems. The usage of resources, parts and semi-finished goods by commodity chains is an indication of the type of freight being transported. Consequently, transport systems must adapt to answer the needs of commodity chains, which encourages diversification. Within a commodity chain, freight transport services can be categorized by:

- **Management of shipments**. Refers to cargo transported by the owner, the manufacturer or by a third party. The tendency has been for corporations to subcontract their freight operations to specialized providers who offer more efficient and cost-effective services.
- **Geographical coverage**. Covers a wide variety of scales ranging from intercontinental, within economic blocs, national, regional or local. Each of these scales often involves specific modes of transport services and the use of specific terminals.
- **Time constraint**. Freight services can have a time element ranging from express, where time is essential, to the lowest cost possible, where time is secondary. There is also a direct relationship between transport time and the level of inventory that has to be maintained in the supply chain. The shorter the time, the lower the inventory level, which can result in significant savings.
- **Consignment size**. Depending on the nature of production, consignments can be carried in full loads, partial loads (less than truck load; LTL), as general cargo, as container loads or as parcels.
- **Cargo type**. Unitized cargo (containers, boxes or pallets) or bulk cargo requires dedicated vehicles, vessels and transshipment and storage infrastructures.
- **Mode**. Cargo can be carried on a single mode (sea, rail, road or air) or in a combination of modes through intermodal transportation.
- **Cold chain**. A temperature-controlled supply chain linked to the material, equipment and procedures used to maintain specific cargo shipments within an appropriate temperature range. Commonly relates to the distribution of food and pharmaceutical products.

Globalization is also concomitant – a by-product – of a post-Fordist environment where just-in-time (JIT) and synchronized flaws are becoming the norm in production and distribution systems. International transportation is shifting to meet the increasing needs of organizing and managing its flows through logistics. In spite of the diversity of transport services supporting various commodity chains, containerization is adaptable enough to cope with a variety of cargo and time constraints.

Concept 3 – Logistics and freight distribution

Authors: Jean-Paul Rodrigue and Markus Hesse

The nature of logistics

The growing flows of freight have been a fundamental component of contemporary changes in economic systems at the global, regional and local scales. These changes are not merely quantitative with more freight in circulation, but structural and operational. Structural changes mainly involve manufacturing systems with their geography of production, while operational changes mainly concern freight transportation with its geography of distribution. As such, the fundamental question does not necessarily reside in the nature, origins and destinations of freight movements, but how this freight is moving. New modes of production are concomitant with new modes of distribution, which brings forward the realm of logistics, the science of physical distribution.

> **Logistics** involves a wide set of activities dedicated to the transformation and distribution of goods, from raw material sourcing to final market distribution as well as the related information flows. Derived from Greek *logistikos* (to reason logically), the word is polysemic. In the nineteenth century the military referred to it as the art of combining all means of transport, revictualing and sheltering of troops. Today it refers to the set of operations required for goods to be made available on markets or to specific locations.

The application of logistics enables a greater efficiency of movements with an appropriate choice of modes, terminals, routes and scheduling. The implied purpose of logistics is to make available goods, raw materials and commodities, fulfilling four major requirements related to order, delivery, quality and cost fulfillment (Figure 5.7).

- **Order fulfillment**. The transaction between the supplier and the customer is satisfied with the specified product provided in the agreed quantity.
- **Delivery fulfillment**. The order must also be delivered at the right location and at the right time. Both involve the scheduling of transportation and freight distribution activities.
- **Quality fulfillment**. The order must be provided intact (in good condition), implying that any form of damage must be avoided during transport and delivery. This is

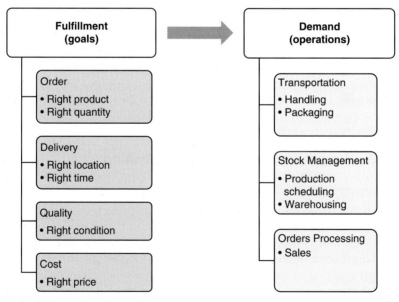

Figure 5.7 Logistics goals and operations

particularly important for products that are fragile, perishable or sensitive to temperature fluctuations.

- **Cost fulfillment**. The final costs of the order, including manufacturing and distribution costs, must be competitive. Otherwise, other options will be considered.

All operations related to logistics aim at ensuring that a demand is satisfied (fulfilling goals), whether it is a part made available to a manufacturer or a good on a store's shelf. There are three major categories of logistics operations:

- **Purchase orders processing**. Operations related to the transactional procurement of goods.
- **Stock management**. Operations related to the physical procurement of goods.
- **Transportation**. Operations related to the physical distribution of goods.

Logistics is thus a multidimensional value-added activity including production, location, time and control of elements of the supply chain. It represents the material and organizational support of globalization. Activities comprising logistics include physical distribution; the derived transport segment, and materials management; and the induced transport segment.

> **Physical distribution** is the collective term for the range of activities involved in the movement of goods from points of production to final points of sale and consumption. It must ensure that the mobility requirements of supply chains are entirely met. Physical distribution includes all the functions of movement and handling of goods, particularly transportation services (trucking, freight rail, air freight, inland waterways, marine shipping and pipelines), transshipment and warehousing services (e.g. consignment, storage, inventory management), trade, wholesale and, in principle, retail. Conventionally, all these activities are assumed to be derived from materials management demands.
>
> **Materials management** considers all the activities related in the manufacturing of commodities in all their stages of production along a supply chain. It includes production and marketing activities such as production planning, demand forecasting, purchasing and inventory management. Materials management must ensure that the requirements of supply chains are met by dealing with a wide array of parts for assembly and raw materials, including packaging (for transport and retailing) and, ultimately, recycling discarded commodities. All these activities are assumed to be inducing physical distribution demands.

The close integration of physical distribution and materials management through logistics is blurring the reciprocal relationship between the derived transport demand function of physical distribution and the induced demand function of materials management. This implies that distribution, as always, is derived from materials management activities (namely production), but also, that these activities are coordinated within distribution capabilities. The functions of production, distribution and consumption are difficult to consider separately, thus recognizing the integrated transport demand role of logistics. Distribution centers are the main facilities from which logistics are coordinated.

> **Distribution center**. Facility or a group of facilities that perform consolidation, warehousing, packaging, decomposition and other functions linked with handling

freight. Their main purpose is to provide value-added services to freight. DCs are often in proximity to major transport routes or terminals. They can also perform light manufacturing activities such as assembly and labeling.

Since it would be highly impractical to ship goods directly from producers to retailers, distribution centers essentially act as a buffer where products are assembled, sometimes from other distribution centers, and then shipped in batches. Distribution centers commonly have a market area in which they offer a service window defined by delivery frequency and response time to order. This structure looks much like a hub-and-spoke network.

The array of activities involved in logistics, from transportation to warehousing and management, have respective costs. Once compiled, they express the burden that logistics impose on distribution systems and the economies they support, which is known as the total logistics costs. Costs are, however, not the only consideration in supply chain management since supply chains can also be differentiated by time, reliability and risk level. The nature and efficiency of distribution systems is strongly related to the nature of the economy in which they operate. Worldwide logistics expenditures represent about 10–15 percent of the total world GDP. In economies dependent on the extraction of raw materials, logistical costs are comparatively higher than for service economies since transport costs account for a larger share of the total added value of goods. For the transport of commodities, logistics costs are commonly in the range of 20 to 50 percent of their total costs.

Driving forces in supply chain management

The emergence of logistics in contemporary supply chains is based upon continuous improvements in transport and inventory management costs, which commonly result in lower lead times: the time it takes for an order to be fulfilled. Lean supply chains, as a managerial concept, is often labeled as seminal in the emergence of modern supply chains where inventory levels are kept at a minimum and where a large share of the inventory is in constant circulation. Typically the manufacturing sector has 6 to 8 inventory turns per year. In the electronics sector, this can even be faster with 10 to 20 inventory turns per year. During the 1980s, the application of flow control permitted inventories to be reduced in time-sensitive manufacturing activities from several days to several hours. Much of these efforts initially took place within the factory, while supply and output flowed as batches from suppliers and to distributors. In the 1990s, with the convergence of logistics and information and communication technologies (ICT), this principle was increasingly applied to the whole supply chain, particularly to the function of distribution.

Another important requirement was containerization, which conferred substantial flexibility to production systems in addition to the container being its own storage unit. The expansion of standard transport infrastructure such as highways, terminals and airports was also essential for the development of modern logistics. Logistics and integrated transport systems are therefore related, particularly because of the container which has concomitantly become a unit of load (transport), production and distribution. Thus, the physical as well as the ICT elements of technological change are being underlined as it helps strengthen the level of control distributors have over the supply chain. The technological dimension of logistics can thus be considered from five perspectives:

Transportation modes. Modes have been the object of very limited technological changes in recent decades. In some cases, modes have adapted to handle containerized operations such as road and rail (e.g. double-stacking). It is maritime shipping that has experienced the most significant technological change, requiring the construction of an entirely new class of ships and the application of economies of scale to maritime container shipping. In this context, a global network of maritime shipping servicing large gateways has emerged.

Transportation terminals. The technological changes have been very significant with the construction of new terminal facilities operating on a high turnover basis. Better handling equipment led to improvements in the velocity of freight at terminals; these are among the most significant technological changes brought by logistics in materials movements. In such a context, the port has become one of the most significant terminals supporting global logistics. Port facilities are increasingly being supported by an array of inland terminals connected by high capacity corridors.

Distribution centers and distribution clusters. Technological changes impacted over the location, design and operation of distribution centers: the facilities handling the requirements of modern distribution. They serve different purposes depending on the combination of fabrication, storage and distribution functions they perform within their supply chains. Modern distribution centers tend to consume more space, both from the site they occupy and the building area. From a locational standpoint, distribution centers mainly rely on trucking, implying a preference for suburban locations with good road accessibility supporting a constant traffic. They service regional markets with a 48-hour service window on average, implying that replenishment orders from their customers are met within that time. They have become one-floor facilities designed more for throughput than for warehousing with specialized loading and unloading bays and sorting equipment. Cross-docking distribution centers represent one of the foremost expressions of a facility that handles freight in a time-sensitive manner. Another tendency has been the setting of freight distribution clusters where an array of distribution activities agglomerate to take advantage of shared infrastructures and accessibility. This tends to expand the added value performed by logistics.

Load units. Since logistics involves improving the efficiency of flows, load units have become particularly important. They are the basic physical management unit in freight distribution and take the form of pallets, swap bodies, semi-trailers and containers. Containers are the privileged load unit for long distance trade, but the growing complexity of logistics required a more specific level of load management. The use of bar codes and increasingly of RFID (Radio Frequency Identification Device) enables a high level of control of the load units in circulation.

Information technologies/e-commerce. Consider the vast array of information processing changes brought by logistics. The commodity chain is linked with physical flows as well as with information flows, notably through Electronic Data Interchange (EDI). Producers, distributors and consumers are embedded in a web of reciprocal transactions. While these transactions mostly take place virtually, their outcomes are physical flows. E-commerce offers advantages for the whole commodity chain, from consumers being exposed to better product information to manufacturers and distributors being able to adapt quickly to changes in the demand. The outcome is often more efficient production and distribution planning with the additional convenience of tracking shipments and inventories.

For logistics, ICT is particularly a time and embeddedness issue: because of ICT, freight distribution is within a paradigm shift from inventory-based logistics (push) to replenishment-based logistics (pull). Demand, particularly in the retailing sector, is very difficult to anticipate accurately. A closer integration between supply and demand enables a more efficient production system with fewer wastes in terms of unsold inventory. Logistics is thus a fundamental component of efficiency improvements in a market economy.

Distribution systems

In a broader sense distribution systems are embedded in a changing macro- and microeconomic framework, which can be roughly characterized by the terms of flexibility and globalization:

Flexibility implies a highly differentiated, strongly market- and customer-driven mode of creating added value. Contemporary production and distribution is no longer subject to single-firm activity, but increasingly practiced in networks of suppliers and subcontractors. The supply chain bundles together all this by information, communication, cooperation and, last but not least, by physical distribution.

Globalization means that the spatial frame for the entire economy has been expanded, implying the spatial expansion of the economy, more complex global economic integration, and an intricate network of global flows and hubs.

The flow-oriented mode affects almost every single activity within the entire process of value creation. The core component of materials management is the supply chain, the time- and space-related arrangement of the whole goods flow between supply, manufacturing, distribution and consumption. Its major parts are the supplier, the producer, the distributor (e.g. a wholesaler, a freight forwarder, a carrier), the retailer, the end-consumer, all of whom represent particular interests. Compared with traditional freight transport systems, the evolution of supply chain management and the emergence of the logistics industry are mainly characterized by three features:

Integration. A fundamental restructuring of goods merchandising by establishing integrated supply chains with integrated freight transport demand. According to macro-economic changes, demand-side oriented activities are becoming predominant. While traditional delivery was primarily managed by the supply side, current supply chains are increasingly managed by the demand.

Time mitigation. Whereas transport was traditionally regarded as a tool for overcoming space, logistics is concerned with mitigating time. Due to the requirements of modern distribution, the issue of time is becoming increasingly important in the management of commodity chains. Time is a major issue for freight shipping as it imposes inventory holding and depreciation costs, which become sensitive for tightly integrated supply chains.

Specialization. This was achieved by shifts towards vertical integration, namely subcontracting and outsourcing, including the logistical function itself. Logistics services are becoming complex and time-sensitive to the point that many firms are now subcontracting parts of their supply chain management to what can be called third-party logistics providers (3PL; asset based). More recently, a new category of providers, called fourth-party logistics providers (4PL; non-asset based) have emerged.

While many manufacturing corporations may have in-house transportation departments, increasingly the complex needs of the supply chain are being contracted out to third parties. Depending on the strategy and costs corporations can outsource in whole or in part their transport and supply chain operations. Third-party logistics providers (3PL) have emerged from traditional intermediaries such as the forwarders, or from transport providers such as FedEx or Maersk. Both groups have been at the forefront of the inter-modal revolution that is now assuming more complex organizational forms and importance. In offering door-to-door services, the customer is no longer aware or necessarily concerned with how the shipment gets to its destination, namely the modes used and the routing selected (Photo 5.1). The preoccupation is with cost, reliability and level of service. This produces a paradox: for the customer of intermodal services geographic space becomes meaningless; but for the intermodal providers routing, costs and service frequencies have significant geographical constraints. The effectiveness of intermodal transport systems is thus masking the importance of transportation to its users.

Photo 5.1   Third-party less-than-truckload service, FedEx

Many freight transport companies are offering less-than-truckload (LTL) services to their customers. Such shipments are generally composed of items of various sizes filling up specific orders that are usually not recurrent. Here, a FedEx (second largest LTL carriers in the United States) truck is picking up cargo at a concrete foundation mold manufacturer in Kansas City. This shipment is likely bound for a specific construction project.

Geography of freight distribution

Logistics has a distinct geographical dimension, which is expressed in terms of flows, nodes and networks within the supply chain. Space/time convergence – a well-known concept in transport geography where time is simply considered as the amount of space that could be traded with a specific amount of time, including travel and transshipment – is being transformed by logistics. Activities that were not previously considered fully in space/time relationships, such as distribution, are being integrated. This implies an organization and synchronization of flows through nodes and network strategies.

The traditional arrangement of **goods flow** included the processing of raw materials to manufacturers, with a storage function usually acting as a buffer (Figure 5.8). The flow continued via wholesaler and/or shipper to retailer, ending at the final customer. Delays were very common on all segments of this chain and accumulated as inventories in warehouses. There was a limited flow of information from the consumer to the supply chain, implying the producers were not well informed (often involving a time lag) about the extent of consumption of their outputs. This procedure is now changing, mainly by eliminating one or more of the costly operations in the supply chain organization. Reverse flows are also part of the supply chain, namely for recycling and product returns. An important physical outcome of supply chain management is the concentration of storage or warehousing in one facility, instead of several. This facility is increasingly being designed as a flow- and throughput-oriented distribution center, instead of a warehouse holding cost-intensive large inventories.

Due to new corporate strategies, a concentration of logistics functions in certain facilities at strategic locations is prevalent. Many improvements in freight flows are achieved at terminals. Facilities are much larger than before, the locations being characterized by a particular connection of regional and long distance relations. Traditionally, freight distribution has been located at major places of production, for instance in the manufacturing belt at the North American east coast and in the Midwest, or in the

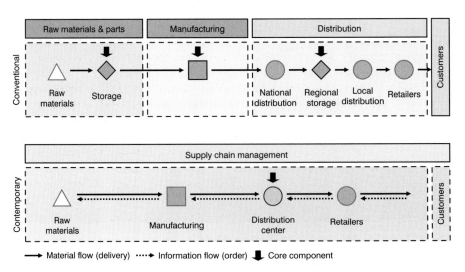

Figure 5.8 Conventional and contemporary arrangement of goods flow

Source: Adapted from Hesse, M. and J-P. Rodrigue (2004) "The transport geography of logistics and freight distribution", *Journal of Transport Geography*, 12(3): 171–84.

old industrialized regions of England and continental Europe. Today, particularly the large-scale goods flows are directed through major gateways and hubs, mainly large ports and major airports, also highway intersections with access to a regional market. The changing geography of manufacturing and industrial production has been accompanied by a changing geography of freight distribution taking advantages of intermediary locations.

The spatial structure of contemporary **transportation networks** is the expression of the spatial structure of distribution. The setting of networks leads to a shift towards larger distribution centers, often serving significant transnational catchments. However, this does not mean the demise of national or regional distribution centers, with some goods still requiring a three-tier distribution system, with regional, national and international distribution centers. The structure of networks has also adapted to fulfill the requirements of an integrated freight transport demand, which can take many forms and operate at different scales. Most freight distribution networks, particularly in retailing, are facing the challenge of the "last mile" which is the final leg of a distribution sequence, commonly linking a distribution center and a customer (store).

Since cities are at the same time zones of production, distribution and consumption, the realm of city logistics is of growing importance. This issue is made even more complex by a growing dislocation between production, distribution and consumption, brought by globalization, global production networks, and efficient freight transport systems and logistics. This dislocation has prompted a growing emphasis on issues related to supply chain integration so that in spite of acute geographical separation physical and managerial processes have minimal friction. How successful individual countries are perceived to be in the setting and management of supply chains can be assessed, by the Logistics Performance Index (LPI; Figure 5.9). The LPI is a composite index based on proxy measures for transport and information infrastructure, supply chain management (SCM) and trade facilitation capabilities, which are calculated based on a world survey of international freight forwarders and express carriers. It underlines that logistical costs in developing countries tend to be higher, which undermines economic development for the main following reasons:

The **regulatory complexity** of distributing goods in developing countries involves higher logistical costs and distributors tend to maintain higher inventory levels to cope with uncertainty. Custom regulations are complex and prone to delays and road transportation can be subject to arbitrary tolls and inspections. This is reflected in higher final goods or component prices that are assumed directly or indirectly by consumers.
Labor and infrastructure productivity in developing countries tend to be lower, which in many cases doubles logistics costs. The advantages of cheap labor can often be counterbalanced by lower levels of productivity. This also impacts the reliability of freight distribution with unreliable lead times and deliveries.
Modal and intermodal capacity is inconsistent. While several terminal facilities, particularly ports, are modern with capacity on a par with global standards, hinterland transportation can be problematic with road segments unable to effectively handle trucks of standard capacity.

In such a context, reforms have been advocated to promote the effectiveness of logistics services and therefore break a vicious cycle in which several developing countries are entangled. This involves a series of reform, pending the capacity to overcome political constraints and the inertia (and commonly rent-seeking behavior) of established

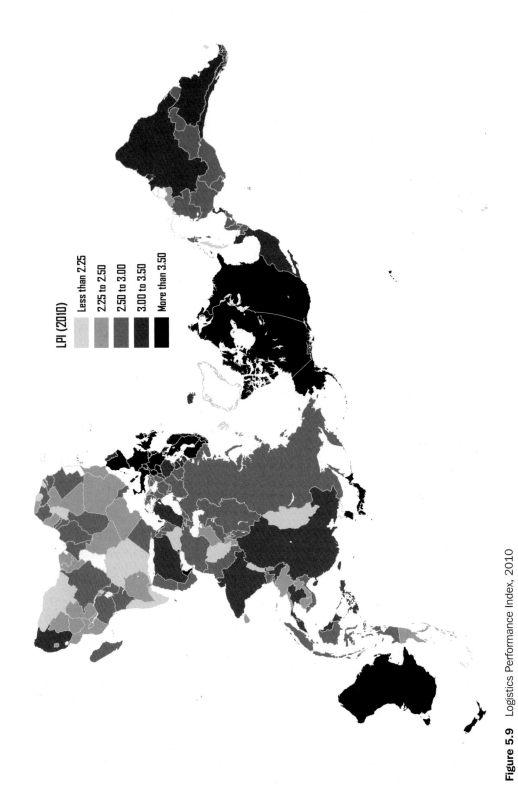

Figure 5.9 Logistics Performance Index, 2010

Source: Adapted from World Bank, Logistics Performance Index.

stakeholders, concerning service providers, infrastructure investment as well as the administrative and regulatory environment.

CASE STUDY   Commodity chain analysis: the cold chain

The structure of commodity chains

Commodity chains reveal much about the global structure of production, the global economy, and thus represent a notable field of investigation that has yet to be fully considered by transport geographers. Understanding the significance of commodity chains requires a comprehensive approach since they include much more than a simple transport consideration; a multitude of activities are involved.

> **Commodity chain analysis**. The identification of the actors and processes that contribute to the origination of a product that is consumed by a market, such as raw materials, products or consumption goods. Thus, a commodity chain includes a sequence of operations ranging from the extraction of raw materials, the assembly of intermediate goods, to the distribution to consumption markets. Commodity chain analysis can also consider only a specific segment related to a single product (or group of products).

The analysis of such a complex chain of agents and processes considers several perspectives:

- **Transactional perspective**. Identification of the flows and of the transactions that create them. This particularly concerns the decision-making process in the establishment and management of commodity chains.
- **Comparative perspective**. Assesses the relative competitiveness of the elements of the commodity chains in terms of added value.
- **Functional perspective**. Identifies the physical processes involved in the circulation of goods, including the capacity constraints in distribution, namely modal, intermodal and terminal effectiveness.

The cold chain

While globalization has made the relative distance between regions of the world much smaller, the physical separation of these same regions is still a very important reality. The greater the physical separation, the more likely freight can be damaged in one of the complex transport operations involved. Some goods can be damaged by shocks while others can be damaged by undue temperature variations. For a range of goods labeled as perishables, particularly food, their quality degrades with time; this can be mostly mitigated with lower temperatures. It takes time and coordination to efficiently move a shipment and every delay can have negative consequences, notably if this cargo is perishable. To ensure that cargo does not become damaged or compromised throughout this process, businesses in the pharmaceutical, medical and food industries are increasingly relying on cold chain technology.

> The **cold chain** refers to the transportation of temperature-sensitive products along a supply chain through thermal and refrigerated packaging methods and the logistical planning to protect the integrity of these shipments.

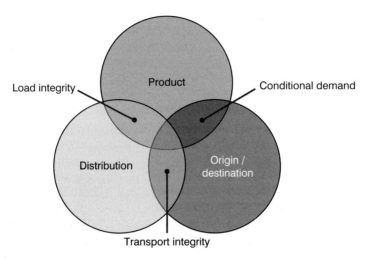

Figure 5.10 Elements of the cold chain

Operational conditions within the cold chain must be consistent so that the load and transport integrity of the shipments are maintained. A cold chain can functionally be considered as the close interaction between three elements (Figure 5.10):

> **Product**. A product is characterized by physical attributes requiring specific temperature and humidity conditions. These conditions dictate its transport, which takes place in a manner that does not undermine its physical attributes to an extent that is judged acceptable. It mostly relates to how perishable and fragile a product can be.
> **Origin/destination**. A function of the respective locations where a temperature-sensitive product is produced and consumed. It is indicative of the potential difficulty of making a product available at a market. Because of advances in cold chain logistics, it became possible to use increasingly distant sourcing strategies.
> **Distribution**. The methods and infrastructures available to transport a product in a temperature-controlled environment. It can involve temperature-controlled containers (reefers), trucks and warehousing facilities.

Specialization has led many companies to not only rely on major shipping service providers such as the United Parcel Service (UPS) and FedEx, but also more focused industry specialists that have developed a niche logistical expertise around the shipping of temperature-sensitive products. The potential to understand local rules, customs and environmental conditions as well as an estimation of the length and time of a distribution route make them an important factor in global trade. As a result, the logistics industry is experiencing a growing level of specialization and segmentation of cold chain shipping in several potential niche markets within global commodity chains. Whole new segments of the distribution industry have been very active in taking advantage of the dual development of the spatial extension of supply chains supported by globalization and the significant variety of goods in circulation. From an economic development perspective, the cold chain enables many developing countries to take part in the global perishable products market. From a geographical perspective, the cold chain has the following impacts:

- **Global**. Specialization of agricultural functions permitting the transport of temperature-sensitive food products to distant markets. Enables the distribution of vaccines and other pharmaceutical or biological products from single large facilities.
- **Regional**. Can support the specialization of production and economies of scale in distribution. This could involve specialized laboratories exchanging temperature-sensitive components or large cold storage facilities servicing regional grocery markets.
- **Local**. Timely distribution to the final consumer of perishables, such as grocery stores and restaurants.

Perishable or temperature-sensitive items are carried in refrigerated containers (called "reefers"), that account for a growing share of the refrigerated cargo being transported around the world. While in 1980, 33 percent of the refrigerated transport capacity in maritime shipping was containerized, this share rapidly climbed to 47 percent in 1990, 68 percent in 2000 and 90 percent in 2010. About 1.69 million TEUs of reefers were being used by 2009. All reefers are painted white to increase the albedo (share of the incident light being reflected; high albedo implies less solar energy absorbed by the surface) with the dominant size being 40-foot high-cube containers.

Food transportation

There is a variety of methods for the transport of food products; the banana accounts for the world's most significant commodity transported in the food cold chain with 20 percent of all seaborne reefers trade. Land, sea and air modes all have different structures for keeping food fresh throughout the transport chain. Innovations in packaging, fruit and vegetable coatings, bioengineering (controlled ripening), and other techniques reducing the deterioration of food products have helped shippers extend the reach of perishable products. For food products such as fruits and vegetables, time has a direct impact on their shelf life and therefore on the potential revenue a consignment may generate. Concomitantly, new transport technologies have permitted the shipment of perishable products over longer distances. For instance, improved roads and intermodal connections along the African coast reduced food transport time to European markets from 10 to 4 days.

Certain domestic or transnational supply chains may only require one transportation mode, but often ground shipments are one link in a combination of transport modes. This makes intermodal transfer critical for the cold chain. Intermodal shipments typically use either 20- or 40-footer refrigerated containers that are capable of holding up to 26 tons of food. The container makes loading and unloading periods shorter and less susceptible to damage. The environments in these containers are currently controlled electronically by either plugging into a generator or power source on the ship or truck, but early food shipments would cycle air from stores of wet or dry ice to keep the food refrigerated. The efficiency of cold chain logistics permitted the consolidation of cold storage facilities (Photo 5.2).

Moving away from ice refrigeration has allowed for much greater distances to be traveled and has greatly increased the size of the global food market, enabling many developing countries to capture new opportunities. Another efficient mode for transporting foodstuffs is air travel. While this is a preferred form of travel for highly perishable and valuable goods due to its ability to move much faster over longer distances, it does lack the environment control and transfer ease of the ground and sea transports. Also, during the flight the cargo is stored in a 15–20°C environment, but

Photo 5.2 Grocery chain cold storage facility, Regina, Saskatchewan

The growing efficiency and reliability of cold chain logistics has enabled major grocery store chains to establish large cold storage distribution facilities servicing large market areas. This strategy confers economies of scale (lower operating costs) as well as the ability to supply a wider variety of cold chain retail goods. This photo depicts a cold storage facility of Loblaw Companies, the largest food retailer in Canada, located at the Global Transportation Hub inland port in the vicinity of Regina, Saskatchewan. It services most of the grocery stores in Western Canada. Several domestic 53-foot reefer containers are being loaded with the orders of specific stores ("no name" refers to the generic store brand).

close to 80 percent of the time the package is exposed to exterior weather while waiting to be loaded onto the plane or being moved to and from the airfield. This is troubling considering the value of the food and the importance placed behind quality and freshness. In order for this form of food transport to experience growth among market users, more uncompromising strategies and regulations will have to be embraced and enacted.

Food transportation is an industry that has fully adapted to the cold chain and can, despite the problems with air transport, be considered the most resilient, particularly since a large majority of food products have a better tolerance to temporary variations of transport temperatures. As a result, small errors can be compounded without the concern of irreversible damage. For instance, for the transportation of produce, for every hour of delay in the pre-cooling of shipments, an equivalent one day loss of shelf life must be accounted. The usage of refrigerated containers has particularly helped, since they account for more than 50 percent of all the refrigerated cargo transported in the world. Source loading can be an important factor extending the shelf life of a cold chain product since it is loaded in a reefer directly at the place of production without additional handling. For instance, source loading into a reefer can expand the shelf life of chilled meat by about 25 days (from 30–35 days to 55–60 days) from conventional methods and thus considerably expand the market potential of the product.

The efficiency and reliability of temperature-controlled transportation has reached a point which allows the food industry to take advantage of global seasonable variations, meaning that during the winter the southern hemisphere can export perishable goods to the northern hemisphere while an opposite trade, generally of smaller scale, takes place during the summer. Countries such as Chile have substantially benefited from this and have developed an active agricultural and food transformation industry mainly servicing the North American market during the winter, but also with several niche markets such as wine. A similar issue concerns some African countries such as Kenya that have developed fresh produce and flower industries catering to the European market. The fast food industry is also an active user of cold chain logistics as every outlet can be considered a factory, with dozens of workers with schedules and shifts, inventory management and the supply chain of components (many of which are temperature sensitive), and which are assembly lines producing quality controlled and high volume products.

Bibliography

Barke, M. (1986) *Transport and Trade*, Edinburgh: Oliver and Boyd.

Bernstein, W.J. (2008) *A Splendid Exchange: How Trade Shaped the World*, New York: Atlantic Monthly Press.

Braudel, F. (1982) *The Wheels of Commerce. Civilization and Capitalism 15th–18th Century*, Vol. II, New York: Harper and Row.

Brooks, M. (2008) *North American Freight Transportation: The Road to Security and Prosperity*, Cheltenham, UK: Edward Elgar.

Coyle, W., W. Hall and N. Ballenger (2001) "Transportation technology and the rising share of U.S. perishable food trade", Economic Research Service/USDA, Changing Structure of Global Food Consumption and Trade/WRS-01-1.

Daniels, J.D., L.H. Radebaugh and D. Sullivan (2010) *International Business: Environments and Operations*, 13th edition, New York: Prentice Hall.

Dicken, P. (2007) *Global Shift: Mapping the Changing Contours of the World Economy*, 5th edition, London: Sage Publications.

Fujita, M., P. Krugman and A.J. Venables (1999) *The Spatial Economy: Cities, Regions and International Trade*, Cambridge, MA: MIT Press.

Gave, C., A. Kaletsky and L.V. Gave (2005) *Our Brave New World*, New York: GaveKal Research.

Hesse, M. (2008) *The City as a Terminal: The Urban Context of Logistics and Freight Transport*, Aldershot, Hampshire: Ashgate.

Hesse, M. and J-P. Rodrigue (2004) "The transport geography of logistics and freight distribution", *Journal of Transport Geography*, 12(3): 171–84.

Hummels, D. (2001) "Time as a trade barrier", GTAP Working Paper No. 18, WestLafayette, IN: Center for Global Trade Analysis.

Lakshmanan, T.J., U. Subramanian, W.P. Anderson and F.A. Léautier (2001) *Integration of Transport and Trade Facilitation: Selected Regional Case Studies*, Washington, DC: World Bank.

O'Connor, K. (2010) "Global city regions and the location of logistics activity", *Journal of Transport Geography*, 18(3): 354–62.

Tallec, F. and L. Bockel (2005) *Commodity Chain Analysis: Constructing the Commodity Chain Functional Analysis and Flow Charts*, Food and Agriculture Organization of the United Nations.

Thompson, J.F., P.E. Brecht, T. Hinsch, and A.A. Kader (2000) "Marine container transport of chilled perishable produce", Publication 21595, University of California, Division of Agriculture and Natural Resources.

World Trade Organization (2010) *International Trade Statistics 2010*, Geneva: WTO.

—— (2008) *World Trade Report 2008: Trade in a Globalizing World*, Geneva: WTO.

⑥ Urban transportation

Urbanization has been one of the dominant contemporary processes as a growing share of the global population lives in cities. Considering this trend, urban transportation issues are of foremost importance to support the passengers and freight mobility requirements of large urban agglomerations. Transportation in urban areas is highly complex because of the modes involved, the multitude of origins and destinations, and the amount and variety of traffic. Traditionally, the focus of urban transportation has been on passengers as cities were viewed as locations of utmost human interactions with intricate traffic patterns linked to commuting, commercial transactions and leisure/cultural activities. However, cities are also locations of production, consumption and distribution, activities linked to movements of freight. Conceptually, the urban transport system is intricately linked with urban form and spatial structure. Urban transit is an important dimension of mobility, notably in high density areas.

Concept 1 – Transportation and the urban form

Global urbanization

No discussion about the urban spatial structure can take place without an overview of urbanization, which has been one of the dominant trends of economic and social change of the twentieth century, especially in the developing world.

> **Urbanization**. The process of transition from a rural to a more urban society. Statistically, urbanization reflects an increasing proportion of the population living in settlements defined as urban, primarily through net rural to urban migration. The level of urbanization is the percentage of the total population living in towns and cities while the rate of urbanization is the rate at which it grows.

This transition is expected to go on well into the second half of the twenty-first century, a trend reflected in the growing size of cities and in the increasing proportion of the urbanized population. Urban mobility problems have increased proportionally, and in some cases exponentially, with urbanization since it concentrates mobility demands over a specific area. Since 1950, the world's urban population has more than doubled, to reach nearly 3.5 billion in 2010, about 50.6 percent of the global population. This is the outcome of three main demographic trends:

> **Natural increase**. It is simply the outcome of more births than deaths in urban areas, a direct function of the fertility rate as well as the quality of healthcare systems (lower mortality rates, particularly for infants). Phases in the demographic transition

are commonly linked with urbanization rates. Although this factor played an important role in the past, it is of much lesser importance today as fertility rates in many developed countries have dropped significantly, in some cases such as Western Europe, Japan and South Korea below replacement rate.

Rural to urban migrations. This has been a strong factor of urbanization, particularly in the developing world where migration accounted for 40–60 percent of the urban growth. Such a process has endured since the beginning of the industrial revolution in the nineteenth century, first in the developed world and then in the developing world. The reasons for urban migration are numerous and may involve the expectation to find employment, improved agricultural productivity which frees rural labor or political and environmental problems where populations are constrained to leave the countryside.

International migration. The growth in international migration has been an important factor in the urbanization of major gateway cities, such as Los Angeles, Miami, New York, London and Paris. This process has a tendency to take place in the largest cities, but there is a trickle down to cities of smaller size.

Current global trends indicate a growth of about 50 million urbanites each year, roughly a million a week. More than 90 percent of that growth occurs in developing countries which places intense pressures on urban infrastructures, particularly transportation, to cope. By 2050, 6.4 billion people, about two-thirds of humanity, are likely to be urban residents. What can be considered as urban includes a whole continuum of urban spatial structures, ranging from small towns to large urban agglomerations. This also raises the question about optimal city size since technical limitations (road, utilities) are not much of an impediment in building very large cities. Many of the world's largest cities (Figure 6.1) can be labeled as dysfunctional mainly because as city

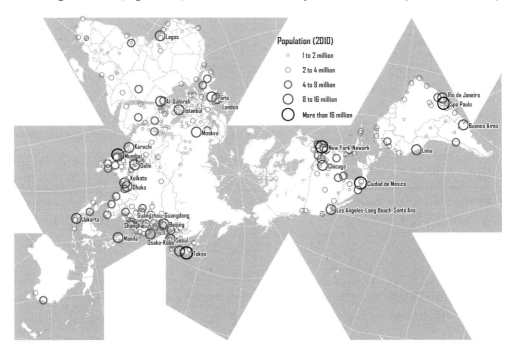

Figure 6.1 World's largest cities

size increases the rising complexities are not effectively coped with by managerial expertise.

Global urbanization and the emergence of large cities has been a significant process with far-reaching economic and social impacts:

- Urbanization involves a much higher level of **concentration** of the global population, which used to be more dispersed. Higher levels of concentration are easier to service from a market perspective, but are also prone to congestion and diseconomies.
- Several **mega-cities** of more than 8 million have emerged. These cities command a large share of the global wealth creation and dominate their respective national economies. Many act as global cities.
- **Urban residents** have different activities, lifestyles and levels of consumption than rural residents, particularly in developing countries.

The urban form

Demographic and mobility growth have been shaped by the capacity and requirements of urban transport infrastructures, such as roads, transit systems or simply walkways. Consequently, there is a wide variety of urban forms, spatial structures and associated urban transportation systems.

> **Urban form**. Refers to the spatial imprint of an urban transport system as well as the adjacent physical infrastructures. Jointly, they confer a level of spatial arrangement to cities.
>
> **Urban (spatial) structure**. Refers to the set of relationships arising out of the urban form and its underlying interactions of people, freight and information. It tries to evaluate to what extent specific urban structures can be achieved with specific transport systems.

In light of transport developments, the urban spatial structure can be categorized by its level of centralization and clustering (Figure 6.2):

> **Centralization**. Refers to the setting of activities in relation to the whole urban area. A centralized city has a significant share of its activities in its center while a decentralized city does not. Large employers such as financial institutions are the main drivers of centralization.
>
> **Clustering**. Refers to the setting of activities in relation to a specific part of the urban area. A cluster of activities is therefore a concentration around a specific focal point, which tends to be transport infrastructures such as a highway interchange, a transit terminal or a smaller town that has been absorbed by the expansion of the metropolis.

Although the four types of urban spatial structures shown in Figure 6.2 are possible, the most significant trend that has impacted urban spatial structures has been decentralization while maintaining a high level of clustering (Type C). This is reflective of a multicentric city. The decentralization of activities resulted in two opposite effects. First, commuting time remained relatively stable in duration. Second, commuting increasingly tends to be longer and made by using the automobile rather than by public

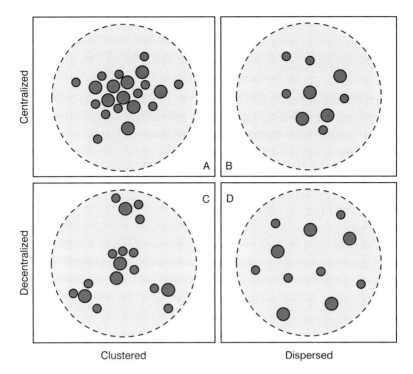

Figure 6.2 Types of urban spatial structures

transit. Most transit and road systems were developed to facilitate suburb-to-city, rather than suburb-to-suburb, commuting. As a result, suburban highways are often as congested as urban highways.

Dispersed urban land development patterns have been dominant in North America over the last 50 years, where land is abundant, transportation costs were low, and where the economy became dominated by tertiary and quaternary activities. Under such circumstances, it is not surprising to find that there is a strong relationship between urban density and automobile use. For many cities their built-up areas have grown at a faster rate than their populations. In addition, commuting became relatively inexpensive compared with land costs, so households had an incentive to buy lower-priced housing at the urban periphery. Similar patterns can be found in many European cities, but this change is occurring at a slower pace and involving a smaller range.

Even if the geographical setting of each city varies considerably, the urban form and its spatial structure are articulated by two structural elements:

Nodes. These are reflected in the centrality of urban activities, which can be related to the spatial accumulation of economic activities or to accessibility to the transport system. Terminals, such as ports, train station, rail yards and airports, are important nodes around which activities agglomerate at the local or regional level. Nodes have a hierarchy related to their importance and contribution to urban functions, with high order nodes such as management and retailing and lower order nodes such as production and distribution.

> **Linkages**. These are the infrastructures supporting flows from, to and between nodes. The lowest level of linkages includes streets, which are the defining elements of the urban spatial structure. There is a hierarchy of linkages moving up to regional roads and railways and international connections by air and maritime transport systems.

Depending on their nature, urban nodes and linkages provide for a functional connectivity, implying interdependent urban functions related to trade, production and telecommunications. Urban transportation is thus associated with a spatial form which varies according to the modes being used. What has not changed much is that cities tend to opt for a grid street pattern. This was the case for many Roman cities built in the first century as it was for American cities built in the twentieth century. The reasons behind this permanence are relatively simple: a grid pattern jointly optimizes accessibility and available real estate. Obviously, many cities are not organized as a grid. They correspond to older cities, many former fortified towns, as well as cities which grew from a constrained location such as an island or a river junction. Local geographical and historical characteristics remain important influences on the urban form.

In an age of motorization and personal mobility, an increasing number of cities are developing a spatial structure that increases reliance on motorized transportation, particularly the privately owned automobile. This has caused a shift from a grid pattern towards curvilinear and cul-de-sac patterns that are commonly found in suburban areas (Figure 6.3).

The conventional street grid is mostly the outcome of the streetcar suburbs that emerged in the early part of the twentieth century. It conferred optimal accessibility and use of available space. The diffusion of the automobile was a driver in the shift in the street network towards a more curvilinear pattern. This implied a reduction in the level of connectivity as well as the density of land use. This was part of a paradox: while the automobile was becoming the dominant support of urban mobility it was also increasingly associated with local disturbances. Planners responded by developing cul-de-sac suburban patterns with the goal to reduce and even eliminate through movements on a large number of residential streets and having them taking place on main arterials. By the 1950s, the conventional cul-de-sac pattern became prevalent in suburban

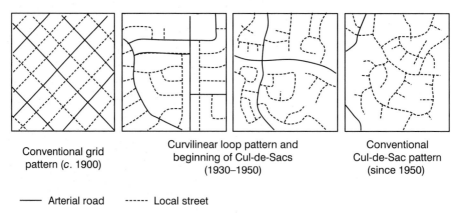

| Conventional grid pattern (c. 1900) | Curvilinear loop pattern and beginning of Cul-de-Sacs (1930–1950) | Conventional Cul-de-Sac pattern (since 1950) |

—— Arterial road   ------ Local street

Figure 6.3 Street network types

Source: Adapted from Marshall, W.E. and N.W. Garrick (2010) "Street network types and road safety," *Urban Design International*, April 21.

developments. Although this pattern minimizes non-local circulation, it also generates more movements and energy consumption.

Dispersion, or urban sprawl, is taking place in many different types of cities, from dense, centralized European metropolises such as Madrid, Paris (Photo 6.1) and London, to rapidly industrializing metropolises such as Seoul, Shanghai and Buenos Aires, to those experiencing recent, fast and uncontrolled urban growth, such as Mumbai and Lagos. Recent urban expansion is consequently almost all geared towards the automobile.

Evolution of transportation and urban form

Historically, movements within cities tended to be restricted to walking, which made medium and long distance urban linkages rather inefficient and time-consuming. Thus, activity nodes tended to be agglomerated and urban forms compact. Many modern cities have inherited an urban form created under such circumstances, even though they are no longer prevailing. The dense urban cores of many European, Japanese and Chinese cities, for example, enable residents to make between one-third and two-thirds of all trips by walking and cycling. At the other end of the spectrum, the dispersed urban forms of most Australian, Canadian and American cities, which were built relatively recently, encourage automobile dependency and are linked with high levels of mobility. Many major cities are also port cities with maritime accessibility playing an enduring role not only for the economic vitality but also in the urban spatial structure

Photo 6.1 High density structured urban form, Paris, France

Paris, like many European cities, has a prominent core, high land use density and high levels of accessibility to urban transit. Still, it remains highly motorized with limited parking space. Photo: Claude Comtois.

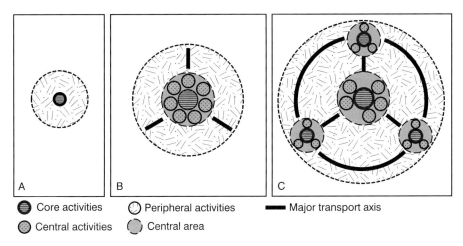

Core activities   Peripheral activities   ▬▬ Major transport axis
Central activities   Central area

Figure 6.4   Evolution of the spatial structure of a city

with the port district being an important node. Airport terminals have also been playing a growing role in the urban spatial structure as they can be considered as cities within cities.

The evolution of transportation has generally led to changes in urban form. The more radical the changes in transport technology have been, the more the alterations on the urban form. Among the most fundamental changes in the urban form is the emergence of new clusters expressing new urban activities and new relationships between elements of the urban system. In many cities, the central business district (CBD), once the primary destination of commuters and serviced by public transportation, has been changed by new manufacturing, retailing and management practices. Whereas traditional manufacturing depended on centralized workplaces and transportation, technological and transportation developments have rendered modern industry more flexible. In many cases, manufacturing relocated in a suburban setting, if not altogether to entirely new low cost locations offshore. Retail and office activities are also suburbanizing, producing changes in the urban form. Concomitantly, many important transport terminals, namely port facilities and rail yards, have emerged in suburban areas following new requirements in modern freight distribution brought in part by containerization. The urban spatial structure shifted from a nodal to a multi-nodal character (Figure 6.4).

The urban spatial structure basically considers the location of different activities as well as their relationships. Core activities are those of the highest order in the urban spatial structure, namely tertiary and quaternary activities involved in management (finance and insurance) and consumption (retailing). Central activities are concerned by production and distribution with activities such as warehousing, manufacturing, wholesaling and transportation. Peripheral activities are predominantly residential or servicing local needs. A central area refers to an agglomeration of core and/or central activities within a specific location. The emergence of a CBD (in the central area of a city) is the result of a historical process, often occurring over several centuries (depending on the age of a city), that has changed the urban form and the location of economic activities. Obviously, each city has its own history, but it is possible to establish a general common process:

Pre-industrial era. For cities that existed before the industrial revolution, the CBD was limited to a small section of the city generally near the waterfront, the market and/or a site of religious or political importance. These were locations where major transactions took place and thus required financial, insurance, warehousing and wholesale services.

Industrial revolution. With the industrial revolution came mass production and mass consumption. This permitted the emergence of a distinct retailing and whole-saling part of the CBD while manufacturing located outside the core. Major terminal facilities, such as ports and rail yards, were also located in proximity to the city core. Managing these expanding activities also created an increasing need for office space that located nearby traditional places of financial interaction. As the industrial revolution matured, major transportation axes spurred from the central area towards the periphery.

Contemporary era. After the Second World War, industries massively relocated away from central areas to suburban areas, leaving room for the expansion of admin-istrative and financial activities. The CBD was thus the object of an important accu-mulation of financial and administrative activities, particularly in the largest cities as several corporations became multinational enterprises. These activities were even more willing to pay higher rents than retailing, thereby pushing some retail activities out of the CBD. New retailing sub-centers emerged in suburban areas because of road accessibility and because of the need to service these new areas. Warehousing and transportation, no longer core area activities, have also relocated to new periph-eral locations close to modern terminal facilities such as container terminals and airports. The spatial structure of many cities became increasingly multi-nodal (or multicentric).

Initially, suburban growth mainly took place adjacent to major road corridors, leaving plots of vacant or farm land in between. Later, intermediate spaces were gradually filled up, more or less coherently. Highways and ring roads, which circled and radiated from cities, favored the development of suburbs and the emergence of important sub-centers that compete with the central business district for the attraction of economic activities. As a result, many new job opportunities have shifted to the suburbs (if not to entirely new locations abroad) and the activity system of cities has been considerably modified. Different parts of a city have a different dynamism depending on its spatial pattern. These changes have occurred according to a variety of geographical and historical contexts, notably in North America and Europe as each subsequent phase of urban transportation developments led to different spatial structures. Sometimes, particularly when new modern urban road infrastructures are built, the subsequent changes in the urban form can be significant.

Although transportation systems and travel patterns have changed considerably over time, one enduring feature remains that most people travel for between 30 and 40 minutes in one direction. Globally, people are spending about 1.2 hours per day commuting, wherever this takes place in a low or a high mobility setting. Different transport technologies, however, are associated with different travel speeds and capacity. As a result, cities that rely primarily on non-motorized transport tend to be different than auto-dependent cities. Transport technology thus plays a very important role in defining urban form and the spatial pattern of various activities. Still, the evolution of the urban form is path-dependent, implying that the current spatial structure is obviously the outcome of past developments, but that those developments were strongly related

to local conditions involving the setting, physical constraints and investments in infrastructures and modes.

The spatiality of urban transportation

The amount of urban land allocated to transportation is often correlated with the level of mobility. In the pre-automobile era, about 10 percent of the urban land was devoted to transportation which was simply roads for a predominantly pedestrian traffic. As the mobility of people and freight increased, a growing share of urban areas was allocated to transport and the infrastructures supporting it. Large variations in the spatial imprint of urban transportation are observed between different cities as well as between different parts of a city, such as between central and peripheral areas. The major components of the spatial imprint of urban transportation are:

Pedestrian areas. Refer to the amount of space devoted to walking. This space is often shared with roads as sidewalks may use between 10 and 20 percent of a road's right of way. In central areas, pedestrian areas tend to use a greater share of the right of way and in some instances whole areas are reserved for pedestrians. However, in a motorized context, most pedestrian areas are for servicing people's access to transport modes such as parked automobiles.

Roads and parking areas. Refer to the amount of space devoted to road transportation, which has two states of activity: moving or parked. In a motorized city, on average 30 percent of the surface is devoted to roads while another 20 percent is required for off-street parking. This implies for each car about 2 off-street and 2 on-street parking spaces. In North American cities, roads and parking lots account for between 30 and 60 percent of the total surface.

Cycling areas. In a disorganized form, cycling simply shares access to pedestrian and road space. However, many attempts have been made to create spaces specifically for bicycles in urban areas, with reserved lanes and parking facilities (Photo 6.2). The Netherlands has been particularly proactive over this issue making biking paths part of the urban transport system: 27 percent of the total amount of commuting is by cycling.

Transit systems. Many transit systems, such as buses and tramways, share road space with automobiles, which often impairs their respective efficiency. Attempts to mitigate congestion have resulted in the creation of road lanes reserved for buses either on a permanent or temporary (during rush hour) basis. Other transport systems such as subways and rail have their own infrastructures and, consequently, their own rights of way.

Transport terminals. Refer to the amount of space devoted to terminal facilities such as ports, airports, transit stations, rail yards and distribution centers. Globalization has increased the mobility of people and freight, both in relative and absolute terms, and consequently the amount of urban space required to support those activities. Many major terminals are located in the peripheral areas of cities, which are the only locations where sufficient amounts of land are available.

The spatial importance of each transport mode varies according to a number of factors, density being the most important. Further, each transport mode has unique performance and space consumption characteristics. The most relevant example is the automobile.

Photo 6.2   Bicycle pool, Paris, France

The pooling of vehicles for short-term rent is an option that is increasingly being considered and applied. This photo shows the "Velib" (Velo Libre) initiative in Paris, France, where bicycles are offered for rental for less than 24 hours, and ideally for less than two hours. The system is composed of 1,450 bicycle rental stations that have an average separation of 300 meters with a pool of 20,000 bicycles. The pooling system turned out to be very popular but also had some problems, notably a high level of vandalism and the need to reposition bicycles every night because of commuting patterns.

It requires space to move around (roads) but it also spends 98 percent of its existence stationary in a parking space. Consequently, a significant amount of urban space must be allocated to accommodate the automobile, especially when it does not move and is thus economically and socially useless. In large urban agglomerations almost all the available street parking space in areas of average density and above is occupied throughout the day. At an aggregate level, measures reveal a significant spatial imprint of road transportation among developed countries. In the United States, more land is thus used by the automobile than for housing. In Western Europe, roads account for between 15 percent and 20 percent of the urban surface while for developing countries this figure is about 10 percent (6 percent on average for Chinese cities but growing fast due to motorization).

Concept 2 – Urban land use and transportation

The land use – transport system

Urban land use comprises two elements: the nature of land use, which relates to which activities are taking place where, and the level of spatial accumulation, which indicates intensity and concentration of activities. Central areas have a high level of spatial accumulation and corresponding land uses, such as retail, while peripheral areas have lower levels of accumulation. Most economic, social or cultural activities imply a multitude of functions, such as production, consumption and distribution. These functions take place at specific locations and are part of an activity system. Some are routine activities, because they occur regularly and are thus predictable, such as commuting and shopping. Others are institutional activities that tend to be irregular. Others are production activities that are related to manufacturing and distribution, whose linkages may be local, regional or global. The behavioral patterns of individuals, institutions and firms have an imprint on land use in terms of their locational choice. The representation of this imprint requires a typology of land use, which can be formal or functional:

> **Formal land use** representations are concerned with qualitative attributes of space such as its form, pattern and aspect and are descriptive in nature.
> **Functional land use** representations are concerned with the economic nature of activities such as production, consumption, residence and transport, and are mainly a socioeconomic description of space.

Within an urban system each activity occupies a suitable, but not necessarily optimal location, from which it derives rent. Transportation and land use interactions mostly consider the retroactive relationships between activities, which are land use related, and accessibility, which is transportation related. These relationships have often been described as a classic "chicken-and-egg" problem since it is difficult to identify the triggering cause of change: do transportation changes precede land use changes or vice versa? There is a scale effect at play as large infrastructure projects tend to precede and trigger land use changes while small-scale transportation projects tend to complement the existing land use pattern. Further, the expansion of urban land uses takes place over various circumstances such as infilling (near the city center) or sprawl (far from the city center) and where transportation plays a different role in each case.

Urban transportation aims at supporting transport demands generated by the diversity of urban activities in a diversity of urban contexts. A key for understanding urban entities thus lies in the analysis of patterns and processes of the transport/land use system. This system is highly complex and involves several relationships between the transport system, spatial interactions and land use:

- **Transport system**. Considers the set of transport infrastructures and modes that support urban movements of passengers and freight. It generally expresses the level of accessibility.
- **Spatial interactions**. Consider the nature, extent, origins and destinations of the urban movements of passengers and freight. They take into consideration the attributes of the transport system as well as the land use factors that are generating and attracting movements.
- **Land use**. Considers the level of spatial accumulation of activities and their associated levels of mobility requirements. Land use is commonly linked with demographic and economic attributes.

Urban activities such as retail or manufacturing have spatial locations from which a land use pattern is derived and influenced by the existing urban form and spatial structure (Figure 6.5). This form is strongly related to the types of activities that can roughly be divided into three major classes:

> **Routine activities** occur regularly and are thus predictable. They involve journey to work (residential to industrial/commercial/administrative) and shopping (residential to retailing). The land use pattern generated is thus stable and coherent. Generally, these activities are zonal and links are from areas to areas.

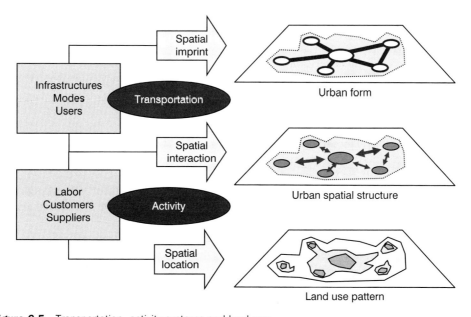

Figure 6.5 Transportation, activity systems and land use

Institutional activities. Most institutions are located at specific points and generally have links with individuals. This activity system is linked to an urban environment where links occur irregularly and according to lifestyle (students, sports, leisure, etc.) or special needs (health).

Production activities involve a complex network of relationships between firms, such as control, distribution, warehousing and subcontracting. This activity system can be linked to a specific urban environment, but also to a region, nation or even the world. Some activities are strongly linked to the local urban area, while others are far more linked to the global economy. The land use pattern of an activity may thus be linked to an external (international) process.

A conundrum concerns the difficulties of linking a specific mode of transportation with specific land use patterns. While public transit systems tend to be associated with higher densities of residential and commercial activities and highways with lower densities, the multiplicity of modes available in urban areas, including freight distribution, conveys an unclear and complex relationship.

Urban land use models

The relationships between transportation and land use are rich in theoretical representations that have contributed much to regional sciences. Since transportation is a distance–decay altering technology, spatial organization is assumed to be strongly influenced by the concepts of location and distance. Several descriptive and analytical models of urban land use have been developed over time, with increased levels of complexity. All involve some consideration of transport in the explanations of urban land use structures.

Von Thunen's regional land use model is the oldest representation based on a **central place, the market town, and its concentric impacts** on surrounding land uses (Figure 6.6). It was initially developed in the early nineteenth century (1826) for the analysis of agricultural land use patterns in Germany. It used the concept of economic rent to explain a spatial organization where different agricultural activities are competing for the usage of land. The underlying principles of this model have been the foundation of many others where economic considerations, namely land rent and distance–decay, are incorporated. The core assumption of the model is that agricultural land use is patterned in the form of concentric circles around a market that consumes all the surplus production, which must be transported. The relative costs of transporting different agricultural commodities to the central market determined the agricultural land use around a city. The most productive activities will thus compete for the closest land to the market and activities not productive enough will locate further away. The relationships between agricultural land use and market distance are very difficult to establish in the contemporary context. However, a strong relationship between the transport system and regional agricultural land use patterns can be acknowledged at the continental level in North America.

The **Burgess concentric model** was among the first attempts to investigate spatial patterns at the urban level (1925). Although the purpose of the model was to analyze social classes, it recognized that transportation and mobility were important factors behind the spatial organization of urban areas (Figure 6.7). The formal land use representation of this model is derived from commuting distance from the central business district, creating concentric circles. Each circle represents a specific socioeconomic

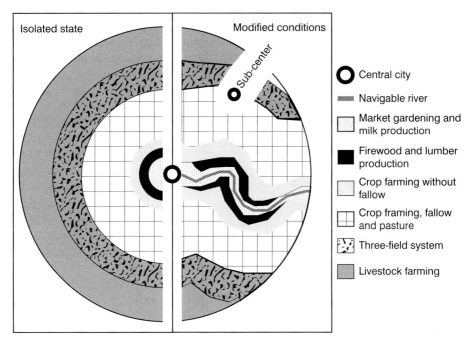

Figure 6.6 Von Thunen's regional land use model

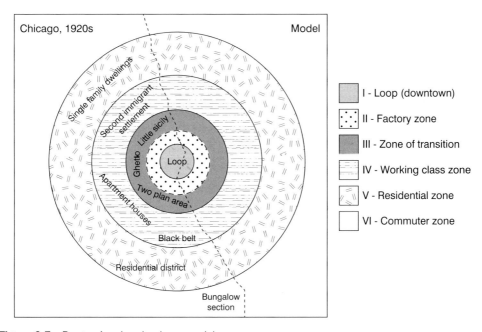

Figure 6.7 Burgess's urban land use model

urban landscape. This model is conceptually a direct adaptation of the Von Thunen's model to urban land use since it deals with a concentric representation. Even close to one century after the concentric urban model was designed, spatial changes in Chicago are still reflective of such a process.

Sector and multiple nuclei land use models were developed to take into account numerous factors overlooked by concentric models, namely the influence of transport axes (Hoyt, 1939) and multiple nuclei (Harris and Ullman, 1945) on land use and growth. Both representations consider the emerging impacts of motorization on the urban spatial structure. Such representations also considered that transportation infra-structures, particularly terminals such as rail stations or ports, occupy specific locations and can be considered as land uses.

Moving further, **hybrid models** are an attempt to include the concentric, sector and nuclei behavior of different processes in explaining urban land use. They try to integrate the strengths of each approach since none of these appears to provide a completely satisfactory explanation. Thus, hybrid models consider the concentric effect of central locations (CBDs and sub-centers) and the radial effect of transport axes, all overlaid to form a land use pattern. Also, hybrid representations are suitable to explain the evolution of the urban spatial structure as they combine different spatial impacts of transportation on urban land use, whether concentric or radial, and this at different points in time.

Land rent theory was also developed to explain land use as a market where different urban activities are competing for land usage at a location. It is strongly based in the market principle of spatial competition where actors are bidding to secure and maintain their presence at a specific location. The more desirable a location is, the higher its rent value (Figure 6.9). By overlapping the bid rent curves of all the urban economic

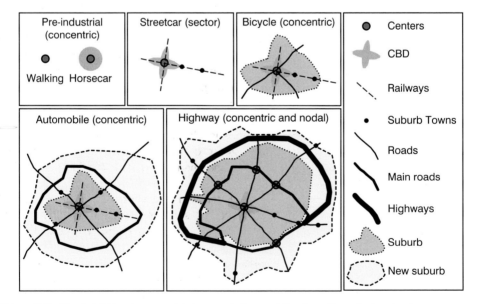

Figure 6.8 The hybrid land use model: transportation and the formation of urban landscapes

Source: Adapted from Taaffe, E.J., H.L. Gauthier and M.E. O'Kelly (1996) *Geography of Transportation*, 2nd edition, Upper Saddle River, NJ: Prentice Hall.

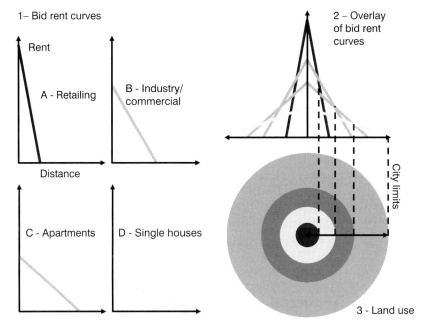

Figure 6.9 Land rent and land use

activities a concentric land use pattern is created with retailing in the CBD, industry/commercial on the next ring, apartments farther on and then single houses. This representation considers an isotropic space. In the real world a set of physiographic (waterfront, hills, etc.), historical (tourism) and social (race, crime, perception) attributes will influence bid rent curves.

Transportation, through accessibility and distance–decay, is a strong explanatory factor on the land rent and its impacts on land use. However, conventional representations of land rent leaning on the concentric paradigm are being challenged by structural modifications of contemporary cities, including:

The downtown area is **not necessarily the most accessible location**. The rapid extension of metropolitan areas involves new locations far from the CBD, notably in suburbia. This has favored the emergence of sub-centers having a concentration of retailing, commercial, distribution and industrial activities, mainly aimed at servicing a growing population.

Improvements in transportation and telecommunications have made several activities **far more tolerant to distance**, but still dependent on accessibility. The urban land use pattern thus tends to be far less coherent, more specialized and dispersed.

A significant share of the land, notably nearby central areas, is captured and **not available on real estate markets**. Governments, institutions, parks, industries and transport infrastructures occupy a large part of most central areas and this ownership can last for several decades (if not several centuries for historical landmarks). This caused an imbalance in the price fixing mechanism in central areas with less land available (thus higher prices) that has favored urban sprawl.

Cellular automata are dynamic land use models developed on the principle that space can be represented as a grid where each cell is a discrete land use unit (Figure 6.10). Cell states thus symbolize land uses and transition rules express the likelihood of a change from one land use state to another. Because cells are symbolically connected and inter-related (e.g. adjacency), models can be used to investigate the dynamics, evolution and self-organization of cellular automata land use systems. Cellular approaches achieve a high level of spatial detail (resolution) and realism, as well as linking the simulation directly to visible outcomes on the regional spatial structure. They are also readily implementable since Geographic Information Systems are designed to work effectively with grid-based spatial representations.

Cellular automata paradigm improves upon most transportation–land use models that are essentially static as they explain land use patterns, but they do not explicitly consider the processes that are creating or changing them. The applicability and dynamics of land use models is related to issues such as the age, size and the locational setting of a city. For instance, concentric cities are generally older and of smaller size, while polycentric cities are larger and relate to urban development processes that took place more recently. While most of the conceptual approaches related to transportation and land use relation-ships have been developed using empirical evidence related to North America and Western Europe, this perspective does not necessarily apply to other parts of the world. A dualism in land uses has been observed in cities in developing countries where on one hand processes such as economic development and motorization are creating an urban landscape that is common in advanced economies. On the other hand an informal land-scape of shantytowns represents a land use structure that is not effectively captured by conventional land use models. It remains to be seen to what extent globalization will favor a convergence of land use patterns across the world's cities.

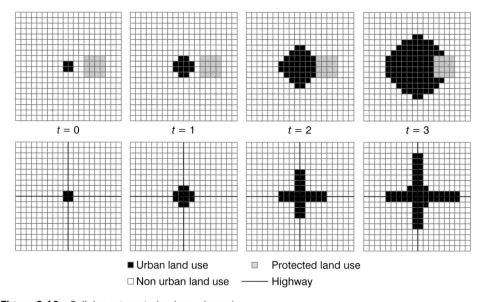

Figure 6.10 Cellular automata land use dynamics

Source: Adapted from Kim, D. and M. Batty (2011) "Calibrating cellular automata models for simulating urban growth: comparative analysis of SLEUTH and Metronamica", UCL Center for Advanced Spatial Analysis, Paper 176.

Transportation and urban dynamics

Both land use and transportation are part of a dynamic system that is subject to external influences. Each component of the system is constantly evolving due to changes in technology, policy, economics, demographics and even culture or values. As a result, the interactions between land use and transportation are played out as the outcome of the many decisions made by residents, businesses and governments. The field of urban dynamics has expanded the scope of conventional land use models, which tended to be descriptive, by trying to consider relationships behind the evolution of the urban spatial structure. This has led to a complex modeling framework including a wide variety of components. Among the concepts supporting urban dynamic representations are retroactions, whereby one component changes its influence on others. The changes will influence the initial component back, either positively or negatively. The most significant components of urban dynamics are:

Land use. This is the most stable component of urban dynamics, as changes are likely to modify the land use structure over a rather long period of time. This comes as little surprise since most real estate is built to last at least several decades. The main impact of land use on urban dynamics is its function as a generator and attractor of movements.

Transport network. This is also considered to be a rather stable component of urban dynamics, as transport infrastructures are built for the long term. This is particularly the case for large transport terminals and subway systems that can operate for a very long period of time. For instance, many railway stations are more than one hundred years old. The main contribution of the transport network to urban dynamics is the provision of accessibility. Changes in the transport network will impact accessibility and movements.

Movements. The most dynamic component of the system since movements of passengers or freight reflect changes almost immediately. Movements thus tend more to be the outcome of urban dynamics than a factor shaping them.

Employment and workplaces. They account for significant inducement effects over urban dynamics since many models often consider employment as an exogenous factor. This is specifically the case for employment that is categorized as basic, or export-oriented, which is linked with specific economic sectors such as manufacturing. Commuting is a direct outcome of the number of jobs and the location of workplaces.

Population and housing. They act as the generators of movements, because residential areas are the sources of commuting. Since there are a wide array of incomes, standards of living, preferences and ethnicity, this diversity is reflected in the urban spatial structure.

The issue about how to articulate these relations remains, particularly in the current context of interdependency between local, regional and global processes. Globalization has substantially blurred the relationships between transportation and land use as well as its dynamics. The main paradigm is concerned with factors once endogenous to a regional setting that have become exogenous. Consequently, many economic activities that provide employment and multiplying effects, such as manufacturing, are driven by forces that are global in scope and may have little to do with regional dynamics. For instance, capital investment could come from external sources and the bulk of the output could be bound for international markets. In such a context it would be difficult to explain urban development processes taking place in coastal Chinese cities, or in a region such as the Pearl River Delta, since export-oriented strategies are among the most significant driving forces. Looking at the urban dynamics of such a system from an endogenous perspective would fail to capture driving forces that are predominantly exogenous.

Concept 3 – Urban mobility

Urban mobility and its evolution

Urban transportation is organized in three broad categories of collective, individual and freight transportation. While passenger movements are the outcome of numerous individual decisions based on different rationales, freight movements are decided in tandem between the cargo owners (producers and customers) and the transportation service providers. In several instances, passengers and freight movements are complementary to one another, but sometimes they may be competing for passengers, the usage of available land and transport infrastructures:

> **Collective transportation** (public transit). The purpose of collective transportation is to provide publicly accessible mobility over specific parts of a city. Its efficiency is based upon transporting large numbers of people and achieving economies of scale. It includes modes such as tramways, buses, trains, subways and ferryboats.
>
> **Individual transportation**. Includes any mode where mobility is the outcome of a personal choice and means such as the automobile, walking, cycling and the motorcycle. The majority of people walk to satisfy their basic mobility, but this number varies according to the city considered. For instance, walking accounts for 88 percent of all movements inside Tokyo while this figure is only 3 percent for Los Angeles.
>
> **Freight transportation**. As cities are dominant centers of production and consumption, urban activities are accompanied by large movements of freight. These movements are mostly characterized by delivery trucks moving between industries, distribution centers, warehouses and retail activities as well as from major terminals such as ports, rail yards, distribution centers and airports. The mobility of freight within cities tends to be overlooked.

Rapid urban development occurring across much of the globe implies increased quantities of passengers and freight moving within urban areas. Movements also tend to involve longer distances, but evidence suggests that commuting times have remained relatively similar through the last hundred years, approximately 1 to 1.2 hours per day. This means that commuting has gradually shifted to faster transport modes and consequently greater distances could be traveled using the same amount of time. Different transport technologies and infrastructures have been implemented, resulting in a wide variety of urban transport systems around the world. In developed countries, there have been three general eras of urban development, and each is associated with a different form of urban mobility:

> **The walking-horsecar era (1800–90)**. Even during the onslaught of the industrial revolution, the dominant means of getting around was on foot. Cities were typically less than 5 kilometers in diameter, making it possible to walk from the downtown to

the city edge in about 30 minutes. Land use was mixed and density was high (e.g. 100 to 200 people per hectare). The city was compact and its shape was more or less circular. The development of the first public transit in the form of omnibus services extended the diameter of the city but did not change the overall urban structure. The railroad facilitated the first real change in urban morphology. These new developments, often referred to as trackside suburbs, emerged as small nodes that were physically separated from the city itself and from one another. The nodes coincided with the location of rail stations and stretched out a considerable distance from the city center, usually up to a half-hour train ride. Within the city proper, rail lines were also laid down and horsecars introduced mass transit.

The electric streetcar or transit era (1890–1920s). The invention of the electric traction motor created a revolution in urban travel. The first electric trolley line opened in 1888 in Richmond. The operating speed of the electric trolley was three times faster than that of horse-drawn vehicles. The city spread outward 20 to 30 kilometers along the streetcar lines, creating an irregular, star-shaped pattern. The urban fringes became areas of rapid residential development. Trolley corridors became commercial strips. The city core was further entrenched as a mixed-use, high density zone. Overall densities were reduced to between 50 and 100 people per hectare. Land use patterns reflected social stratification where suburban outer areas were typically middle class while the working class continued to concentrate in the central city. As street congestion increased in the first half of the twentieth century, the efficiency of streetcar systems deteriorated and fell out of favor; many were abandoned.

The automobile era (1930 onward). The automobile was introduced in European and North American cities in the 1890s, but only the wealthy could afford this innovation. From the 1920s, ownership rates increased dramatically, with lower prices made possible by Henry Ford's revolutionary assembly-line production techniques. As automobiles became more common, land development patterns changed. Developers were attracted to greenfield areas located between the suburban rail axes, and the public was attracted to these single-use zones, thus avoiding many inconveniences associated with the city, mainly pollution, crowding and lack of space. Transit companies ran into financial difficulties and eventually transit services throughout North America and Europe became subsidized, publicly owned enterprises. As time went on, commercial activities also began to suburbanize. Within a short time, the automobile was the dominant mode of travel in all cities of North America. The automobile reduced the friction of distance considerably which led to urban sprawl.

In many areas of the world where urbanization is more recent, the above synthetic phases did not took place. In the majority of cases fast urban growth led to a scramble to provide transport infrastructure in an inadequate fashion. Each form of urban mobility, be it walking, the private car or urban transit, has a level of suitability to fill mobility needs. Motorization and the diffusion of personal mobility has been an ongoing trend linked with substantial declines in the share of public transit in urban mobility.

A taxonomy of urban mobilities

Movements are linked to specific urban activities and their land use. Each type of land use involves the generation and attraction of a particular array of movements. This relationship is complex, but is linked to factors such as recurrence, income, urban form, spatial accumulation, level of development and technology. Urban movements are either obligatory, when they are linked to scheduled activities (such as home-to-work movements), or voluntary, when those generating it are free to decide their scheduling (such as leisure). The most common types of urban movements are:

- **Pendulum movements**. These are obligatory movements involving commuting between locations of residence and work. They are highly cyclical since they are predictable and recurring on a regular basis, most of the time a daily occurrence, thus the term pendulum.
- **Professional movements**. These are movements linked to professional, work-based, activities such as meetings and customer services, mainly taking place during work hours.
- **Personal movements**. These are voluntary movements linked to the location of commercial activities, which includes shopping and recreation.
- **Touristic movements**. Important for cities having historical and recreational features they involve interactions between landmarks and amenities such as hotels and restaurants. They tend to be seasonal in nature or occurring at specific moments. Major sports events such as the World Cup or the Olympics are important generators of urban movements during their occurrence.
- **Distribution movements**. These are concerned with the distribution of freight to satisfy consumption and manufacturing requirements. They are mostly linked to transport terminals, distribution centers and retail outlets.

The consideration of urban movements involves their generation, the modes and routes used and their destination:

- **Trip generation**. On average, an urban resident undertakes between 3 and 4 trips per day. Moving in an urban area is usually done to satisfy a purpose such as employment, leisure or access to goods and services. Each time a purpose is satisfied, a trip is generated. Important temporal variations of the number of trips by purpose are observed with the most prevalent pattern being pendulum movements.
- **Modal split**. Implies which transportation mode is used for urban trips and is the outcome of a modal choice. Modal choice depends on a number of factors such as technology, availability, preference, travel time (distance) and income. For instance, everything within five minutes of walking is considered to be readily accessible to pedestrians.
- **Trip assignment (routing)**. Involves which routes will be used for journeys within the city. For instance, a commuter driving a car has most of the time a fixed route. This route may be modified if there is congestion or if another activity (such as shopping) is linked with that trip; a practice often known as trip chaining. Several factors influence trip assignment, the two most important being transport costs and availability.
- **Trip destination**. Changes in the spatial distribution of economic activities in urban areas have caused important modifications to the destination of movements, notably those related to work. Activity-based considerations are important since each economic activity tends to be associated with a level of trip attraction. Retail, public administration, entertainment and restoration are the activities that attract the most movements per person employed. The central city used to be a major destination for

movements, but its share has substantially declined in most areas and suburbs now account for the bulk of urban movements.

Mobility is also a social equity issue. The share of the automobile in urban trips varies in relation to location, social status, income, quality of public transit and parking availability. Mass transit is often affordable, and several social groups, such as students, the elderly and the poor, are a captive market. There are important variations in mobility according to age, income, gender and disability. The gender gap in mobility is the outcome of socioeconomic differences as access to individual transportation is mostly a matter of income. Consequently, in some instances modal choice is more a modal constraint linked to economic opportunities.

Central locations generally have the most urban mobility options because private and public transport facilities are present. However, this does not mean that mobility is easier since central areas are congested. In locations outside the central core, a share of the population not having access to the automobile faces a level of isolation, or at least a more limited access to amenities and employment opportunities. Limited public transit and high automobile ownership costs have created a class of spatially constrained (mobility deprived) people.

Urban transit

Transit is almost exclusively an urban transportation mode, particularly in large urban agglomerations. The urban environment is particularly suitable for transit because it provides conditions fundamental to its efficiency, namely high density and significant short distance mobility demands. Since transit is a shared service, it potentially benefits from economies of agglomeration related to high densities and from economies of scale related to high mobility demands. The lower the density in which a transit system is operating, the lower the demand, with the greater likelihood that it will be run at a loss. In fact, most public transit systems are not financially sound and have to be subsidized. Transit systems are made up of many types of services, each suitable to a specific set of market and spatial contexts. Different modes are used to provide complementarity services within the transit system and in some cases between the transit system and other transport systems.

There are approximately 184 subway (metro) systems in operation around the world, with several carrying more than 1 billion passengers per year (Figure 6.11). The construction and setting of subway systems has accelerated in recent years, particularly in cities in developing economies where mass transit becomes a clear strategy to improve urban mobility and mitigate congestion. The level of ridership is linked with several geographical and economic considerations. Some cities with high ridership such as Moscow, Beijing and Shanghai have a tradition of subsidized public transportation and transit-oriented urban planning. Many high density cities have a level of subway ridership proportional to their population (London, Hong Kong, Osaka, Paris, Seoul and Tokyo) indicating a consistent level of ridership where public transit accounts for between 25 and 50 percent of commuting. Cities where ridership is proportionally less than the population have either a significant portion of their population too poor to afford the subway (Sao Paulo and Mexico), or wealthy enough to prefer the automobile (New York).

Contemporary transit systems tend to be publicly owned, implying that many decisions related to their development and operation are politically motivated. This is a sharp contrast to what took place in the past as most transit systems were private and

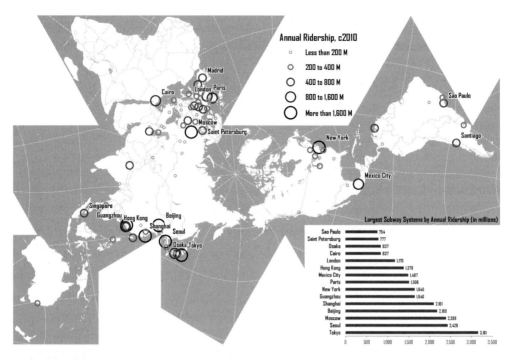

Figure 6.11 World's main subway systems, c. 2010

Source: Adapted from Metrobits.org, World Metro Database.

profit-driven initiatives. With the fast diffusion of the automobile in the 1950s, many transit companies faced financial difficulties, and the quality of their service declined as in a declining market there were limited incentives to invest. Gradually, they were purchased by public interests and incorporated into large agencies, mainly for the sake of providing mobility. As such, public transit often serves more a social function of public service and a tool of social equity than having any sound economic role. Transit has become dependent on government subsidies, with little if any competition permitted as wages and fares are regulated. As a result, they tend to be disconnected from market forces and subsidies are constantly required to keep a level of service. With suburbanization transit systems tend to have even less of a relationship with economic activities.

Government-owned public transit systems are facing financial difficulties for three main reasons. The first is that they are often designed to service taxpayers, not necessarily potential customers. Because of the funding base, transit systems may be spread into neighborhoods that do not provide a significant customer base. The second is that transit unions were able to extract significant advantages in terms of wages and social benefits, increasing labor costs. The third concerns a technology fixation which pushes investment in high cost transit (e.g. light rail) while low cost solutions (buses) would have been sufficient to many transit systems, particularly in lower density areas.

Reliance on urban transit as a mode of urban transportation tends to be high in Asia (Photo 6.3), intermediate in Europe and low in North America. Since their inception in the early nineteenth century, comprehensive urban transit systems had significant impacts on the urban form and spatial structure, but this influence is receding. Three major classes of cities can be found in terms of the relationships they have with their transit systems (Figure 6.12):

- **Adaptive cities**. Represent true transit-oriented cities where urban form and urban land use developments are coordinated with transit developments. While central areas are adequately serviced by a metro system and are pedestrian friendly, peripheral areas are oriented along transit rail lines.
- **Adaptive transit**. Represents cities where transit plays a marginal and residual role and where the automobile accounts for the dominant share of movements. The urban form is decentralized and of low density.
- **Hybrids**. Represent cities that have sought a balance between transit development and automobile dependency. While central areas have an adequate level of service, peripheral areas are automobile-oriented.

Contemporary land development tends to precede the introduction of urban transit services, as opposed to concomitant developments in earlier phases of urban growth. Thus, new services are established once a demand is deemed to be sufficient, often after being the subject of public pressure. Transit authorities operate under a service warrant and are often running a recurring deficit as services are becoming more expensive to provide. This has led to a set of considerations aimed at a higher integration of transit in the urban planning process, especially in North America, where such a tradition is not well established. Still, in spite of decades of investment, North American public transit

Photo 6.3 BTS Skytrain, Bangkok

Many developing countries experienced a fast-paced urbanization but were not serviced by a subway system. By the 1980s, the needs to develop a rail transit system in Bangkok became urgent as the city's roadway system became increasingly congested. The Bangkok Mass Transit System, which is an elevated rail system (Skytrain) began operations in 1999. It is composed of 30 stations and carries around 600,000 people per day.

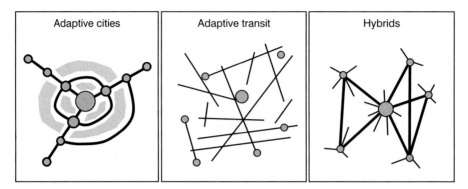

Figure 6.12 Transit and urban form

Source: Adapted from Cervero, R. (1998) *The Transit Metropolis*, Washington, DC: Island Press.

ridership roughly remained the same throughout the 1980s and 1990s, but has increased since then.

From a transportation perspective, the potential benefits of a better integration between transit and local land uses are reduced trip frequency and increased use of alternative modes of travel (i.e. walking, biking and transit). Evidence is often lacking to support such expectations as the relative share of public transit ridership is declining across the board. Community design can consequently have a significant influence on travel patterns. Land use initiatives should be coordinated with other planning and policy initiatives to cope with automobile dependence. However, there is a strong bias against transit in the general population because of negative perceptions, especially in North America, but increasingly globally. As personal mobility is a symbol of status and economic success, the users of public transit are perceived as the least successful segment of the population. This bias may undermine the image of transit use within the general population.

Concept 4 – Urban transport problems

Challenges facing urban transportation

Cities are locations having a high level of accumulation and concentration of economic activities and are complex spatial structures that are supported by transport systems. The larger the city, the greater its complexity and the potential for disruptions, particularly when this complexity is not effectively managed. The most important transport problems are often related to urban areas and take place when transport systems, for a variety of reasons, cannot satisfy the numerous requirements of urban mobility. Urban productivity is highly dependent on the efficiency of its transport system to move labor, consumers and freight between multiple origins and destinations. Additionally, important transport terminals such as ports, airports and rail yards are located within urban areas, contributing to a specific array of problems. Some problems are ancient, like congestion (which plagued cities such as Rome), while others are new like urban freight distribution or environmental impacts. Among the most notable urban transport problems are:

Traffic congestion and parking difficulties. Congestion is one of the most prevalent transport problems in large urban agglomerations, usually above a threshold of about one million inhabitants. It is particularly linked with motorization and the diffusion of the automobile, which has increased the demand for transport infrastructures. However, the supply of infrastructures has often not been able to keep up with the growth of mobility. Since vehicles spend the majority of the time parked, motorization has expanded the demand for parking space, which has created space consumption problems particularly in central areas; the spatial imprint of parked vehicles is significant. Congestion and parking are also interrelated since looking for a parking space (called "cruising") creates additional delays and impairs local circulation. In central areas of large cities cruising may account for more than 10 percent of the local circulation as drivers can spend 20 minutes looking for a parking spot. This practice is often judged more economically effective than using a paying off-street parking facility as the time spent looking for a free (or low cost) parking space is compensated by the monetary savings. Also, many delivery vehicles will simply double-park at the closest possible spot to unload their cargo.

Longer commuting. On par with congestion people are spending an increasing amount of time commuting between their residence and workplace. An important factor behind this trend is related to residential affordability as housing located further away from central areas (where most of the employment remains) is more affordable. Therefore, commuters are trading time for housing affordability. However, long commuting is linked with several social problems, such as isolation, as well as poorer health (obesity).

Public transport inadequacy. Many public transit systems, or parts of them, are either over- or under-used. During peak hours, crowdedness creates discomfort for users as the system copes with a temporary surge in demand. Low ridership makes many services financially unsustainable, particularly in suburban areas. In spite of significant subsidies and cross-financing (e.g. tolls) almost every public transit system cannot generate sufficient income to cover its operating and capital costs. While in the past deficits were deemed acceptable because of the essential service public transit was providing for urban mobility, its financial burden is increasingly controversial.

Difficulties for non-motorized transport. These difficulties are the outcome of intense traffic, where the mobility of pedestrians, bicycles and vehicles is impaired, and also because of a blatant lack of consideration for pedestrians and bicycles in the physical design of infrastructures and facilities.

Loss of public space. The majority of roads are publicly owned and free of access. Increased traffic has adverse impacts on public activities which once crowded the streets such as markets, agoras, parades and processions, games and community interactions. These have gradually disappeared to be replaced by automobiles. In many cases, these activities have shifted to shopping malls while in other cases they have been abandoned altogether. Traffic flows influence the life and interactions of residents and their usage of street space. More traffic impedes social interactions and street activities. People tend to walk and cycle less when traffic is high.

Environmental impacts and energy consumption. Pollution, including noise, generated by circulation has become a serious impediment to the quality of life and even the health of urban populations. Further, energy consumption by urban transportation has dramatically increased and so the dependency on petroleum. Yet, peak oil considerations are increasingly linked with peak mobility expectations where high energy prices cause a shift towards more efficient and sustainable forms of urban transportation, namely public transit.

Accidents and safety. Growing traffic in urban areas is linked with a growing number of accidents and fatalities, especially in developing countries. Accidents account for a significant share of recurring delays. As traffic increases, people feel less safe to use the streets.

Land consumption. The territorial imprint of transportation is significant, particularly for the automobile. Between 30 and 60 percent of a metropolitan area may be devoted to transportation, an outcome of the over-reliance on some forms of urban transportation. Yet, this land consumption also underlines the strategic importance of transportation in the economic and social welfare of cities.

Freight distribution. Globalization and the materialization of the economy have resulted in growing quantities of freight moving within cities. As freight traffic commonly shares infrastructures with the circulation of passengers, the mobility of freight in urban areas has become increasingly problematic. City logistics strategies can be established to mitigate the variety of challenges faced by urban freight distribution.

Many dimensions to the urban transport challenge are linked with the dominance of the automobile.

Automobile dependency

Automobile use is obviously related to a variety of advantages such as on-demand mobility, comfort, status, speed and convenience. These advantages jointly illustrate why automobile ownership continues to grow worldwide, especially in urban areas. When given the choice and the opportunity, most individuals will prefer using an automobile. Several factors influence the growth of the total vehicle fleet, such as sustained economic growth (increase in income and quality of life), complex individual urban movement patterns (many households have more than one automobile), more leisure time and suburbanization. Therefore, rising automobile mobility can be perceived as a positive consequence of economic development. The acute growth in the total number of vehicles also gives rise to congestion at peak traffic hours on major thoroughfares, in business districts and often throughout the metropolitan area.

Cities are important generators and attractors of movements, which have created a set of geographical paradoxes that are self-reinforcing. For instance, specialization leads to additional transport demands while agglomeration leads to congestion. Over time, a state of automobile dependency has emerged which results in a diminution in the role of other modes, thereby limiting still further alternatives to urban mobility. In addition to the factors contributing to the growth of driving, two major factors contributing to automobile dependency are:

Underpricing and consumer choices. Most road infrastructures are subsidized as they are considered a public service. Consequently, drivers do not bear the full cost of automobile use. Like the "Tragedy of the Commons", when a resource is free of access (road), it tends to be overused and abused (congestion). This is also reflected in consumer choice, where automobile ownership is a symbol of status, freedom and prestige, especially in developing countries. Single home ownership also reinforces automobile dependency.

Planning and investment practices. Planning and the ensuing allocation of public funds aim towards improving road and parking facilities in an ongoing attempt to avoid congestion. Other transportation alternatives tend to be disregarded. In many cases, zoning regulations impose minimum standards of road and parking services and de facto impose a regulated automobile dependency.

There are several levels of automobile dependency, ranging from low to acute, with their corresponding land use patterns and alternatives to mobility. Among the most relevant indicators of automobile dependency are the level of vehicle ownership, per capita motor vehicle mileage and the proportion of total commuting trips made using an automobile. A situation of high automobile dependency is reached when more than three-quarters of commuting trips are done using the automobile. For the United States, this proportion has remained around 88 percent over the recent decades. Automobile dependency is also served by a cultural and commercial system promoting the automobile as a symbol of status and personal freedom, namely through intense advertising and enticements to purchase new automobiles. Not surprisingly, many developing countries perceive motorization as a condition for development. Even if the term automobile dependency is often negatively perceived and favored by market distortions such as the provision of roads, its outcome reflects the choice of individuals who see the automobile more as an advantage then an inconvenience.

There is a growing body of evidence underlining that a peak level of car mobility is unfolding, at least in developed countries. Higher energy prices, congestion and the general aging of the population are all countervailing forces to car dependency. For instance, since 2006 the amount of vehicle-miles traveled in the United States has peaked, a process associated with higher energy prices and a strong recession. There are many alternatives to automobile dependency such as intermodality (combining the advantages of individual and collective transport), carpooling (strengthened by policy and regulation by the US government) or non-motorized transportation (walking and cycling). These alternatives can only be partially implemented as the automobile remains, in the short and medium term, the prime choice for providing urban mobility.

Congestion

Congestion occurs when transport demand exceeds transport supply at a specific point in time and in a specific section of the transport system. Under such circumstances, each vehicle impairs the mobility of others.

Congestion can be perceived as an unavoidable consequence of the usage of scarce transport resources, particularly if they are not priced. The last decades have seen the extension of roads in rural but particularly in urban areas, most of them free of access. Those infrastructures were designed for speed and high capacity, but the growth of urban circulation occurred at a rate higher than often expected. Investments came from diverse levels of government with a view to providing accessibility to cities and regions. There were strong incentives for the expansion of road transportation by providing high levels of transport supply. This has created a vicious circle of congestion which supports the construction of additional road capacity and automobile dependency. Urban congestion mainly concerns two domains of circulation, often sharing the same infrastructures:

Passengers. In many regions of the world incomes have significantly increased to the point that one automobile per household or more is common. Access to an automobile conveys flexibility in terms of the choice of origin, destination and travel time. The automobile is favored at the expense of other modes for most trips, including commuting.

Freight. Several industries have shifted their transport needs to trucking, thereby increasing the usage of road infrastructure. Since cities are the main destinations for freight flows (either for consumption or for transfer to other locations) trucking adds to further congestion in urban areas. The "last mile" problem remains particularly prevalent for freight distribution in urban areas. Congestion is commonly linked with a drop in the frequency of deliveries tying additional capacity to ensure a similar level of service.

It is important to underline that congestion in urban areas is predominantly caused by commuting patterns and little by truck movements. On average, infrastructure provision was not able to keep up with the growth in the number of vehicles, even less with the total number of vehicles-km. During infrastructure improvement and construction, capacity impairment (fewer available lanes, closed sections, etc.) favors congestion. Important travel delays occur when the capacity limit is reached or exceeded, which is the case in almost all metropolitan areas. In the largest cities such as London, road traffic is actually slower than it was 100 years ago. Marginal delays are thus increasing and driving speed becomes problematic as the level of population density increases. Once a population threshold of about one million is reached, cities start to experience recurring congestion problems. This observation must be nuanced by numerous factors related to the urban setting, modal preferences and the quality of existing urban transport infrastructures. Still, large cities have become congested most of the day, and congestion is getting more acute. Another important consideration concerns parking, which consumes large amounts of space and provides limited economic benefit. In automobile-dependent cities, this can be very constraining as each economic activity has to provide an amount of parking space proportional to their level of activity. Parking has become a land use that greatly inflates the demand for urban land (Photo 6.4).

Urban mobility also reveals congestion patterns. Daily trips can be either "mandatory" (workplace–home) or "voluntary" (shopping, leisure, visits). The former is often performed within fixed schedules while the latter complies with variable and discretionary schedules. Correspondingly, congestion comes in two major forms:

Recurrent congestion. The consequence of factors that cause regular demand surges on the transportation system, such as commuting, shopping or weekend trips. However, even recurrent congestion can have unforeseen impacts in terms of its duration and severity. Mandatory trips are mainly responsible for the peaks in circulation flows, implying that about half the congestion in urban areas is recurring at specific times of the day and on specific segments of the transport system.

Non-recurrent congestion. The other half of congestion is caused by random events such as accidents and unusual weather conditions (rain, snowstorms, etc.), which are unexpected and unplanned. Non-recurrent congestion is linked to the presence and effectiveness of incident response strategies. As far as accidents are concerned, their randomness is influenced by the level of traffic as the higher the traffic on specific road segments the higher the probability of accidents.

Photo 6.4   Parking in a public park, Brussels

Many cities are facing acute difficulties to provide parking space, especially in central areas. This problem is prevalent in Western Europe. In North American cities, suburbanization has often resulted in the abandonment of buildings in central areas. Many of these facilities have been torn down and converted into parking space (either as a vacant lot or as a dedicated multistoried facility), while awaiting a better use. This process was not so common in Western Europe where the real estate of most central areas has remained relatively unchanged and includes real estate of historical and architectural significance. To provide parking space, many European cities have been forced to build underground facilities that are expensive and rarely able to meet the demand. This photo was taken in downtown Brussels where motorists desperate for parking space have parked their vehicles along a walkway in a public space, which has been severely degraded.

Behavioral and response time effects are also important. In a system running close to capacity, a vehicle simply braking suddenly may trigger what is known as a backward traveling wave, that is as vehicles are forced to stop, the bottleneck moves up from the location where the braking initially took place, often leaving drivers puzzled about the cause. The spatial convergence of traffic causes a surcharge on transport infrastructures up to the point where congestion can lead to the total immobilization of traffic. Not only does the massive use of the automobile have an impact on traffic circulation and congestion, but it also leads to the decline in public transit efficiency when both are sharing the same roads.

The urban transit challenge

As cities continue to become more dispersed, the cost of building and operating public transportation systems increases. For instance, as of 2012 only about 184 urban agglomerations have a subway system, the great majority of them being in developed countries. Furthermore, dispersed residential patterns characteristic of automobile-dependent cities makes public transportation systems less convenient to support urban mobility. In many cities additional investments in public transit did not result in significant additional ridership. Unplanned and uncoordinated land development has led to rapid expansion of the urban periphery. Residents, by selecting housing in outlying areas, restrict their potential access to public transportation. Over-investment (when investments do not appear to imply significant benefits) and under-investment (when there is a substantial unmet demand) in public transit are both complex challenges.

Urban transit is often perceived as the most efficient transportation mode for urban areas, notably large cities. However, surveys reveal a stagnation of public transit systems, especially in North America. The economic relevance of public transit is being questioned. Most urban transit developments had little if any impact on alleviating congestion in spite of mounting costs and heavy subsidies. This paradox is partially explained by the spatial structure of contemporary cities which are oriented along servicing the needs of the individual, not necessarily the needs of the collectivity. Thus, the automobile remains the preferred mode of urban transportation. In addition, public transit is publicly owned, implying that it is a politically motivated service that provides limited economic returns. Even in transit-oriented cities such as in Europe, transit systems depend massively on government subsidies. Little or no competition is permitted as wages and fares are regulated, undermining any price adjustments to changes in ridership. Thus, public transit often serves the purpose of a social function ("public service") as it provides accessibility and social equity, but with limited relationships with economic activities. Among the most difficult challenges facing urban transit are:

Decentralization. Public transit systems are not designed to service low density and scattered urban areas that are increasingly dominating the landscape. The greater the decentralization of urban activities, the more difficult and expensive it becomes to serve urban areas with public transit. Additionally, decentralization promotes long distance trips on transit systems causing higher operating costs and revenue issues for flat fare transit systems.

Fixity. The infrastructures of several public transit systems, notably rail and subway systems, are fixed, while cities are dynamical entities, even if the pace of change can take decades. This implies that travel patterns tend to change and that a transit system built for servicing a specific pattern may eventually face "spatial obsolescence".

Connectivity. Public transit systems are often independent from other modes and terminals. It is consequently difficult to transfer passengers from one system to the other. This leads to a paradox between the preference of riders to have direct connections and the need to provide a cost-efficient service network that involves transfers.

Competition. In view of cheap and ubiquitous road transport systems, public transit faced strong competition and loss of ridership in relative terms and in some cases in absolute terms. The higher the level of automobile dependency, the more inappropriate the public transit level of service. The public service being offered is simply outpaced by the convenience of the automobile. However, changes in energy prices are likely to impose a new equilibrium in this relationship.

Financing and fare structures. Most public transit systems have abandoned a distance-based fare structure to a simpler flat fare system. This had the unintended consequence of discouraging short trips for which most transit systems are well suited, and encouraging longer trips that tend to be more costly per user than the fares they generate. Information systems offer the possibility for transit systems to move back to a more equitable distance-based fare structure.

Legacy costs. Most public transit systems employ unionized labor that have consistently used strikes (or the threat of a strike) and the acute disruptions they create as leverage to negotiate favorable contracts, including health and retirement benefits. Since public transit is subsidized these costs were not well reflected in the fare systems. In many transit systems, additional subsidies went into compensation or to cover past debt, and not necessarily into performance improvements or additional infrastructure. As most governments are facing stringent budgetary constraints because of unsustainable social welfare commitments, public transit agencies are being forced to reassess their budgets through an unpopular mix of higher fares, deferred maintenance and the breaking of labor contracts.

There are indications that public transit is reassessing its role in societies with high levels of automobile dependency. The rise in petroleum prices since 2006 has increased the cost of vehicle ownership and operation. A younger generation perceives the automobile as a less attractive proposition than prior generations and is more willing to use public transit and live in higher density areas. Electronic fare systems are also making the utilization of public transit more convenient. A recent trend concerns the usage of incentives, such as point systems (e.g. air miles with purchase of a monthly pass) to further promote the use of public transit and to influence consumer behavior.

CASE STUDY City logistics

Authors: Jean-Paul Rodrigue and Laetitia Dablanc

Freight and the city

Freight transportation maintains a set of core relations with urban areas since a city is an entity where production, distribution and consumption activities are using scarce land. The global urban and economic system has also become functionally specialized, permitting a global division of production and its associated freight volumes. City logistics, as a strategy ensuring efficient freight movements and innovative responses to urban customer and business demands, is an emerging field of investigation that was created by the challenges of the commodification and massification of production and consumption.

> **City logistics/urban logistics**. The means over which freight distribution can take place in urban areas as well as the strategies that can improve its overall efficiency, such as mitigating congestion and environmental externalities.

Urban economies are evolving rapidly towards a higher level of material intensiveness. Store inventory levels have shrunk and businesses are increasingly supplied on a just-in-time basis. The number of different products sold has increased considerably, and inventories change several times a year. With the rise of the service economy, the demand for express transport and courier services is also soaring. These factors have made urban economies more dependent on transportation systems, with more frequent and customized deliveries. All this promotes a greater intensity and frequency of urban freight distribution and correspondingly improved forms, organization and management. There is a complex relationship between the spatial and functional structure of city logistics where the organization and density of land uses interact with various forms of transport infrastructure influences on the location of logistics activities (Figure 6.13).

City logistics has a spatial and functional structure relating to the impacts of a city's organization of activities and transport infrastructure on freight distribution. Suburbanization has impacted a large share of the global urban landscape and characterizes a specific context in which freight distribution takes place. Although suburbia is functionally integrated to the central city, it is also a distinct space with its own consumption patterns. Suburban logistics differs from city logistics over two fundamental issues:

Spatial structure. The urban spatial structure is multipolar, of lower density and commonly involves higher consumption patterns (higher income levels). It is in suburbia where large freight terminals such as ports, airports and intermodal yards tend to be located. This implies that suburbia handles the majority of the interface between the metropolitan area and national as well as global freight distribution systems. At accessible locations (such as a highway interchange) suburban centers with commercial and office activities have emerged. They have become new nexuses of freight distribution, particularly if large scale commercial activities such as shopping malls are concerned. Yet, the spatial structure is prone to diseconomies as lower population densities and a more disorganized land use pattern are associated with longer trips.

Functional structure. Suburbia faces less congestion than the central city, implying that "last mile" constraints are less acute; parking difficulties are rarer and full truck lengths/loads (e.g. 53-foot trailers in North America) are able to circulate on most of the major roads. Suburbia is thus an environment highly conducive to logistics as it offers accessibility to markets (the central city as well as neighboring suburban areas), the availability of land as well as lower congestion levels. Through logistics

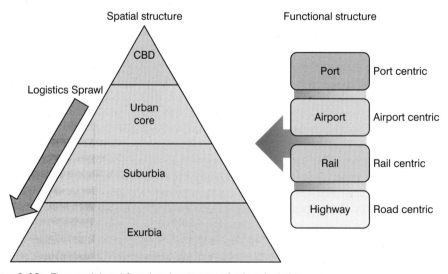

Figure 6.13 The spatial and functional structure of urban logistics

sprawl, terminal and warehousing activities that conventionally were located close to the city center have been replaced by terminal and modal specific clustering of logistics activities. Port centric (A) and airport centric (B) activities tend to support an interaction between global and city logistics. Road centric (C) and highway centric (E) activities involve a variety of supply chains and seek accessible locations with affordable land. The growth of intermodal rail transportation, particularly in relation with port container traffic, has been prone to the setting of rail centric logistical activities.

Therefore, although city logistics appears to be an issue taking place at the local (urban metropolitan) level, a comprehensive understanding of its drivers and dynamics requires the following:

Global supply chains. A salient issue relates to urban freight distribution in the context of global supply chains as global processes are imposing local forms of adaptation to ensure that freight is delivered in a timely and reliable fashion. Offshoring has also contributed to the setting of supply chains where freight distribution activities taking place within an urban area cannot be effectively explained by the regional economic structure.

Global freight distribution. Since the distances involved in supporting global supply chains have increased, the function of distribution has taken a new significance, particularly with the setting of large terminal facilities such as ports, airports, rail yards and distribution centers. They are handling movements originating from, bound to or simply passing through a metropolitan area. With containerization as a tool supporting the bulk of international trade, intermodal terminals have become a notable element of the urban landscape. With the growth of valuable cargo carried over long distances, airports are also active nodes interacting with urban freight distribution. Along with their attached freight distribution facilities (e.g. transloading facilities and warehouses) large terminals form a fundamental element of the interface between global distribution and city logistics.

Urbanization. Global urbanization is compounding the challenges of city logistics since the share and the level of concentration of the global population living in cities is increasing. Historically, the production and consumption of freight have predominantly taken place in cities, but with the industrial revolution and subsequently with globalization this share has increased. Cities also present a variety of forms and levels of density, each associated with specific city logistics patterns. Socioeconomic factors, such as rising income and consumer preferences, should also not be neglected.

Most of the early applications of city logistics were undertaken in Japan and Western Europe as these cities were more constrained by the lack of available land and had an established tradition pertaining to urban planning. Up to the twenty-first century the consideration of urban freight distribution within the planning discipline remained limited. This implies that urban planning generally does not pay much attention to issues related to urban freight distribution. Yet these issues are linked with externalities and growing concerns by private and public interests to address them:

Private concerns. Actors directly involved in urban freight distribution are mainly concerned by its constraints such as congestion since it imposes additional costs and delays in their operations. Also, restrictions on street access by trucks as well as for pickups and deliveries, provide additional challenges. Urban logistics becomes a competitive factor that needs to be addressed since it impairs the cost and efficiency of the "last mile" (or first) of global supply chains.

Public concerns. Actors involved in the oversight of urban freight distribution as well as urban residents in general are concerned by its externalities. Congestion and noise impacts residents, particularly their commuting and social interactions, as well as the general livability of a city. Passengers and freight flows are therefore subject to conflicts. Environmental concerns such as air pollution are recurrent in many cities. City logistics is becoming a salient urban sustainability issue.

The diversity of urban freight distribution

A city is supplied by an impressive variety of supply chains servicing a wide array of economic activities such as grocery stores, retail, restaurants, office supplies, raw materials and parts, construction materials and wastes. The level of economic development is linked with the level of urban freight activity as income and consumption levels are interdependent. Because of the divergence in built environments and the diversity of urban economic activities, each city around the world has different freight transport and logistics activities and level of intensity. This raises the issue of the specific size threshold after which urban freight distribution problems, such as delays and congestion, become more prevalent, which requires a concerted approach. Using the United States as evidence, congestion starts to be a serious issue once a threshold of about one million inhabitants is reached. For cities of less than one million, city logistics is less likely to be a problem and may be localized to specific areas such as the downtown or the port or other terminal areas.

The unique and often non-replicable conditions of each city influence the nature and intensity of congestion in its urban freight distribution system. The share of public transit use, land use pattern and density and income levels are common factors relatively unique to each city. Considering the growing level of material intensiveness related to the functions of production, distribution and consumption, cities above four million inhabitants should have planning and circulation management schemes where urban freight distribution is preeminent. Cities of smaller size can also proactively be involved in supporting specific and localized urban freight distribution activities.

The intensity of urban freight distribution depends on local economic, geographic and cultural characteristics, which leads to different objectives and preoccupations in urban freight distribution. For instance, each world city has a diverse array of concerns:

- **Paris** aims to limit the environmental footprint of freight distribution so that the quality of life of its residents can be maintained and improved. The city's status as one of the world's leading cultural and touristic hubs has a notable impact on the strategies and priorities accorded to urban freight distribution.
- **Mexico** tries to cope with the contradictory demands related to the dual presence of both modern (motorized) and traditional forms of urban distribution in terms of infrastructure provision and regulations. Modern logistics services are as vital to the urban economies of developing countries as are more basic freight activities serving street vendors or home-based manufacturing workshops.
- **Chicago** aims at maintaining its role as a major rail hub and freight distribution platform for North America. The metropolitan area is the point of convergence for the rail lines of the Class I carriers, but the different terminal facilities are in separate

parts of the city and are not well connected. This creates truck congestion as containers need to be carried from one terminal to the other.

- **Los Angeles** is facing congestion and environmental issues such as noise and air pollution. The city is facing conflicts between its function as a major commercial gateway for the East Asian trade, and other functions linked with touristic and cultural activities. Recent initiatives involve trucking associated with the main port facilities as well as nearby major import-based distribution centers.
- **Shanghai**. The megacity has become the largest cargo port in the world and acts as the major transport hub supporting China's export-oriented strategies. A significant share of the freight circulating within the city is therefore linked with global distribution processes.

All urban freight distribution systems involve a wide array of supply chains, each of varying importance depending on the urban setting and the level of development, but broadly coming under two main functional classes. The first involves consumer-related distribution:

- **Independent retailing**. Urban areas have a notable variety of retailing activities, many of which define the commercial and social character of neighborhoods. More often than not they are small single owner stores and in developing countries these retailing activities are often complemented by informal street markets and stalls.
- **Chain retailing**. In the contemporary commercial landscape large retail stores (such as "Big box" stores) have become an important element. While they used to be located in central areas and were a defining element of urban centrality, they have also emerged in suburban and peri-urban areas. Shopping malls, many quite large, are set on the principle of economies of agglomeration and the provision of ample parking space. Chain retailing tends to rely on the expertise of third party logistics service providers to mitigate urban freight distribution challenges, but mostly to organize complex multinational sourcing strategies sourcing mass retailers. Large stores are commonly accessed through dedicated delivery bays where they are resupplied on a daily basis through their own regional warehousing facilities.
- **Food deliveries**. Since most food products are perishable, a specialized form of urban distribution has been set to supply outlets such as grocery stores and restaurants. The outdoor market (central markets offer enclosed facilities) also take an important role in supplying urban populations with perishables, particularly in developing countries. This may be linked with informal forms of distribution where food producers deliver their production to urban markets. Limited information is available about urban food consumption levels but high levels of spoilage are observed, in the range of 50 percent of all the food consumed.
- **Parcel and home deliveries**. Globalization and the setting of advanced services, such as insurance, finance or corporate management (head or regional offices), are linked with a growth in the movement of parcels. While some are serviced by local companies, large parcel carriers have established services covering the majority of the world's main commercial cities. They maintain a network of strategically located distribution centers where shipments are consolidated or deconsolidated. International shipments are often taken care of by parent companies, namely air freight integrators. Online shopping (for retail goods, but also for groceries) has grown significantly, a trend which is associated with the growth of home deliveries.

The second functional class of city logistics is related to producer-related distribution:

- **Construction sites**. Urban infrastructures, from roads and residences to office and retail spaces, are constantly being constructed, renovated and repaired. Such activities are intensive in material use and must be supplied on an irregular basis, both in terms of the time and location of the deliveries.
- **Waste collection and disposal**. Urban activities generate large quantities of waste, namely paper, paperboard, food, plastics, metals and glass. These materials must be collected and carried to recycling or disposal sites. In particular, recycling has become an important activity taking place in urban areas and involves specialized vehicles and dedicated pick-up tours. As standards of living are increasing across the world, the amount of waste generated by cities has grown accordingly.
- **Industrial and terminal haulage**. Cities are zones of production as well as gateways for the circulation of goods. Large transportation terminals such as ports, airports and railyards are dominant elements of the urban landscape, including logistics zones where freight is distributed to extensive markets. Transport terminals and logistics zones are also generators of goods movements that may impact urban circulation (the last mile). Gate access at large intermodal terminals, such as ports, can lead to congestion (queuing) and local disruptions.

Two actors, private and common carriers, handle commercial freight transportation. Private carriers are at the same time beneficial cargo owners (manufacturers or retailers) using their own transportation assets (fleet and workforce). They can also subcontract this function to an independent carrier. Common carriers service any customer on a contractual basis, which also leads to the opportunity to consolidate cargo and deliveries, which can benefit smaller users. The share of private carriers is dominant for urban freight distribution in developing countries while in developed countries common carriers account for about half of urban deliveries.

The issue of dualism remains prevalent in urban freight distribution as it underlines different modes of operation between distribution systems that are integrated to globally oriented supply chains and distribution systems linked with informal activities that are more related to the local or regional economy. This is best represented by owner-drivers, or small independent truckers acting as subcontractors to large carriers for the final distribution of goods in urban areas. Dualism is therefore illustrative of a coexistence of modern and traditional means of freight distribution within the same metropolitan area. Another aspect of dualism is related to an active informal transportation sector that supplies the needs of lower income segments of the population, a very important component of city logistics services in developing countries.

Bibliography

Barry, M. (1991) *Through the Cities: The Revolution in Light Rail Transit*, Dublin: Frankfort Press.

Batty, M. and Y. Xie (1994) "From cells to cities", *Environment and Planning B*, 21: 531–48.

Boudoin, D. (2011) "Les Espaces Logistiques Urbains (ELU): une réponse aux problèmes de circulation des marchandises en ville", *Rail, Ville et Transport*, 515: 87–91.

Browne, M., J. Allen, A. Woodburn and M. Piotrowska (2007) *Urban Freight Transport – Literature Review*, Transport Studies Group, University of Westminster.

Camagni, R., M.C. Gibelli and P. Rigamonti (2002) "Urban mobility and urban form: the social and environmental costs of different patterns of urban expansion", *Ecological Economics*, 40: 199–216.

Carter, H. (1995) *The Study of Urban Geography*, 4th edition, London: Arnold.

Cervero, R. (1998) *The Transit Metropolis: A Global Inquiry*, Washington, DC: Island Press.

Crawford, J.H. (2005) "A brief history of urban form: street layout through the ages", first published on Carfree.com.

Dablanc, L. (2009) "Freight transport, a key for the new urban economy", Freight Transport for Development: A Policy Toolkit, Washington, DC: World Bank.

Dimitriou, H. (1993) *Urban Transport Planning*, New York: Routledge.

Docherty, I., G. Giuliano and D. Houston (2008) "Connected cities", in R.D. Knowles, J. Shaw and I. Docherty (eds) *Transport Geographies: Mobilities, Flows and Spaces*, London: Blackwell, pp. 83–101.

Gwilliam, K. (ed.) (2001) "Cities on the move: a World Bank Urban Transport Strategy Review", Strategy Paper, Washington, DC: World Bank.

Hanson, S. and G. Giuliano (eds) (2004) *The Geography of Urban Transportation*, 3rd edition, New York: Guilford Press.

Harvey, J. (1996) *Urban Land Economics*, Houndmills: Macmillan.

Hesse, M. (1995) "Urban space and logistics: on the road to sustainability?", *World Transport Policy and Practice*, 1(4): 39–45.

Isard, W. (1956) *Location and Space-Economy*, Cambridge, MA: MIT Press.

Kaplan, D., J. Wheeler and S. Holloway (2009) *Urban Geography*, 2nd edition, New York: Wiley.

Kauffman, R.J. (2001) "Paving the planet: cars and crops competing for land", *Alert*, Washington, DC: Worldwatch Institute.

Kenworthy, J., F. Laube, P. Newman, P. Barter, T. Raad, C. Poboon and B. Guia (1999) *An International Sourcebook of Automobile Dependence in Cities, 1960–1990*, Boulder: University Press of Colorado.

Litman, T. (2010) *Evaluating Transportation Land Use Impacts*, Victoria, BC: Victoria Transport Policy Institute.

Marchetti, C. (1994) "Anthropological invariants in travel behaviour", *Technical Forecasting and Social Change*, 47(1): 75–8.

Meyer, M.D. and E.J. Miller (2000) *Urban Transportation Planning: A Decision-Oriented Approach*, New York: McGraw-Hill.

Moore, T. and P. Thorsnes (2007) *The Transportation/Land Use Connection*, Washington, DC: American Planning Association. Report no. 448/449.

7 Transportation and the economy

Transport systems are closely related to socioeconomic changes. The mobility of people and freight and levels of territorial accessibility are at the core of this relationship. Economic opportunities are likely to arise where transportation infrastructures are able to answer mobility needs and ensure access to markets and resources. From the industrial revolution in the nineteenth century to globalization and economic integration processes of the late twentieth and early twenty-first centuries, regions of the world have been affected differently by economic development. International, regional and local transportation systems alike have become fundamental components of economic activities. A growing share of the wealth is thus linked to trade and distribution. However, even if transportation has positive impacts on socioeconomic systems, there are also negative consequences such as congestion, accidents and mobility gaps. Transportation is also a commercial activity derived from operational attributes such as transportation costs, capacity, efficiency, reliability and speed. Transportation systems are evolving within a complex set of relationships between transport supply – mainly the operational capacity of the network – and transport demand – the mobility requirements of an economy.

Concept 1 – Transportation and economic development

Authors: Jean-Paul Rodrigue and Theo Notteboom

The economic importance of transportation

Like many economic activities that are intensive in infrastructure, the transport sector is an important component of the economy impacting on development and the welfare of populations. When transport systems are efficient, they provide economic and social opportunities and benefits that result in positive multiplier effects such as better accessibility to markets, employment and additional investments. When transport systems are deficient in terms of capacity or reliability, they can have an economic cost such as reduced or missed opportunities. Efficient transportation reduces costs, while inefficient transportation increases costs. The impacts of transportation are not always intended, and can have unforeseen or unintended consequences such as congestion. Transport also carries an important social and environmental load, which cannot be neglected.

The added value and employment effects of transport services usually extend beyond employment and added value generated by that activity; indirect effects are salient. For instance, transportation companies purchase a part of their inputs from local suppliers. The production of these inputs generates additional value added and employment in the local economy. The suppliers in turn purchase goods and services from other local firms.

There are further rounds of local re-spending which generate additional value added and employment. Similarly, households that receive income from employment in transport activities spend some of their income on local goods and services. These purchases result in additional local jobs and added value. Some of the household income from these additional jobs is in turn spent on local goods and services, thereby creating further jobs and income for local households. As a result of these successive rounds of re-spending in the framework of local purchases, the overall impact on the economy exceeds the initial round of output, income and employment generated by passenger and freight transport activities. Thus, from a general standpoint the economic impacts of transportation can be direct, indirect and related:

Direct impacts (also known as induced): the outcome of accessibility changes where transport enables employment, added value, larger markets, and time and cost savings.

Indirect impacts: the outcome of the economic multiplier effects where the price of commodities, goods or services drop and/or their variety increases. Indirect value-added and jobs are the result of local purchases by companies directly dependent upon transport activity. Transport activities are responsible for a wide range of indirect value added and employment effects, through the linkages of transport with other economic sectors (e.g. office supply firms, equipment and parts suppliers, maintenance and repair services, insurance companies, consulting and other business services).

Related impacts: the outcome of economic activities and firms partly relying on efficient transport services for both passengers and freight. For instance, the steel industry requires cost-efficient import of iron ore and coal for blast furnaces, and export activities for finished products such as steel booms and coils. Manufacturers and retail outlets and distribution centers handling imported containerized cargo rely on efficient transport and seaport operations.

Mobility is one of the most fundamental and important characteristics of economic activity as it satisfies the basic need of going from one location to the other, a need shared by passengers, freight and information. All economies and regions do not share the same level of mobility as most are in a different stage in their mobility transition towards motorized forms of transport. Economies that possess greater mobility are often those with better opportunities to develop than those with scarce mobility. Reduced mobility impedes development while greater mobility is a catalyst for development. Mobility is thus a reliable indicator of development. Providing this mobility is an industry that offers services to its customers, employs people and pays wages, invests capital and generates income. The economic importance of the transportation industry can thus be assessed from both a macroeconomic and a microeconomic perspective:

At the **macroeconomic level** (the importance of transportation for a whole economy), transportation and the mobility it confers are linked to a level of output, employment and income within a national economy. In many developed countries, transportation accounts for between 6 and 12 percent of GDP.

At the **microeconomic level** (the importance of transportation for specific parts of the economy) transportation is linked to producer, consumer and production costs. The importance of specific transport activities and infrastructure can thus be assessed for each sector of the economy. Transportation accounts on average for between 10 and 15 percent of household expenditures while it accounts for around 4 percent of the costs of each unit of output in manufacturing, but this figure varies greatly according to subsectors.

Transportation links together the factors of production in a complex web of relationships between producers and consumers. The outcome is commonly a more efficient division of production by an exploitation of geographical comparative advantages, as well as the means to develop economies of scale and scope. The productivity of space, capital and labor is thus enhanced with the efficiency of distribution and personal mobility. It is acknowledged that economic growth is increasingly linked with transport developments, namely in infrastructures, but also managerial expertise is crucial for logistics. The following impacts can be assessed:

- **Networks**. Setting of routes enabling new or existing interactions between economic entities.
- **Performance**. Improvements in cost and time attributes for existing passenger and freight movements.
- **Reliability**. Improvement in the time performance, notably in terms of punctuality, as well as reduced loss or damage.
- **Market size**. Access to a wider market base where economies of scale in production, distribution and consumption can be improved.
- **Productivity**. Increases in productivity from access to a larger and more diverse base of inputs (raw materials, parts, energy or labor) and broader markets for diverse outputs (intermediate and finished goods).

Transportation and economic opportunities

Transportation developments that have taken place since the beginning of the industrial revolution have been linked to growing economic opportunities (Figure 7.1). At each stage of human societal development, a particular transport mode has been developed or adapted. However, it has been observed that throughout history no single transport has been solely responsible for economic growth. Instead, modes have been linked with the function and the geography in which growth was taking place. The first trade routes established a rudimentary system of distribution and transactions that would eventually be expanded by long distance maritime shipping networks and the setting of the first multinational corporations. Major flows of international migration that occurred since the 18th century were linked with the expansion of international and continental transport systems that radically shaped emerging economies such as in North America and Australia. Transport has played a catalytic role in these migrations, transforming the economic and social geography of many nations. Concomitantly, transportation has been a tool of territorial control and exploitation, particularly during the colonial era where resource-based transport systems supported the extraction of commodities in the developing world and forwarded them to the industrializing nations of the time. More recently, port development, particularly container ports, has been of strategic interest as a tool of integration to the global economy as the case of China illustrates.

The development of the mass production system at the beginning of the twentieth century increasingly relied on the commercial opportunities introduced by road transportation, particularly the automobile. Later in the twentieth century, globalization benefited from the joint synergy of maritime transportation, roadways, railways, air and

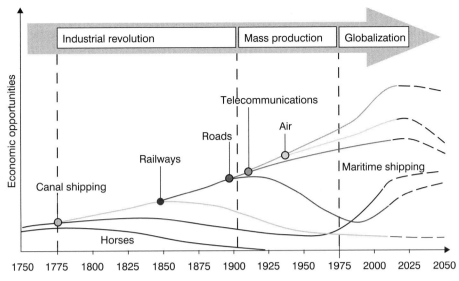

Figure 7.1 Cumulative modal contribution to economic opportunities

Source: Adapted from HOP Associates (2005) "Time, mobility and economic growth".

telecommunications, all of which supported integrated transport systems and supply chain management. Economic opportunities became global in scale and scope, particularly because of the capacity to maintain an intricate network of trade and transactions through transport systems. More recently, new opportunities arose with the convergence of telecommunications and information technologies, supporting a higher level of management of production, consumption and distribution, as well as a more efficient mobility of passengers. It is expected that such a process, building upon the advantages conferred by other transportation modes, will account for a significant share of economic opportunities in the first half of the twenty-first century.

While some regions benefit from the development of transport systems, others are often marginalized by a set of conditions in which inadequate transportation plays a role. Transport by itself is not a sufficient condition for development. However, the lack of transport infrastructures can be seen as a constraining factor on development. In developing countries, the lack of transportation infrastructures and regulatory impediments are jointly impacting economic development by conferring higher transport costs, but also delays rendering supply chain management unreliable. A poor transport service level can negatively affect the competitiveness of regions and corporations and thus have a negative impact on the regional added value and employment.

The standard assumption is that transportation investments tend to be more wealth-producing as opposed to wealth-consuming investments, such as services. Still, several transportation investments can be wealth consuming if they merely provide convenience, such as parking and sidewalks, or service a market size well below any possible economic return, with for instance projects labeled "bridges to nowhere". In such a context, transport investment projects can be counterproductive by draining the resources of an economy instead of creating wealth and additional opportunities. Efficient and sustainable transport markets and systems play a key role in regional development although the direction of causality between transport and wealth generation is not always clear. In a number of regions around the world, transport markets and related transport infrastructure networks are seen as key drivers in the promotion of a more balanced and sustainable development of the region or even the entire continent, particularly by improving accessibility and the situation of weaker regions and disadvantaged social groups.

There is also a tendency for transport investments to have declining marginal returns. While initial infrastructure investments tend to have a high return since they provide an entirely new range of mobility options, the more the system is developed the more likely additional investment would result in lower returns. At some point, the marginal returns can be close to zero or even negative, implying a shift of transport investments from wealth producing to wealth consuming. A common fallacy is assuming that additional transport investments will have a similar multiplying effect to that of the initial investments, which can lead to capital misallocation. This means quite understandably that the economic impacts of transport investments tend to be significant where infrastructures were previously nonexistent, or deficient and marginal where an extensive network is already present. Therefore, each development project must be considered independently.

Types of transport impacts

The relationship between transportation and economic development is difficult to formally establish and has been debated for many years. There are several layers of activity that transportation can valorize, from a suitable location that experiences the development of its accessibility through infrastructure investment to a better usage of existing transport assets through management. This is further nuanced by the nature, scale and scope of possible impacts:

> **Timing of the development** varies as the impacts of transportation can precede, occur during or take place after economic development. The lag, concomitant and lead impacts make it difficult to separate the specific contributions of transport to development. Each case study appears to be specific to a set of timing circumstances that are difficult to replicate elsewhere.
>
> **Types of impacts** vary considerably. The spectrum of impacts ranges from the positive through the permissive to the negative. In some cases transportation impacts can promote, in others they may hinder economic development in a region. In many cases, few, if any, direct linkages could be clearly established.

Cycles of economic development provide a revealing conceptual perspective about how transport systems evolve in time and space as they include the timing and the nature of the transport impact on economic development (Figure 7.2). This perspective underlines that after a phase of introduction and growth, a transport system will eventually reach a phase of maturity through geographical and market saturation. There is also the risk of over-investment when economic growth is credit driven, which can lead to significant misallocations of capital, including in the transportation sector. The outcome is a surplus capacity in infrastructures and modes creating deflationary pressures that undermine profitability. In periods of recession that commonly follow periods of expansion, transportation activities may experience a setback, namely in terms of lower demand and a scarcity of capital investment.

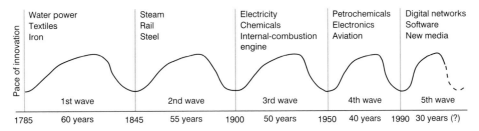

Figure 7.2 Long wave cycles of innovation

Source: Adapted from *The Economist*, February 20, 1999.

Technological innovation and economic growth are closely related and can be articulated within the concept of cycles or waves. Each wave represents a diffusion phase of technological innovations creating entirely new economic sectors, and thus opportunities for investment and growth. Since the beginning of the industrial revolution in the late eighteenth century, five waves have been identified:

1st wave (1785–1845). Leaned on innovations such as water power, textiles and iron. The beginning of the industrial revolution was mainly focusing on simple commodities such as clothes and tools. The conventional maritime technology relying on sailships was perfected, supporting the creation of large colonial/trading empires, mainly by Great Britain, France, the Netherlands and Spain. Significant inland waterway systems were also constructed. The costs of production and transportation were significantly reduced.

2nd wave (1845–1900). Involved the massive application of coal as a source of energy, mainly through the steam engine. This induced the development of rail transport systems, opening new markets and giving access to a wider array of resources. The steamship had a similar impact for maritime transportation and permitted expanded commercial opportunities in global trade.

3rd wave (1900–50). Electrification was a major economic change as it permitted the usage of a variety of machines and appliances and permitted the development of urban transit systems (subways and tramways). Another significant improvement was the internal combustion engine, around which the whole automotive industry was created and expanded the mobility of passengers and freight.

4th wave (1950–90). The post-Second World War period represented significant industrial changes with new materials such as plastics (petrochemicals) and new sectors such as electronics (television). The jet engine expanded the aviation industry towards the mass market and mobility could be realized nationally and globally.

5th wave (1990–2020?). The current wave mainly relies on information systems, which have tremendously modified the transactional environment with new methods of communication and more efficient forms of management of production and distribution systems (logistics). This spawned new industries related to personal computing devices, mainly computer manufacturing and software programming, and more recently e-commerce as information processing converged with telecommunications.

As time progressed, the lapse between each wave got shorter. For instance, the first wave lasted 60 years while the fourth wave lasted 40 years. This reflects a growing potential for innovation and the capacity of economic systems to derive commercial opportunities from an innovation once it has been adopted. Innovations are no longer the result of individual efforts, but are organized and concerted actions whose results are rapidly diffused. It is thus expected that the fifth wave will last about 30 years.

Transport, as a technology, typically follows a path of experimentation, introduction, adoption and diffusion and, finally, obsolescence, each of which has an impact on the rate of economic development. In this cyclic behavior a high level of benefits and productivity is realized in the early phase while later phases face diminishing returns. Containerization is a relevant example of such a diffusion behavior. As most innovations are eventually abandoned, many technologies go through what can be called a "hype phase" with unrealistic expectations. In addition, transport modes and infrastructures are depreciating assets that constantly require maintenance and upgrades. At some point, their useful lifespan is exceeded and the vehicle must be retired or the infrastructure rebuilt. Thus, transport investments for their amortization must consider the lifespan of the concerned mode or infrastructure.

Transport as a factor of production

Contemporary trends have underlined that economic development has become less dependent on relations with the environment (resources) and more dependent on relations across space. While resources remain the foundation of economic activities, the commodification of the economy has been linked with higher levels of material flows of all kinds. Concomitantly, resources, capital and even labor have shown increasing levels of mobility. This is particularly the case for multinational firms that can benefit from transport improvements in two significant markets:

- **Commodity market**. Improvement in the efficiency with which firms have access to raw materials and parts as well as to their respective customers. Thus, transportation expands opportunities to acquire and sell a variety of commodities necessary for industrial and manufacturing systems.
- **Labor market**. Improvement in the access to labor and a reduction in access costs, mainly by improved commuting (local scale) or the use of lower cost labor (global scale).

A common fallacy in assessing the importance and impact of transportation on the economy is to focus only on transportation costs, which tend to be relatively low (5 to 10 percent of the value of a good). Transportation is an economic factor of production of goods and services, implying that relatively small changes can have substantial impacts on costs, locations and performance. An efficient transport system with modern infrastructures favors many economic changes, most of them positive. It provides market accessibility by linking producers and consumers. The major impacts of transport on economic processes can be categorized as follows:

Geographic specialization. Improvements in transportation and communication favor a process of geographical specialization that increases productivity and spatial interactions. An economic entity tends to produce goods and services with the most appropriate combination of capital, labor and raw materials. A given area will thus tend to specialize in the production of goods and services for which it has the greatest advantages (or the least disadvantages) compared to other areas as long as appropriate transport is available for trade (Figure 7.3). Through geographic speciali-zation supported by efficient transportation, economic productivity is promoted. This process is known in economic theory as comparative advantages.

Large-scale production. An efficient transport system offering cost, time and reliability advantages permits goods to be transported over longer distances. This facilitates mass production through economies of scale because larger markets can be accessed. The concept of "just-in-time" has further expanded the productivity of production and distribution with benefits such as lower inventory levels and better responses to shifting market conditions. Thus, the more efficient transportation becomes, the larger the markets that can be serviced and the larger the scale of production.

Increased competition. When transport is efficient, the potential market for a given product (or service) increases, and so does competition. A wider array of goods and services becomes available to consumers through competition which tends to

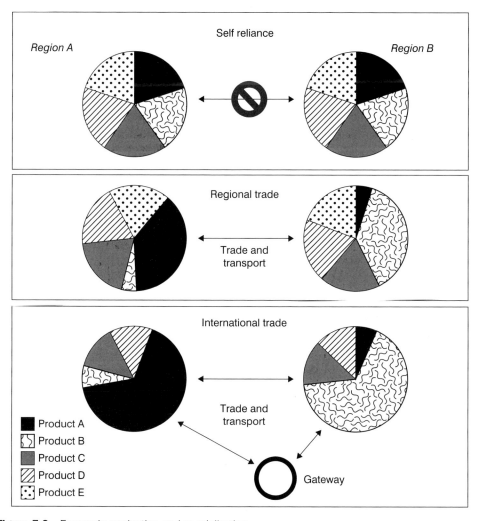

Figure 7.3 Economic production and specialization

reduce costs and promote quality and innovation. Globalization has clearly been associated with a competitive environment that spans the world.

Increased land value. Land which is adjacent to or serviced by good transport systems generally has greater value due to the utility it confers to many activities. In some cases, the opposite can be true if related to residential activities. Land located near airports and highways, near noise and pollution sources, will thus suffer from corresponding diminishing land value.

The evolution of transport systems impacts the regional economy in terms of spatial specialization. Figure 7.3 represents a simplified example of how changes in transport may impact on the specialization of regional economies, which a priori have similar environmental endowments. In a situation of self-reliance there is no apparent transport link between two regions. They are isolated from one another and must satisfy their own needs. Each region thus tends to be similar in terms of economic output. While regions have different environmental endowments, they must still provide for every basic necessity such as food. Quantities produced depend on the demand and the industrial capacity.

With a transport link between two regions, specialization can take place. Each region develops its respective potential; Product A for the first region and Product B for the second, assuming that they respectively have a comparative advantage for these two products. If Product A is cheaper to produce in the first region, it becomes more efficient to lessen the production of other products and concentrate on Product A. Respectively, the second region can do so for Product B. Therefore, the first region can dedicate more resources for the production of Product A, of which it can then sell the surplus (minus local consumption) to the second region. The key of this specialization becomes the difference between transport costs and production costs of a product. If the unit costs savings resulting from specialization exceed the unit transports costs, then specialization can take place.

Regional specialization is greatly expanded with international trade. By having access to a larger market through a gateway, namely a seaport, region A and B can specialize even more in the production for which they have a respective comparative advantage. They can even cease production in a specific array of products, which are now imported. Under such circumstances, the reliance on transportation increases, even if its relative costs may be declining.

Transport also contributes to economic development through job creation and its derived economic activities. Accordingly, a large number of direct (freighters, managers, shippers) and indirect (insurance, finance, packaging, handling, travel agencies, transit operators) employment are associated with transport. Producers and consumers take economic decisions on products, markets, costs, location, prices which are themselves based on transport services, their availability, costs and capacity.

Socioeconomic impacts

While many of the economic impacts of transportation are positive, there are also significant negative impacts that are assumed by individuals or by the society in one way or another. Among the most significant are:

Mobility gaps. Since mobility is one of the fundamental components of the economic benefits of transportation, its variations are likely to have substantial impacts on the opportunities of individuals. Mobility needs do not always coincide due to several factors, namely the lack of income, lack of time, lack of means and the lack of access. People's mobility and transport demands thus depend on their socioeconomic situation. The higher the income, the higher the mobility, which may give rise to substantial mobility gaps between different population groups. Gender gaps exist in mobility as women tend to have lower incomes. Mobility gaps are particularly prevalent for long distance travel. With the development of air transport, a segment of the global population has achieved a very high level of mobility for their business and leisure activities, while the great majority of the global population has little mobility. This issue is expected to become more acute as the population of many advanced economies is aging rapidly, which implies that access to mobility will not be an income issue but an age issue. By 2020, about 10 percent of the global population (719 million) will be over 65 while by 2050 it will be 16 percent (1,492 million).

Cost differences. Locations that have low levels of accessibility, such as landlocked countries, tend to have higher costs for many goods (sometimes basic necessities such as food) as most have to be imported, often over long distances. The resulting higher transport costs inhibit the competitiveness of such locations and limits opportunities. Consumers and industries will pay higher prices, impacting on their welfare (disposable income) and competitiveness.

Congestion. With the increased use of transport systems, it has become common for parts of the network to be used above design capacity. Congestion is the outcome of such a situation with its associated costs, delays and waste of energy. Distribution systems that rely upon on-time deliveries are particularly susceptible to congestion.

Accidents. The use of transport modes and infrastructure is never entirely safe. Every motorized vehicle contains an element of danger and nuisance. Due to human errors and various forms of physical failures (mechanical or infrastructural) injuries, damages and even death occur. Accidents tend to be proportional to the intensity of use of transport infrastructures which means the more traffic the higher the probability for an accident to occur. They have important socioeconomic impacts including healthcare, insurance, damage to property and the loss of life. The respective level of safety depends on the mode of transport and the speed at which an accident occurs. No mode is completely safe but the road remains the most dangerous medium for transportation, accounting for 90 percent of all transport accidents on average. At the global level about 1.3 million people died in road accidents in 2010 in addition to 50 million injured. China has one of the highest car accident death rates in the world, with more than 110,000 fatalities per year (300 per day), a factor mainly due to recent growth in vehicle ownership.

The emission of pollutants related to transport activities has a wide range of environmental consequences that have to be assumed by society, more specifically on four elements:

Air quality. Atmospheric emissions from pollutants produced by transportation, especially by the internal combustion engine, are associated with air pollution and, arguably, global climate change. Some pollutants (nitrogen oxides, carbon monoxides, ozone, volatile organic compounds – NOx, CO, O_3, VOC, etc.) can produce respiratory troubles and aggravate cardiovascular illnesses. In urban regions, about 50 percent of all air pollution emanates from automobile traffic.

Noise. A major irritant, noise can impact on human health and most often human welfare. Noise can be manifested on three levels depending on emissions intensity: psychological disturbances (perturbations, displeasure), functional disturbances (sleep disorders, loss of work productivity, speech interference) or physiological disturbances (health issues such as fatigue, and hearing damage). Noise and vibration associated with trains, trucks and planes in the vicinity of airports are major irritants.

Water quality. Accidental and nominal runoff of pollutants from transport such as oil spills, are sources of contamination for both surface water and groundwater.

Land take. Transport is a large consumer of space when all of its supporting infrastructure and equipment are considered. Furthermore, the planning associated with these structures does not always consider aesthetic values as is often the case in the construction of urban highways. These visual impacts have adverse consequences on the quality of life of nearby residents.

Concept 2 – Transport costs

Authors: Jean-Paul Rodrigue and Theo Notteboom

Transport costs and rates

Transport systems face requirements to increase their capacity and to reduce the costs of movements. All users (e.g. individuals, enterprises, institutions, governments, etc.) have to negotiate or bid for the transfer of goods, people, information and capital because supplies, distribution systems, tariffs, salaries, locations, marketing techniques as well as fuel costs are changing constantly. There are also costs involved in gathering information, negotiating, and enforcing contracts and transactions, which are often referred to as the cost of doing business. Trade involves transaction costs that all agents attempt to reduce since transaction costs account for a growing share of the resources consumed by the economy.

Frequently, enterprises and individuals must take decisions about how to route passengers or freight through the transport system. This choice has been considerably expanded in the context of the production of lighter and high value consumer goods, such as electronics, and less bulky production techniques. It is not uncommon for transport costs to account for 10 percent of the total cost of a product. This share also roughly applies to personal mobility where households spend about 10 percent of their income on transportation, including the automobile which has a complex cost structure. Thus, the choice of a transportation mode to route people and freight between origins and destinations becomes important and depends on a number of factors such as the nature of the goods, the available infrastructures, origins and destinations, technology, and particularly their respective distances. Jointly, they define transportation costs.

Transport costs are a monetary measure of what the transport provider must pay to produce transportation services. They come as fixed (infrastructure) and variable (operating) costs, depending on a variety of conditions related to geography, infrastructure, administrative barriers, energy, and on how passengers and freight are carried. Three major components, related to transactions, shipments and the friction of distance, impact on transport costs.

Transport costs have significant impacts on the structure of economic activities as well as on international trade. Empirical evidence underlines that raising transport costs by 10 percent reduces trade volumes by more than 20 percent. In a competitive environment where transportation is a service that can be bid for, transport costs are influenced by the respective rates of transport companies, the portion of the transport costs charged to users.

> **Rates** are the price of transportation services paid by their users. They are the negotiated monetary cost of moving a passenger or a unit of freight between a specific origin and destination. Rates are often visible to consumers since transport providers must provide this information to secure transactions. They may not necessarily express the real transport costs.

The difference between costs and rates either results in a loss or a profit from the service provider. Considering the components of transport costs previously discussed, rate setting is a complex undertaking subject to constant change. For public transit, rates are often fixed and the result of a political decision where a share of the total costs is subsidized by society. The goal is to provide an affordable mobility to the largest possible segment of the population even if this implies a recurring deficit (public transit systems rarely make any profit). It is thus common for public transit systems to have rates that are lower than costs. For freight transportation and many forms of passenger transportation (e.g. air transportation) rates are subject to competitive pressure. This means that the rate will be adjusted according to the demand and the supply. They either reflect costs directly involved with shipping (cost-of-service) or are determined by the value of the commodity (value-of-service). Since many actors are involved in freight transportation private rates tend to vary, often significantly, but profitability is paramount.

Costs and time components

Transportation offers a spectrum of costs and level of services. The price of a transport service does not only include the direct out-of-the-pocket money costs to the user but also includes time costs and costs related to possible inefficiencies, discomfort and risk (e.g. unexpected delays). However, economic actors often base their choice of a transport mode or route on only part of the total transport price. For example, motorists are biased by short-run marginal costs. They might narrow down the price of a specific trip by car to fuel costs only, thereby excluding fixed costs such as depreciation, insurance and vehicle tax. Many shippers or freight forwarders are primarily guided by direct money costs when considering the price factor in modal choice. The narrow focus on direct money costs is to some extent attributable to the fact that time costs and costs related to possible inefficiencies are harder to calculate and often can only be fully assessed after the cargo has arrived. Among the most significant conditions affecting transport costs and thus transport rates are:

> **Geography**. Its impacts mainly involve distance and accessibility. Distance is commonly the most basic condition affecting transport costs (Figure 7.4). The more it is difficult to trade space for a cost, the more the friction of distance is important. It can be expressed in terms of length, time, economic costs or the amount of energy used. It varies greatly according to the type of transportation mode involved and the

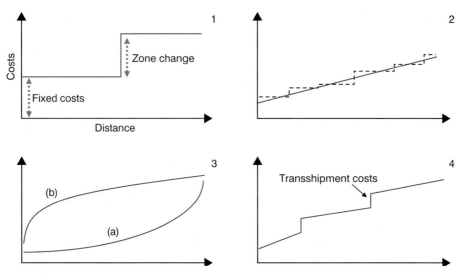

Figure 7.4 Different friction of distance functions

efficiency of specific transport routes. Landlocked countries tend to have higher transport costs, often twice as much, as they do not have direct access to maritime transportation.

There are four major categories of friction of distance functions: **No effects of distance** (1). Uncommon are economic activities on which distance has no effects. However, the distance–cost function telecommunication networks and the virtual space of the Internet have such a cost structure. Telephone calls can be included in this category as well as postal fees and public transit fares. All those activities generally have a fixed cost which is not related to distance, but often to a service zone. Once a new zone is entered (such as for international phone calls) a new cost structure applies. **Linear effects of distance** (2). Transport costs are increasing proportionally to distance. Fuel consumption can be included in this category since it is a direct function of the distance traveled. For reasons of simplicity, a step-wise approach is often used to establish transport rates by using administrative units as distance units. **Non-linear effects of distance** (3). Freight distribution costs are growing in a non-linear fashion with distance from the terminal or the distribution center, particularly because of empty back hauls (a). Inversely, international air transportation costs are not usually much higher than regional air transportation costs because long haul planes have more capacity (b). **Multimodal transport chain** (4). A combination of linehaul and terminal costs where transshipment costs at terminals (e.g. ports and airports) increase the friction of distance as efforts must be spent at loading or unloading when passing from one mode to another.

Type of product. Many products require packaging, special handling, are bulky or perishable. Coal is obviously a commodity that is easier to transport than fruits or fresh flowers as it requires rudimentary storage facilities and can be transshipped using rudimentary equipment. Insurance costs are also to be considered and are commonly a function of the value to weight ratio and the risk associated with the movement. As such, different economic sectors incur different transport costs as they each have their own transport intensity. With containerization the type of product plays little in the transport cost since rates are set per container, but products still need to be loaded or unloaded from the container. For passengers, comfort and amenities must be provided, especially if long distance travel is involved.

Economies of scale. Another condition affecting transport costs is related to economies of scale or the possibilities to apply them as the larger the quantities transported, the lower the unit cost. Bulk commodities such as energy (coal, oil), minerals and grains are highly suitable to obtain lower unit transport costs if they are transported in large quantities. For instance, larger container ships involve lower unit costs.

Energy. Transport activities are large consumers of energy, especially oil. About 60 percent of all the global oil consumption is attributed to transport activities. Transport typically accounts for about 25 percent of all the energy consumption of an economy. The costs of several energy-intensive transport modes, such as air transport, are particularly susceptible to fluctuations in energy prices.

Trade imbalances. Imbalances between imports and exports have impacts on transport costs. This is especially the case for container transportation since trade imbalances imply the repositioning of empty containers that have to be taken into account in the total transport costs. Consequently, if a trade balance is strongly negative (more imports than exports), transport costs for imports tend to be higher than for exports. Significant transport rate imbalances have emerged along major trade routes. The same condition applies at the national and local levels where freight flows are often unidirectional, implying empty back haul movements.

Infrastructures. The efficiency and capacity of transport modes and terminals has a direct impact on transport costs. Poor infrastructures imply higher transport costs, delays and negative economic consequences. More developed transport systems tend to have lower transport costs since they are more reliable and can handle more movements.

Mode. Different modes are characterized by different transport costs, since each has its own capacity limitations and operational conditions. When two or more modes are directly competing for the same market, the outcome often results in lower transport costs. Containerized transportation permitted a significant reduction in freight transport rates around the world.

Competition and regulation. Concerns the complex competitive and regulatory environment in which transportation takes place. Transport services taking place over highly competitive segments tend to be of lower cost than on segments with limited competition (oligopoly or monopoly). International competition has favored concentration in many segments of the transport industry, namely maritime and air modes. Regulations, such as tariffs, cabotage laws, labor, security and safety impose additional transport costs, particularly in developing countries.

Surcharges. Refer to an array of fees, often set in an arbitrary fashion, to reflect temporary conditions that may impact on costs assumed by the transporter. The most common are fuel surcharges, security fees, geopolitical risk premiums and additional baggage fees. The passenger transport industry, particularly airlines, has become dependent on a wide array of surcharges as a source of revenue.

Transport time is also an important dimension in the evaluation of transport costs, particularly since logistics concomitantly involves cost and time management. They

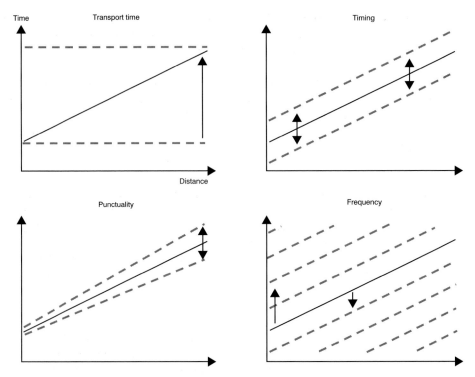

Figure 7.5 Different components of transport time

Source: Adapted from Woxenius, J. (2006) "Temporal elements in the spatial extension of production networks", *Growth and Change*, 37(4): 526–49.

include the transport time, the order time, the timing, the punctuality and the frequency (Figure 7.5):

> **Transport time**. Concerns the real duration of transport, which tends to be easily understood since commonly a proportional function of distance. Geographical constraints such as weather or technical limitations such as operational speed have a direct impact on transport time. Transport time on roads is technically limited to legal speed limits. For maritime and air, the limitation mainly concerns fuel economy and design speed. Although rail can accommodate a variety of speeds, tight schedules impose limited variations in operational speeds.
>
> **Order time** (not shown on Figure 7.5). Almost all transport requires a form of advance preparation, mainly to secure a capacity, an itinerary and a rate. In some cases, the order time is short and a matter of queuing on a first-come-first-served basis, while in other cases orders have to be secured months in advance.
>
> **Timing**. Involves the usage of a specific departure time, which depending on the mode can have a level of flexibility. While for air and rail travel timing is commonly tight due to fixed schedules and access to a terminal capacity (such as a gate and a takeoff time), commuters and trucking have more flexibility. If there is congestion either at the origin, destination or in between, trucking companies may elect to modify their schedule accordingly (earlier or later delivery).
>
> **Punctuality**. Represents the ability to keep a specified schedule, which can be represented as an average deviation from a scheduled arrival time. The longer the distance,

the more likely are potential disruptions that may affect schedule integrity. Some movements may have a level of tolerance to disruptions in punctuality while others, such as heading to a business meeting or flows in a just-in-time supply chain, have limited tolerance.

Frequency. The number of departures for a specific time range. The higher the frequency, the better the level of service. However, a high frequency ties up a larger quantity of vehicles. Distance is also a factor for lower frequency since transport demand tends to decline accordingly. Combining long distance travel and high frequency is an expensive undertaking for transport providers as a greater number of vehicles must be assigned to a specific route, as in the case of maritime container shipping.

For instance, a maritime shipping company may offer a container transport service between a number of North American and Pacific Asian ports. It may take 12 days to service two ports across the Pacific (transport time) and a port call is done every two days (frequency). In order to secure a slot on a ship, a freight forwarder must call at least five days in advance (order time). For a specific port terminal, a ship arrives at 8 a.m. and leaves at 5 p.m. (timing) with the average delay being two hours (punctuality).

Types of transport costs

Mobility tends to be influenced by transport costs. Empirical evidence for passenger vehicle use underlines the relationship between annual vehicle mileage and fuel costs, implying the higher fuel costs are, the lower the mileage. At the international level, doubling of transport costs can reduce trade flows by more than 80 percent. The more affordable mobility is, the more frequent the movements and the more likely they will take place over longer distances. A wide variety of transport costs can be considered.

Terminal costs. Costs that are related to loading, transshipment and unloading. Two major terminal costs can be considered: loading and unloading at the origin and destination, which are unavoidable; and intermediate (transshipment) costs that can be avoided.

Linehaul costs. Costs that are a function of the distance over which a unit of freight or passenger is carried. Weight is also a cost function when freight is involved. They include labor and fuel and commonly exclude transshipment costs.

Capital costs. Costs applying to the physical assets of transportation, mainly infrastructures, terminals and vehicles. They include the purchase or major enhancement of fixed assets, which can often be a one-time event. Since physical assets tend to depreciate over time, capital investments are required on a regular basis for maintenance.

Transport providers make a variety of decisions based on their cost structure, a function of all the above types of transport costs. To simplify transactions and clearly identify the respective responsibilities specific commercial transportation terms have been set. **International Commercial Terms**, or Incoterms, are predefined commercial terms that are used to define the transport component and the share of costs and risks for international commercial transactions. They provide a consistent framework of the expected transport service to be provided, which removes uncertainty, and define legally enforceable responsibilities across international jurisdictions. Although for many transactions it is either the seller's or the buyer's responsibility to carry the cargo, this task is often attributed to a shipper or a third-party logistics provider that will act on their behalf. The most common Incoterms are (Figure 7.6):

EXW (Ex Works). The buyer virtually takes care of all the transport responsibilities and the only obligation of the seller is to have the cargo available at an agreed-upon time and premises (factory, distribution center). It often refers to the factory price since it excludes all costs related to transportation, such as insurance and duties.

FCA (Free Carrier). The seller's responsibility is simply to provide the cargo that has been cleared for export (duties paid) at a specific delivery point. This is common for intermodal transport.

FAS (Free Alongside Ship). Usually used for maritime bulk cargo where the seller provides the cargo at the dock, which is ready to be loaded on a ship. For instance, a grain seller would make available bulk grain at a dockside grain elevator and it would be the buyer's responsibility to charter a bulk cargo ship and load the cargo.

FOB (Free On Board). The seller provides, transports and loads the cargo on board a vessel, which is usually selected by the buyer. This is common for bulk cargo.

CFR (Cost and Freight). The seller brings and unloads the cargo at a port of destination, but the buyer assumes the risk as soon as the cargo is loaded at the port of origin. The buyer has the responsibility of picking up the cargo.

CIF (Cost, Insurance and Freight). Same as above, but the seller also provides insurance for the cargo up to the port of destination. It usually applies to bulk cargo.

CIP (Carriage and Insurance Paid). Common in intermodal transport chains where the seller takes the complete responsibility of bringing the cargo to the point of destination (e.g. the door of a distribution center) as well as providing insurance. The buyer is responsible for unloading the cargo at the point of destination.

While the transport price plays an important role in modal choice, firms using freight transport services are not always motivated by notions of cost minimization. They often

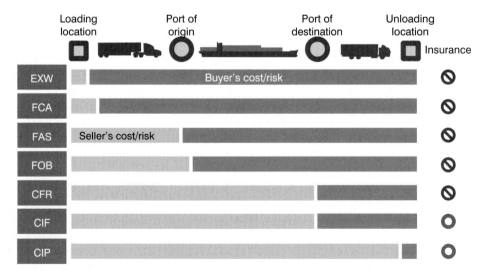

Figure 7.6 Selected International Commercial Terms (Incoterms)

Source: Adapted from International Chamber of Commerce.

show "satisficing behavior" whereby the transport costs need to be below a certain threshold combined with specific requirements regarding reliability, frequency and other service attributes. Such complexities make it more difficult to clearly assess the role of transport price in the behavior of transport users. The role of transport companies has sensibly increased in the general context of the global commercial geography. However, the nature of this role is changing as a result of a general reduction of transport costs but growing infrastructure costs, mainly due to greater flows and competition for land. Each transport sector must consider variations in the importance of different transport costs. While operating costs are high for air transport, terminal costs are significant for maritime transport.

Technological changes and their associated decline in transport costs have weakened the links with transport modes and their terminals. There is less emphasis on heavy industries and more importance given to manufacturing and transport services (e.g. warehousing and distribution). Indeed, new functions are being grafted on to transport activities that are henceforward facilitating logistics and manufacturing processes. Relations between terminal operators and carriers have thus become crucial notably in containerized traffic. They are needed to overcome the physical and time constraints of transshipment, notably at ports.

The requirements of international trade gave rise to the development of specialized and intermediary firms providing transport services. These are firms that do not physically transport the goods, but are required to facilitate the grouping, storage and handling of freight as well as the complex paperwork and financial and legal transactions involved in international trade. Examples include freight forwarders, customs brokers, warehousing, insurance agents and banking, etc. Recently, there has been a trend to consolidate these different intermediate functions, and a growing proportion of global trade is now being organized by multinational corporations that are offering door-to-door logistics services.

Concept 3 – Transport supply and demand

Authors: Jean-Paul Rodrigue and Theo Notteboom

The supply and demand for transportation

Each transport mode shares the common goal of fulfilling a derived transport demand, and each transport mode thus fills the purpose of supporting mobility. Transportation is a service that must be utilized immediately since unlike the resources it often carries, the transport service itself cannot be stored. Mobility must occur over transport infrastructures having a fixed capacity, providing a transport supply. In several instances, transport demand is answered in the simplest means possible, notably by walking. However, in some cases elaborate and expensive infrastructures and modes are required to provide mobility, such as for international air transportation.

Transportation is a market composed of suppliers of transport services and users of these services. Well-functioning transport markets should allow transport supply to meet transport demand so that transport needs for mobility are satisfied. An economic system including numerous activities located in different areas generates movements that must be supported by the transport system. Without movements infrastructures would be useless and without infrastructures movements could not occur, or would not occur in a cost-efficient manner. This interdependency can be considered according to two concepts, which are transport supply and demand:

Transport supply. The capacity of transportation infrastructures and modes, generally over a geographically defined transport system and for a specific period of time. Supply is expressed in terms of infrastructures (capacity), services (frequency) and networks (coverage). Capacity is often assessed in static and dynamic terms, where static capacity represents the amount of space available for transport (e.g. terminal surface) and dynamic capacity is the improvement that can be made through better technology and management. The number of passengers, volume (for liquids or containerized traffic) or mass (for freight) that can be transported per unit of time and space is commonly used to quantify transport supply.

Transport demand. Transport needs, even if those needs are satisfied, fully, partially or not at all. Similar to transport supply, it is expressed in terms of number of people, volume or tons per unit of time and space.

The supply side of the transport market can be divided into two categories:

Third-party transportation. Transport companies offer transport services to users who require such services, often on open markets. Transport users pay for the services delivered according to the terms of the agreed contract. Examples include third-party trucking companies, container shipping lines, railway operators and bus companies. Competitiveness is a key advantage of third-party transportation as providers strive to offer better and lower cost services for their customers. There is also the risk of fluctuating prices due to changing market conditions and that transport capacity may not be available when a customer requires it.

Own account transportation. The transport user deploys his/her own transport means to move freight or to travel (e.g. motorists using private cars or large industrial companies owning a fleet of trucks or rail wagons). The transport user has a direct access to a known capacity, but at the risk of a lower level of asset utilization (e.g. empty movements or idle equipment).

Transport demand is generated by the economy, composed of persons, institutions and industries which generate movements of people and freight. A distinction can be made between consumptive and productive transport needs. Productive transport needs have a clear economic focus. For example, the transport of semi-finished products from one production site to the final production or assembly site creates added value in the production process by benefiting from the locational advantages of each of the production sites. Consumptive transport needs generate less visible added value. For example, a road trip does not really add value in a pure economic sense, but generates subjective utility and satisfaction to the users. A discussion on the functioning of transport markets is particularly relevant where it concerns the fulfillment of productive transport needs, but the consumptive dimension of transport must also be considered.

The location of resources, factories, distribution centers and markets is obviously related to freight movements. Transport demand can vary under two circumstances that are often concomitant: the quantity of passengers or freight increases or the distance over which these passengers or freight are carried increases (Figure 7.7). Geographical considerations and transport costs account for significant variations in the composition of freight transport demand between countries. For the movements of passengers, the location of residential, commercial and industrial areas tells a lot about the generation and attraction of movements.

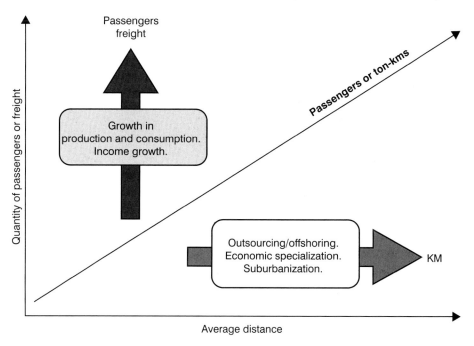

Figure 7.7 Growth factors in transport demand

The realized transport demand, expressed in passenger or ton-km, can increase for two reasons. The first is obviously that more passengers or freight are being carried. This is an outcome of growth in population, production, consumption and income. The second is a growth in the average distance over which passengers or freight loads are being carried. Outsourcing, offshoring, economic specialization (factors linked with globalization) and suburbanization are relevant factors behind this trend. These two factors are often concomitant, creating multiplying effects on the transport demand.

Supply and demand functions

Transport supply and demand have a reciprocal but asymmetric relationship. While a realized transport demand cannot take place without a corresponding level of transport supply, a transport supply can exist without a corresponding transport demand. This is common in infrastructure projects that are designed with a capacity fulfilling an expected demand level, which may or may not materialize, or may take several years to do so. Scheduled transport services, such as public transit or airlines, offer a transport supply that runs even if the demand is insufficient. Infrastructures also tend to be designed at a capacity level higher than the expected base scenario in case that demand turns out to be higher than anticipated. In other cases, the demand does not materialize, often due to improper planning or unexpected socioeconomic changes.

Transport demand that is met by a supply of transport services generates traffic (trucks, trains, ships, airplanes, buses, bicycles, etc.) on the corresponding transport infrastructure networks. The traffic capacity is generally larger than the actual transport demand since the average utilization degree of vehicles rarely reaches 100 percent: e.g. empty hauls of trucks, an underutilized container ship capacity sailing on a shipping route characterized by imbalanced container flows, an underutilized offpeak bus service and the one person per car situation in commuter traffic. There is a simple statistical way to measure transport supply and demand for passengers or freight:

> The **passenger-km** (or passenger-mile) is a common measure expressing the realized passenger transport demand as it compares a transported quantity of passengers with a distance over which it gets carried. The ton-km (or ton-mile) is a common measure expressing the realized freight transport demand. Although both the passenger-km and ton-km are most commonly used to measure realized demand, the measure can equally apply for transport supply.

There are several factors impacting the capacity of transport infrastructure, from the physical characteristics of the network, how it is operated and maintained to the presence of bottlenecks. Transport supply can be simplified by a set of functions representing what are the main variables influencing the capacity of transport systems. These variables are different for each mode. For road, rail and telecommunications, transport supply is often dependent on the capacity of the routes and vehicles (modal supply) while for air and maritime transportation transport supply is strongly influenced by the capacity of the terminals (intermodal supply).

Transport demand tends to be expressed at specific times that are related to economic and social activity patterns. In many cases, transport demand is stable and recurrent, which allows a good approximation in planning services. In other cases, transport demand is unstable and uncertain, which makes it difficult to offer an adequate level of service. For instance, commuting is a recurring and predictable pattern of movements, while emergency response vehicles such as ambulances are dealing with an unpredictable demand that can be expressed as a probability. Transport demand functions vary according to the nature of what is to be transported:

- **Passengers**. For the road and air transport of passengers, demand is a function of demographic attributes of the population such as income, age, standard of living, race and sex, as well as modal preferences.
- **Freight**. For freight transportation, the demand is function of the nature and the importance of economic activities (GDP, commercial surface, number of tons of ore extracted, etc.) and of modal preferences. Freight transportation demand is more complex to evaluate than passengers.
- **Information**. For telecommunications, the demand can be a function of several criteria including the population (telephone calls) and the volume of financial activities (stock exchange). The standard of living and education levels are also factors to be considered.

Supply/demand relationships

Relationships between transport supply and demand continually change, but they are interrelated. From a conventional economic perspective, transport supply and demand interact until an equilibrium is reached between the quantity of transportation the market is willing to use at a given price and the quantity being supplied for that price level (Figure 7.8).

The demand curve assumes that if transport costs are high, demand is low as the consumers of a transport service (either freight or passengers) are less likely to use it. If

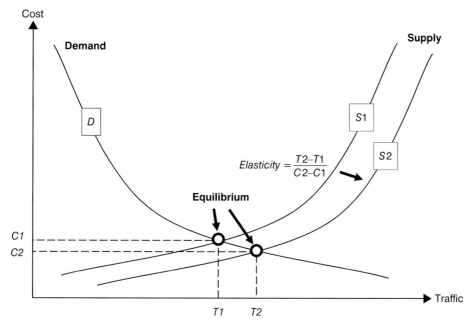

Figure 7.8 Classic transport demand/supply function

transport costs are low, the demand would be high as users would get more services for the same cost. The supply curve behaves inversely. If costs are high, transport providers would be willing to supply high quantities of services since high profits are likely to arise under such circumstances. If costs are low, the quantity of transport services would be low as many providers would see few benefits operating at a loss. The equilibrium point represents a compromise between what users are willing to pay and what providers are willing to offer. Under such circumstances, an amount of traffic T1 would flow at an operating cost C1. If because of an improvement a larger amount of service is possible for the same cost (the supply curve moves from S1 to S2), a new equilibrium will be reached with a quantity of traffic T2 at a price C2.

Many transport systems behave in accordance with supply and demand, which are influenced by cost variations. In line with microeconomic theory, the Law of Demand states that the demand for transport services decreases when the price of this service increases. This is reflected in the transport demand curve which plots the aggregate quantity of a transport service that consumers are willing to buy at different prices, holding constant other demand drivers such as prices of other transport services and goods, the budget or income and quality aspects such as reliability. Any change in another factor that affects the consumers' willingness to pay for the transport service results in a shift in the demand curve for the good.

Price changes not only affect the level of transport demand, but can also lead to shifts of demand to other routes, alternative transport modes and/or other time periods. In the medium or long term structural changes in the pricing of transport can affect location decisions of individuals and businesses. However, several considerations are specific to the transport sector which make supply/demand relationships more complex:

Entry costs. These are the costs incurred to operate at least one vehicle in a transport system. In some sectors, notably maritime, rail and air transportation, entry costs are very high, while in others such as trucking, they are very low. High entry costs imply that transport companies will consider seriously the additional demand before adding new capacity or new infrastructures (or venturing in a new service). In a situation of low entry costs, the number of companies is fluctuating with the demand. When entry costs are high, the emergence of a new player is uncommon while dropping out is often a dramatic event linked to a large bankruptcy. Consequently, transport activities with high entry costs tend to be oligopolistic while transport activities with low entry costs tend to have many competitors.

Public sector. Few other sectors of the economy have seen such a high level of public involvement than transportation, which creates many disruptions in conventional price mechanisms. The provision of transport infrastructures, especially roads, was massively funded by governments, mainly for the sake of national accessibility and regional equity. Transit systems are also heavily subsidized, to provide accessibility to urban populations and especially to the poorest segment judged to be deprived in mobility. As a consequence, transport costs are often considered as partially subsidized. Government control (and direct ownership) was also significant for several modes, such as rail and air transportation in a number of countries. The recent years have, however, been characterized by privatization and deregulation.

Elasticity. The notion of price elasticity is at the core of transport demand and refers to the variation of demand in response to a variation of cost. For example, an elasticity of -0.5 for vehicle use with respect to vehicle operating costs means that an increase of 1 percent in operating costs would imply a 0.5 percent reduction in vehicle mileage or trips. Variations of transport costs have different consequences for different modes, but transport demand has a tendency to be inelastic. While commuting tends to be inelastic in terms of costs, it is elastic in terms of time. For economic sectors where freight costs are a small component of the total production costs, variations in transport costs have limited consequences on the demand. For air transportation, especially the tourism sector, price variations have significant impacts on the demand. There are thus differences among the obtained price elasticities, which raise questions about the transferability of the results to other locations and/or other time periods. Hence, each case is characterized by a specific local environment in terms of modal choice options, budget/income of the transport user, spatial planning, price levels, etc. All these factors combined can make the behavior of transport users somewhat different across regions and settings.

As transport demand is a derived demand from individuals, groups and industries it can be desegregated into a series of partial demands fulfilled by the adaptation and evolution of transport techniques, vehicles and infrastructures to changing needs. Moreover, the growing complexity of economies and societies linked with technological changes force the transport industry to constant changes. This leads to growing congestion, a reduction in transport safety, a degradation of transport infrastructures and concerns about environmental impacts.

CASE STUDY **The cruise industry**

Authors: Jean-Paul Rodrigue and Theo Notteboom

Emergence of the cruise industry

From the mid-nineteenth century liner services supported long distance passenger transportation between continents, particularly between Europe and North America. The need to accommodate a large number of passengers of different socioeconomic status for at least a week led to the emergence of specific ship designs, radically different from cargo ships, where speed and comfort (at least for the elite) were paramount. The emergence of the cruise industry can be traced to the demise of the ocean liner in the 1960s as it was replaced by fast jet services with which it could not compete. It took more than a decade to see the complete demise of liner services with the final realization that long distance travel was now to be assumed by air transport. The last liners became the first cruise ships, especially considering the 30-year lifespan of a liner. The availability of a fleet of liners whose utility was no longer commercially justifiable prompted their reconversion to form the first fleet of cruise ships.

For instance, one of the last purposely designed liners, the SS *France*, operating between 1961 and 1974, was mainly used for the conventional transatlantic service between Le Havre and New York. With rising oil prices and more efficient jet liners, including the 747 (introduced in 1970), the liner was no longer able to effectively compete over the transatlantic route. While a jet plane could link Paris or London to New York in about 8 hours, it took about 4 days for a liner to cross the Atlantic, excluding a train segment between London and Southampton (or Paris and Le Havre). Considering one round trip per day, a 747 could carry about 3,200 passengers across the Atlantic in the time it took the SS *France* to carry 2,000 passengers on a single journey. Unable to generate enough revenue to justify its operating costs the SS *France* was mothballed in 1974 and purchased by the Norwegian Cruise Line (and renamed the SS *Norway*). Its final commercial years between 1980 and 2003 were spent as a cruise ship. However, liners were not particularly suitable to the requirements of the emerging cruise industry. For instance, since many liners were designed to operate on the North Atlantic throughout the year for scheduled passenger services, their outdoor amenities such as boardwalks and swimming pools were limited. Additionally, they were built for speed (which was their trademark) with the related high levels of fuel consumption.

The modern cruise industry began in the late 1960s and early 1970s with the founding of Norwegian Cruise Line (1966), Royal Caribbean International (1968) and Carnival Cruise Lines (1972), which have remained the largest cruise lines. The early goal of the cruise industry was to develop a mass market since cruising was until then an activity for the elite. A way to achieve this was through economies of scale as larger ships are able to accommodate more customers as well as creating additional opportunities for onboard sources of revenue. The first dedicated cruise ships began to appear in the 1970s and could carry about 1,000 passengers. By the 1980s, economies of scale were further expanded with cruise ships that could carry more than 2,000 passengers. The current large cruise ships have a capacity of about 6,000 passengers. The market for the cruise industry was by then established and recognized as a full-fledged touristic alternative directly competing with well-known resort areas such as Las Vegas or Orlando.

Cruise ships tend to have a low draft since they do not carry cargo; they are more volume than weight. This confers the advantage of being able to access a large number of ports and therefore multiplying itinerary options since the setting of a pure cruise port leans on criteria that are different from commercial ports. Cruise ports tend to be located close to either city centers (cultural and commercial amenities) or to natural amenities (e.g. a protected beach). These sites do not have on average very deep drafts and dredging would be socially or environmentally unacceptable. For instance, ships of the Oasis class, which as of 2011 accounted for the largest cruise ship class, have a draft of 31 feet. Comparatively, a container ship of 2,500 TEUs requires a draft of 33 feet, while a sovereign class container ship of 8,400 TEUs requires a draft of 46 feet. Draft issues that have plagued container ports are a much more marginal constraint for cruise shipping. Additionally, cruise ships have the option to anchor and use tendering services, which opens a wide array of ports of call.

Market dynamics

The global cruise industry carried about 19.1 million passengers in 2011, up from 7.2 million in 2000. The global growth rate of the cruise industry has been enduring and stable, at around 7 percent per year in spite of economic cycles of growth and recession. For instance, the financial crisis of 2008–9 has not impacted the demand for cruises. This underlines that the industry has been so far fundamentally supply based; the ships are built and the customers are found to fill them through various marketing and discounting strategies. The possibility for cruise ship operators to successfully follow a supply push strategy makes the cruise industry quite different from other shipping markets, such as container shipping. Hence, in most shipping markets the shipping activity is a clearly derived activity of trade and demand is rather price inelastic. As mentioned earlier, demand in the cruise business is "created" through pricing and branding/marketing. Cruise operators are challenged to develop competitive cruise packages which involve a high quality stay onboard, an array of shore-based activities offering access to a variety of cultures and sites, and easy transfers to/from the vessel.

The construction of cruise ships tends to take place in cycles where several ships are ordered and enter the market within a short time frame. Since the cruise industry is a relatively small segment of the touristic sector, it has so far been very successful at finding customers to fill a greater number of ever larger ships. The cruise product has become diversified to attract new customers and to respond to the preferences of a wide array of customer groups. In view of fulfilling the desires of its guests, the cruise industry has innovated through the development of new destinations, new ship designs, new and diverse onboard amenities, facilities and services, plus wide-ranging shore-side activities. Most cruise ship operators work around specific cruise themes and voyage lengths can vary to meet the changing vacation patterns of customers. Its highest level of market penetration is in North America with about 3 percent of the population taking a cruise each year. This includes people who may take more than one cruise in a year so actual figures are lower.

The market drivers of the cruise industry are similar to those that have fostered the growth of tourism since the Second World War, particularly the rising affluence of the global population and the growing popularity of exotic and resort destinations. The general aging of the population is also a factor in favor of cruise shipping as the main market remains older adults, albeit customers are getting significantly younger. While in 1995 the average age of a cruiser was about 65 years, this figure dropped to 45 years by 2006. What is novel with cruising is that the ship represents in itself the destination, essentially acting as a floating hotel (or a theme park) with all the related facilities (bars, restaurants, theaters, casinos, swimming pools, etc.). This permitted cruise lines to develop a captive market within their ships as well as for shore-based activities (e.g. excursions or facilities entirely owned by subsidiaries of the cruise line).

Some cruise operators go very far in developing new entertainment concepts on board their vessels since onboard services account for between 20 and 30 percent of the total cruise line revenues. The average customer spends about $1,700 for their cruise, including ship and off-ship expenses for goods and services. The majority of these expenses are captured within the cruise ship as passengers spend on average $100 per port of call. The Caribbean has been the dominant deployment market of the cruise industry since its inception, but the Mediterranean cruise market has grown substantially in recent years. Both markets offer a variety of cultures in close proximity and are thus ideally suited. Furthermore, strong niche markets have developed focusing on, for instance, history (Hanseatic cities in northern Europe) or natural amenities (Alaska).

The cruise industry has a very high level of ownership concentration, since the four largest cruise shipping companies account for 96 percent of the market (Carnival Lines, Royal Caribbean, Norwegian Cruise Line and MSC Cruises). High levels of horizontal integration are also observed since most cruise companies have acquired parent companies but kept their individual names for the purpose of product differentiation. For instance, Royal Caribbean Cruises, which is the world's second largest cruise company behind Carnival Lines, accounts for 24 percent of the global market serviced under six different brands such as Celebrity Cruises (which caters to higher end customers) and Azamara Club Cruises (smaller ships servicing more exotic destinations with shore stay options). Cruise shipping is increasingly capital intensive as each new cruise ship class comes with better amenities. A ship of the latest Oasis class, which is able to carry more than 6,000 passengers, costs about 1.2 billion dollars and can take four years to be delivered. Larger ships command higher booking prices since they offer more amenities, but current trends indicate that the cruise industry has no ships larger than the Oasis class in its order books. Optimal economies of scale may have been reached.

Networks and ports of call

The cruise industry sells itineraries, not destinations, implying a greater flexibility in the selection of ports of call. The selection of an itinerary is the outcome of several commercial considerations including potential revenue generation, distance between ports of call (cruise ships can cover 200 nautical miles per night), brand positioning (exotic ports of call for premium services), guest satisfaction (customer-oriented industry), economic trends and market research such as evaluating changes in disposable incomes and the demographics of the customer base. Ships are constantly moving between ports of call and shore leaves are of low duration; 4.3 hours on average in the Caribbean. A standard cruise itinerary is a loop beginning and ending at a hub port (also called a turn port) and typically lasting 7 days with 3 to 5 ports of call depending on their respective proximity. Cruises of 10 to 21 days are also offered but they tend to have lower profit margins as customers are inclined to spend less as the cruise progresses.

The global cruise port system is characterized by a high level of regional concentration as well as a clustering of port visits (Figure 7.9). The observed destination patterns clearly underline the prominence of port visits around the Caribbean and the Mediterranean in line

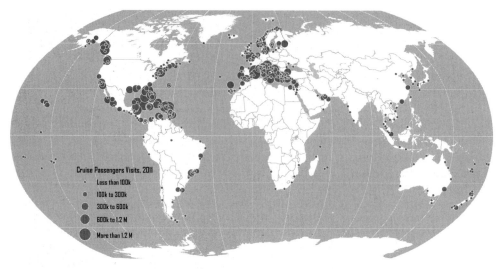

Figure 7.9 The global cruise port system

Source: Data adapted from Cruise Market Watch.

with the operational characteristics of 7-day cruises calling at 3 to 5 ports. Other clusters of significant activity concern the US Northeast and Atlantic Canada, Alaska, Hawaii, Hanseatic ports and the coast of Norway. Limited cruise activity takes place in East and Southeast Asia in spite of the significant economic development processes that have occurred in the region in recent decades. Therefore, the geography of cruise and commercial ports is completely different in terms of the dominant ports and the regions being serviced. It is also indicative that new cruising clusters may emerge to serve a latent demand from a growing middle and upper class in Asia, the Middle East and South America.

Three main types of itineraries can be found:

Perennial. The region covered by the itinerary is serviced throughout the year as the demand remains resilient, which is associated with stable (subtropical) weather conditions as well as stable itineraries. There may be significant seasonal variations in the number of ships deployed but the market remains serviced throughout the year. The Caribbean is the foremost perennial cruise market (summer low season), but the Mediterranean is also serviced year-round with a winter low season.

Seasonal. Weather is the dominant factor explaining seasonality, implying that some regions have a market potential only during a specific period or season. This is particularly the case for Baltic, Norwegian, Alaskan and New England cruises that are serviced during summer months. Inversely, South American and Australian itineraries are serviced during the winter months.

Repositioning. Because of the seasonality of the cruise industry the repositioning of ships between seasons is required. Cruise companies are increasingly using this opportunity to offer customers lower cost cruises for the inconvenience of having to book air travel arrangements for the return trip since the beginning and ending ports of call are not the same. This mainly takes place across the Atlantic as ships move from the winter Caribbean peak season to the summer Mediterranean peak season (and vice versa). The beginning and the end of the Alaska season are also combined

Photo 7.1   Cruise ships at the Port of Barcelona, Spain

The Port of Barcelona on the Mediterranean coast of Spain is the most important cruise shipping hub of the Mediterranean and the second most important in the world after Miami. In 2010, it handled around 2.35 million passengers with the months between April and October being the busiest (summer peak season). The cruise ship in this photo is the MSC *Splendida*, one of the newest ships that entered service in 2009 and has a capacity of 3,900 passengers and a crew of 1,300. The cruise ship in the back is being refueled by a port-based tanker.

with a Hawaiian cruise as ships get repositioned. Barcelona (Photo 7.1) and Dubai are emerging repositioning hubs since the Mediterranean and the Indian Ocean are growing faster than the conventional Caribbean market.

Bibliography

Berry, B.J.L. (1967) *Geography of Market Centers and Retail Distribution*, Englewood Cliffs, NJ: Prentice-Hall.

—— (1991) *Long-wave Rhythms in Economic Development and Political Behavior*, Baltimore, MD: Johns Hopkins University Press.

Brueckner, K. (2003) "Airline traffic and urban economic development", *Urban Studies*, 40(8): 1455–69.

Button, K. (2010) *Transport Economics*, 3rd edition, Aldershot: Edward Elgar.

Dowling, R.K. (ed.) (2006) *Cruise Ship Tourism*, Cambridge, MA: CAB International.

European Conference of Ministers of Transport (2001) *Transport and Economic Development*, Paris: OECD.

Goodbody Economic Consultants (2003) *Transport and Regional Development*, Ballsbridge Park, Ballsbridge, Dublin.

Hall, P. (1984) *The World Cities*, 3rd edition, New York: St Martin's Press.

Harrington, J.W. and B. Warf (1995) *Industrial Location: Principles, Practice and Policy*, London: Routledge.

Henderson, J.V., Z. Shalizi and A.J. Venables (2000) "Geography and development", *Journal of Economic Geography*, 1: 81–106.

Lakshmanan, T.R. (2011) "The broader economic consequences of transport infrastructure investments", *Journal of Transport Geography*, 19(1): 1–12.

Pred, A. (1977) *City Systems in Advanced Economies: Past Growth, Present Processes and Future Development Options*, New York: Wiley.

8 Transport, energy and environment

Transportation systems are linked with a wide range of environmental considerations at all geographical scales, from the global to the local. The nature of these environmental impacts is related to the transport modes themselves, their energy supply systems, their emissions and the infrastructures over which they operate. While consuming large quantities of energy, especially oil, vehicles also emit numerous pollutants such as carbon dioxide and nitrogen oxide as well as noise, and transport infrastructures have damaged many ecological systems. Several of the environmental impacts of transport systems have been externalized, implying that the benefits of mobility are realized by a few while the costs are assumed by the whole society. The spatial structure of economic activities, notably their land use, is also increasingly linked with environmental impacts. The sustainability of transport systems has become one core issue in the provision of mobility.

Concept 1 – The environmental impacts of transportation

The issue of transport and the environment

The issue of transportation and the environment is paradoxical in nature since transportation conveys substantial socioeconomic benefits, but at the same time transportation is impacting environmental systems. From one side, transportation activities support increasing mobility demands for passengers and freight, while on the other, transport activities are associated with growing levels of environmental externalities. This has reached a point where transportation is a dominant source of emission of most pollutants and their multiple impacts on the environment. These impacts fall within three categories:

- **Direct impacts**. The immediate consequence of transport activities on the environment where the cause and effect relationship is generally clear and well understood.
- **Indirect impacts**. The secondary (or tertiary) effects of transport activities on environmental systems. They are often of higher consequence than direct impacts, but the relationships involved are often misunderstood and difficult to establish.
- **Cumulative impacts**. The additive, multiplicative or synergetic consequences of transport activities. They take into account the varied effects of direct and indirect impacts on an ecosystem, which are often unpredicted.

The complexities of the problems have led to much controversy in environmental policy and in the role of transportation. The transportation sector is often subsidized by the public sector, especially through the construction and maintenance of road infrastructure which tend to be free of access. Sometimes, public stakes in transport modes, terminals and infrastructure can be at odds with environmental issues. If the owner and the regulator are the same (different branches of the government), then there is a risk that regulations will not be effectively complied with. It can also lead to another extreme where compliance would lead to inefficient transport systems, with subsidized costs.

Total costs incurred by transportation activities, notably environmental damage, are generally not fully assumed by the users. The lack of consideration of the real costs of transportation could explain several environmental problems. Yet, a complex hierarchy of costs is involved, ranging from internal (mostly operations), compliance (abiding by regulations), contingent (risk of an event such as a spill) to external (assumed by society). For instance, external costs account on average for more than 30 percent of the estimated automobile costs. If environmental costs are not included in this appraisal, the usage of the car is consequently subsidized by society and costs accumulate as environmental pollution. This requires due consideration as the number of vehicles, especially automobiles, is steadily increasing.

The transport–environment link

The relationships between transport and the environment are multidimensional. Some aspects are unknown and some new findings may lead to drastic changes in environmental policies, as it did in regards of acid rain and chlorofluorocarbons in the 1970s and 1980s. The 1990s were characterized by a realization of global environmental issues, epitomized by the growing concerns about anthropogenic effects and climate change. Transportation also became an important dimension of the concept of sustainability, which is expected to become the prime focus of transport activities in the coming decades, ranging from vehicle emissions to green supply chain management practices. These impending developments require a deep understanding of the reciprocal influences between the physical environment and transport infrastructures; yet this understanding is often lacking. The main factors considered in the physical environment are geographical location, topography, geological structure, climate, hydrology, soil, natural vegetation and animal life.

The main environmental dimensions of transportation are related to the causes, the activities, the outputs and the results of transport systems (Figure 8.1):

> **Causes**. Two major factors contribute to the level of transport activities. Economics refers to the general level of development, income and transport supply. An advanced economy is likely to generate more transportation activities per capita than a developing one. Land use refers to the spatial structure and location of transport demand.
> **Activities**. Involve a wide array of factors expressing the usage of transportation infrastructures and all the related services. All these activities have environmental outputs.
> **Outputs**. The first outcome of transportation activities are emissions of all sorts (carbon monoxide, nitrogen oxides, particulates, etc.). According to the geographical characteristics of the area where emissions occur (e.g. wind patterns) ambient pollution levels are created. Once these levels are correlated with population density, a level of exposure to harmful pollutants can be calculated. This exposure is likely to have consequences.
> **End results**. They include all the health, environmental and welfare effects of the exposure to emissions from transportation activities, which are very difficult to measure.

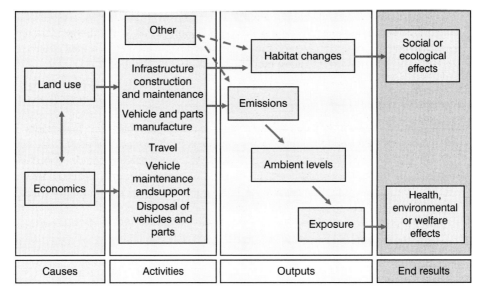

Figure 8.1 Environmental dimensions of transportation

Source: EPA.

Establishing linkages between environmental dimensions is a difficult undertaking. For instance, to what extent are carbon monoxide emissions linked to land use patterns? Furthermore, transportation is embedded in environmental cycles, notably over the carbon cycle. The relationships between transport and the environment are also complicated by two observations:

> First, transport activities contribute among other anthropogenic and natural causes, directly, indirectly and cumulatively to environmental problems. In some cases, they may be a dominant factor, while in others their role is marginal and difficult to establish.
>
> Second, transport activities contribute at different geographical scales to environmental problems, ranging from local (noise and carbon monoxide emissions) to global (climate change), not forgetting continental/national/regional problems (smog and acid rain).

Environmental policies for transportation thus have to take account of the level of contribution and the geographical scale, otherwise some policies may just move the problems elsewhere and have unintended consequences. A noted example is the local/regional policies that have forced the construction of higher chimneys for coal burning facilities (power plants) and induced the continental diffusion of acid rain. Thus, even if an administrative division (municipality, county, state/province) has adequate environmental enforcement policies, the geographical scale of pollutants diffusion (notably air pollutants) obviously goes beyond established jurisdictions.

In addition to the environmental impacts of the network, traffic and modes, economic/industrial processes sustaining the transport system must be considered. These include the production of fuels, vehicles and construction materials, some of which are very energy intensive (e.g. aluminum), and the disposal of vehicles, parts and the provision of infrastructure. They all have a life cycle timing their production, utilization and disposal. Thus, evaluation of the transport–environment link without consideration of cycles in both the environment and product life is likely to convey a limited overview of the situation and may even lead to incorrect appraisal and policies.

Environmental dimensions

Transportation activities support increasing mobility demands for passengers and freight, notably in urban areas. But transport activities have resulted in growing levels of motorization and congestion. As a result, the transportation sector is becoming increasingly linked to environmental problems. The most important impacts of transport on the environment relate to climate change, air quality, noise, water quality, soil quality, biodiversity and land take:

> **Climate change**. The activities of the transport industry release several million tons of gases each year into the atmosphere. These include lead (Pb), carbon monoxide (CO), carbon dioxide (CO_2; not a pollutant), methane (CH_4), nitrogen oxides (NOx), nitrous oxide (N_2O), chlorofluorocarbons (CFCs), perfluorocarbons (PFCs), silicon tetrafluoride (SiF_4), benzene and volatile components (BTX), heavy metals (zinc, chrome, copper and cadmium) and particulate matters (ash, dust). There is an ongoing debate to what extent these emissions are linked to climate change and the role of anthropogenic factors. Some of these gases, particularly nitrous oxide, also participate in depleting the stratospheric ozone (O_3) layer which naturally screens the earth's surface from ultraviolet radiation.
>
> **Air quality**. Highway vehicles, marine engines, locomotives and aircraft are the source of pollution in the form of gas and particulate matters emissions that affects air quality causing damage to human health. Toxic air pollutants are associated with cancer, cardiovascular, respiratory and neurological diseases. Carbon monoxide (CO) when inhaled affects the bloodstream, reduces the availability of oxygen and can be extremely harmful to public health. An emission of nitrogen dioxide (NO_2) from transportation sources reduces lung function, affects the respiratory immune defense system and increases the risk of respiratory problems. The emissions of sulfur dioxide (SO_2) and nitrogen oxides (NOx) in the atmosphere form various acidic compounds that when mixed in cloud water create acid rain. Acid precipitation has detrimental effects on the built environment, reduces agricultural crop yields and causes forest decline. The reduction of natural visibility by smog has a number of adverse impacts on the quality of life and the attractiveness of tourist sites. Particulate emissions in the form of dust emanating from vehicle exhaust as well as from non-exhaust sources such as vehicle and road abrasion have an impact on air quality. The physical and chemical properties of particulates are associated with health risks such as respiratory problems, skin irritations, eyes inflammations, blood clotting and various types of allergies.
>
> **Noise**. Noise represents the general effect of irregular and chaotic sounds. It is traumatizing for the hearing organ and that may affect the quality of life by its unpleasant and disturbing character. Long term exposure to noise levels above 75dB seriously hampers hearing and affects human physical and psychological wellbeing. Transport noise emanating from the movement of transport vehicles and the operations of ports, airports and rail yards affects human health, through an increase in the risk of cardiovascular diseases. Increasing noise levels have a negative impact on the urban environment reflected in falling land values and loss of productive land uses.

Water quality. Transport activities have an impact on hydrological conditions. Fuel, chemical and other hazardous particulates discarded from aircraft, cars, trucks and trains or from port and airport terminal operations, such as de-icing, can contaminate rivers, lakes, wetlands and oceans. Because demand for shipping services is increasing, marine transport emissions represent the most important segment of water quality inventory of the transportation sector. The main effects of marine transport operations on water quality predominantly arise from dredging, waste, ballast waters and oil spills. Dredging is the process of deepening harbor channels by removing sediments from the bed of a body of water. Dredging is essential to create and maintain sufficient water depth for shipping operations and port accessibility. Dredging activities have a twofold negative impact on the marine environment. They modify the hydrology by creating turbidity that can affect the marine biological diversity. The contaminated sediments and water raised by dredging require spoil disposal sites and decontamination techniques. Waste generated by the operations of vessels at sea or at ports cause serious environmental problems, since they can contain a very high level of bacteria that can be hazardous for public health as well as marine ecosystems when discharged in waters. And various types of garbage containing metals and plastic are not easily biodegradable; they can persist on the sea surface for a long time and can be a serious impediment for maritime navigation in inland waterways and at sea and as well as affecting berthing operations. Ballast waters are required to control a ship's stability and draft and to modify the center of gravity in relation to cargo carried and the variance in weight distribution. Ballast waters acquired in a region may contain invasive aquatic species that, when discharged in another region, may thrive in a new marine environment and disrupt the natural marine ecosystem. There are about 100 non-indigenous species recorded in the Baltic Sea. Invasive species have resulted in major changes in nearshore ecosystems, especially in coastal lagoons and inlets. Major oil spills from oil cargo vessel accidents are one of the most serious problems of pollution from maritime transport activities.

Soil quality. The environmental impact of transportation on soil consists of soil erosion and soil contamination. Coastal transport facilities have significant impacts on soil erosion. Shipping activities are modifying the scale and scope of wave actions leading to serious damage in confined channels such as river banks. The removal of the earth's surface for highway construction or lessening surface grades for port and airport developments have led to important loss of fertile and productive soils. Soil contamination can occur through the use of toxic materials by the transport industry. Fuel and oil spills from motor vehicles are washed on road sides and enter the soil. Chemicals used for the preservation of railroad ties may enter the soil. Hazardous materials and heavy metals have been found in areas contiguous to railroads, ports and airports.

Biodiversity. Transportation also influences natural vegetation. The need for construction materials and the development of land-based transportation has led to deforestation. Many transport routes have required draining land, thus reducing wetland areas and driving-out water plant species. The need to maintain road and rail right of way or to stabilize slopes along transport facilities has resulted in restricting growth of certain plants or has produced changes in plants with the introduction of new species different from those which originally grew in the areas. Many animal species are becoming extinct as a result of changes in their natural habitats and reduction of ranges.

Land take. Transportation facilities have an impact on the urban landscape. The development of port and airport infrastructures is a significant feature of the urban and peri-urban built environment. Social and economic cohesion can be severed when new transport facilities such as elevated train and highway structures cut across an existing urban community. Arteries or transport terminals can define urban borders and produce segregation. Major transport facilities can affect the quality of urban life by creating physical barriers, increasing noise levels, generating odors, reducing the urban aesthetic and affecting the built heritage.

Environmental externalities

Externalities are an economic concept that refers to activities of a group that have unintended consequences, positive or negative, on other groups and most importantly that those consequences, particularly if they are negative, are not assumed by those causing them. The impacts are therefore "externalized". A common example of a positive externality concerns technology since it obviously benefits the innovative firm but also the whole economy through various productivity improvements. Negative externalities have a lot of relevance over environmental issues, since many of the negative consequences of pollution are assumed by the whole society.

The environmental externalities of transportation include the consideration of physical measures of environmental damage and the evaluation of involved costs for society. The main fallacy underlined by externalities is that the costs attributed to a few sources (e.g. users of cars) must be carried by many (users and nonusers alike). Knowing the sources of environmental externalities is a relatively easy undertaking, while the evaluation of damage and other costs has not yet reached comparative standards among governmental and non-governmental agencies. The challenge resides over three issues:

Relationships. The nature and extent of the relationships between transport and the environment has to be considered. This is particularly complex as most environmental relationships tend to be indirect and cumulative.

Quantification. Relationships have to be quantified and also a value given to environmental externalities. This is almost impossible as only general figures, much subject to debate, can be assessed. The quantification of economic, social and environmental costs is very difficult but possible if some simplifications and generalizations are assumed.

Policy making. The level and extent of corrective actions that can be taken to alleviate and mitigate environmental externalities linked to transportation so that those contributing bear the consequences of their activities. In view of the two above points attempts at regulation, particularly if they involve a comprehensive framework, can be hazardous.

The costs of environmental externalities can be considered from economic, social and environmental dimensions. The basic types of transportation externalities attributed to the environment fall within air pollution, water pollution, noise, and hazardous materials. Establishing and quantifying environmental externalities is a complex undertaking. Quantification is only at its preliminary stage and many have used this argument to defer the application of several environmental policies (e.g. on acid rain, CFCs and most importantly, climate change) by lobbying governments. Additionally, the wider the geographical scale the more complex the environmental problem becomes, mainly due to cross-jurisdictional issues. Recent attempts to reach a consensus about climate change have underlined that multilateral environmental agreements are close to impossible.

The sources/emitters of pollutants rarely bear the consequences of their impacts. This has several implications. First, when specific sources are concerned, like road transportation, users only take account of the direct costs of modal ownership like a car (vehicle, fuel, insurance, etc.). Ownership is often the only entry and utilization cost for several transportation modes. The society generally assumes the role of providing and maintaining infrastructure and any indirect costs such as damage to structures and infrastructure, losses in productivity (agriculture and labor), cleanup, health services and damage to ecosystems. Second, the geographic separation between sources and recipients is often acute. Acid rain and climate change are obvious examples. On a local level, a community may be affected by noise levels well over its own contribution (notably those near major highways), while another (in the suburbs) may be affected in a very marginal way and still significantly contribute to noise elsewhere during commuting.

There is a tendency towards a shift from direct to indirect consequences for environmental externalities, in terms of total costs involved. For instance, the absolute levels of air pollutants emissions have considerably dropped in developed countries such as the United States. The problem of source reduction by vehicles was addressed because it was a straightforward cause of air pollutants emissions. This has tended to displace problems elsewhere and developed new types of externalities. Thus, the relative share of air pollution impacts is lessening, but not the number of vehicles, investment in infrastructure or noise levels, which have their own externalities. Reductions in the relative importance of one type of externality can redirect the focus on to other types that were less well addressed but probably as important in the overall impacts of transport on the environment.

Transfers and additions of costs are very common attributes of environmental externalities. Trying to lessen economic costs will either lessen or worsen social and environmental costs, depending on the externality. For instance, keeping salt as the main de-icing agent is a cheaper solution for authorities responsible for road maintenance, but this practice transfers economic benefits into environmental costs (damage to the ecosystem). In the context of limited resources, the distribution of economic, social and environmental costs takes an important role in what type of damage is most acceptable and in what proportions. It is clear from past strategies that several economic costs have been minimized, notably for producers and users, while social and environmental consequences were disregarded. This practice is less applicable since society is less willing to bear the costs and consequences of externalities for various reasons (public awareness, high health costs, etc.).

Concept 2 – Transportation and energy

Authors: Jean-Paul Rodrigue and Claude Comtois

Energy

Human activities are closely dependent on the usage of several forms and sources of energy to perform work. Energy is the potential that allows movement and/or the modification of matter (e.g. making steel by combining iron and carbon). The energy content of an energy source is the available energy per unit of weight or volume, but the challenge is to effectively extract and use this energy (Figure 8.2). Thus, the more

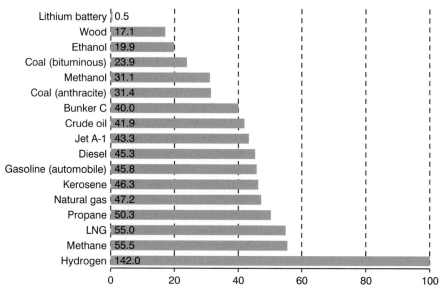

Figure 8.2 Chemical energy content of some fuels (in MJ/kg)

Source: Adapted from Ronneau, C. (2004) *Énergie, pollution de l'air et développement durable*, Louvain-la-neuve: Presses Universitaires de Louvain.

energy consumed the greater the amount of work realized and it comes as no surprise that economic development is correlated with greater levels of energy consumption.

Different fuels have different levels of energy content and the graph in Figure 8.2 relates to energy released through combustion. The higher the energy content the higher the quality of the fuel, which is inversely proportional to its chemical complexity. High quality fuels are gases while low quality fuels are solids, with liquids in between. The fuel which has the highest energy content is hydrogen, which is also the simplest chemical component in existence. Gasoline, which is derived from refining crude oil, contains much more energy than coal (almost twice that of the lower grade bituminous) or wood (nearly three times). Liquefied natural gas (LNG) is almost entirely composed of methane while natural gas has about 85 percent of its mass accounted by methane. Jet A-1 is the standard fuel used by commercial jet planes and is mostly composed of kerosene and a number of additives (antifreeze, antioxidant and antistatic) since the fuel must meet very strict specifications as it will be exposed to high altitudes and low temperatures. Conversely, Bunker C fuel which is the main fuel used for maritime shipping can be considered as one of the lower quality fuels in liquid form, but suitable for vast ship engines.

Although methane and hydrogen both have higher energy content than gasoline, their gaseous form creates storage difficulties. Furthermore, hydrogen must be synthesized, which requires energy. As a comparison, with a conversion rate of 100 percent, it would require 100 hours to capture the solar energy equivalent of 1 kg of gasoline on a surface of one square meter. One of the most efficient energy storage devices, the lithium battery, can only hold about the equivalent of 0.5 MJ per kilogram, underlining the challenge of developing electric vehicles.

There are enormous reserves of energy able to meet the future needs of mankind. Unfortunately, one of the main contemporary issues is that many of these reserves cannot be exploited at reasonable costs, such as solar energy, or are unevenly distributed around the world, such as oil. Through the history of mankind's use of energy, the choice of an energy source depended on a number of utility factors which involved a transition in energy systems from solid, liquid and eventually to gas sources of energy. Since the industrial revolution, efforts have been made for work to be performed by machines, which considerably improves industrial productivity. The development of the steam engine and the generation and distribution of electric energy over considerable distances have also altered the spatial pattern of manufacturing industries by liberating production from a direct connection to a fixed power system. While in the earlier stages of the industrial revolution factories located close to sources of energy (a waterfall or a coal field) or raw materials, mass conveyances and new energy sources (electricity) enabled a much greater locational flexibility.

Industrial development places large demands on fossil fuels. At the turn of the twentieth century, the invention and commercial development of the internal combustion engine, notably in transport equipment, made possible the efficient movement of people, freight and information and stimulated the development of the global trade network. With globalization, transportation is accounting for a growing share of the total amount of energy spent for implementing, operating and maintaining the international range and scope of human activities. Energy consumption has a strong correlation with the level of development. Among developed countries, transportation now accounts for between 20 and 25 percent of the total energy being consumed. The benefits conferred by additional mobility, notably in terms of a better exploitation of comparative advantages, have so far compensated for the growing amount of energy spent to support it. At the beginning of the twenty-first century, the transition in energy systems has reached a stage where fossil fuels, notably petroleum, are dominant. Out of the world's total power production, 87.1 percent is derived from fossil fuels.

Transportation and energy consumption

Transportation and energy is a standard physics application where giving momentum to masses (people, vehicles, cargo, etc.) requires a proportional amount of energy. The relationship between transport and energy is a direct one, but subject to different interpretations since it concerns different transport modes, each having their own performance levels. There is often a compromise between speed and energy consumption, related to the desired economic returns (Figure 8.3).

Within the transport sector there are significant differences between speed, energy costs, mode and type of loads (freight and passengers). Since economies of scale play a crucial role in freight transportation, this is reflected in its lower levels of energy consumption. Overall, transport operators try to seek a compromise between speed (returns in overcoming distance) and energy (costs in overcoming distance). The lowest energy consumption levels are associated with bulk freight traveling at slow speed (like oil) while high levels of energy consumption correspond to passengers or merchandise being carried at high velocities. The first corresponds to a compromise of energy over speed, while the second is a compromise of speed over energy. The compromise for supersonic planes was so disadvantageous that commercial services (Concorde) were abandoned in 2003.

Passengers and high value goods can be transported by fast but energy-intensive modes since the time component of their mobility tends to have a high value, which conveys the willingness to use more energy. Economies of scale, mainly those achieved by maritime transportation, are linked to low levels of energy consumption per unit of mass being transported, but at a lower speed. This fits relatively well freight transport

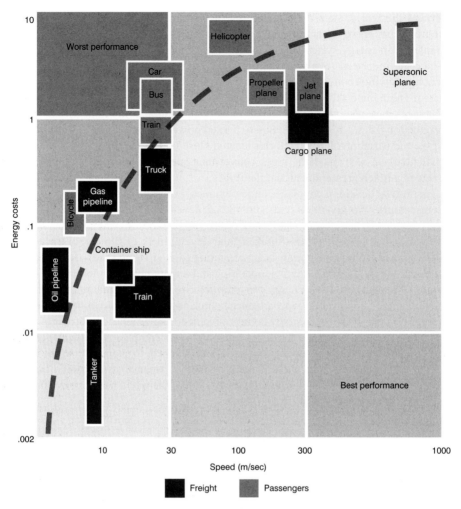

Figure 8.3 Energy efficiency by transportation mode

Source: Adapted from Chapman, J.D. (1989) *Geography and Energy: Commercial Energy Systems and National Policies*, New York: Longman Scientific & Technical.

imperatives, particularly for bulk. Comparatively, air freight has high energy consumption levels linked to high speed services. Transportation markets are particularly impacted by these three energy issues:

- The **price level and volatility** of energy sources which are dependent on the processes used in their production. Stable energy sources are obviously preferred.
- Technological and technical changes in the level of **energy performance** of transport modes and terminals. An important goal is thus to improve this energy performance.
- **Environmental externalities** related to the use of specific modes and energy sources and the goal to reduce them.

A trend that emerged since the 1950s concerns the growing share of transportation in the world's total oil consumption; transportation accounts for approximately 25 percent of world energy demand and for about 61.5 percent of all the oil used each year. Energy consumption has strong modal variations:

Land transportation accounts for the great majority of energy consumption. Road transportation alone consumes on average 85 percent of the total energy used by the transport sector in developed countries. This trend is not however uniform within the land transportation sector itself, as road transportation is almost the sole mode responsible for additional energy demands over the last 25 years. Despite a falling market share, rail transport, on the basis of 1 kg of oil equivalent, remains four times more efficient for passenger movement and twice as efficient for freight movement as road transport. Rail transport accounts for 6 percent of global transport energy demand.

Maritime transportation accounts for 90 percent of cross-border world trade as measured by volume. The nature of water transport and its economies of scale make it the most energy-efficient mode since it uses only 7 percent of all the energy consumed by transport activities, a figure way below its contribution to the mobility of goods.

Air transportation plays an integral part in the globalization of transportation networks. The aviation industry accounts for 8 percent of the energy consumed by transportation. Air transport has high energy consumption levels, linked to high speeds. Fuel is the second most important cost for the air transport industry accounting for 13–20 percent of total expenses. This accounts for about 1.2 million barrels per day. Technological innovations, such as more efficient engines and better aerodynamics, have led to a continuous improvement of the energy efficiency of each new generation of aircrafts.

Further distinctions in the energy consumption of transport can be made between passenger and freight movements:

Passenger transportation accounts for 60 to 70 percent of energy consumption from transportation activities. The private car is the dominant mode but has a poor energetic performance, although this performance has seen substantial improvements since the 1970s, mainly due to growing energy prices and regulations. Only 12 percent of the fuel used by a car actually provides momentum. There is a close relationship between rising income, automobile ownership and distance traveled by vehicle. The United States has one of the highest levels of car ownership in the world with one car for every two people. About 60 percent of all American households owned two or more cars, with 19 percent owning three or more. Another trend has been the increasing rise in ownership of minivans, sport utility vehicles (SUVs) and light-duty trucks for personal use and the corresponding decline in fuel economy. Fuel consumption is however impacted by diminishing returns, implying that higher levels of fuel efficiency involve declining marginal gains in fuel consumption. Also, the growth of vehicle-miles traveled is correlated with changes in energy prices and is entering a phase of maturity in several developed countries.

Freight transportation is dominated by rail and maritime shipping, the two most energy-efficient modes. Coastal and inland waterways also provide an energy-efficient method of transporting passengers and cargoes. A tow boat moving a typical load of 15 barges in tow holds the equivalent of 225 rail car loads or 870 truckloads. The rationale for favoring coastal and inland navigation is based on lower energy consumption rates of shipping and the general overall smaller externalities of water transportation.

Transportation and alternative fuels

All other things being equal, the energy source with the lowest cost will always be sought. The dominance of petroleum-derived fuels is a result of the relative simplicity with which they can be stored and efficiently used in the internal combustion engine vehicle. Other fossil fuels (natural gas, propane and methanol) can be used as transportation fuels but require a more complicated storage system. The main issue concerning the large-scale uses of these alternative vehicle fuels is the large capital investments required in distribution facilities as compared with conventional fuels. Another issue is that in terms of energy density, these alternative fuels have lower efficiency than gasoline and thus require greater volume of on-board storage to cover the equivalent distance as a gasoline propelled vehicle if performance is kept constant.

Alternative fuels in the form of non-crude oil resources are drawing considerable attention as a result of shrinking oil reserves, increasing petroleum costs and the need to reduce emissions of harmful pollutants. The most prevalent alternatives being considered are:

Biogas such as ethanol, methanol and biodiesel can be produced from the fermentation of food crops (sugar cane, corn, cereals, etc.) or wood-waste. Their production, however, requires large harvesting areas that may compete with other types of land use. Besides, it is estimated that one hectare of wheat produces less than 1,000 liters of transportation fuel per year which represents the amount of fuel consumed by one passenger car traveling 10,000 kilometers per year. This limit is related to the capacity of plants to absorb solar energy and transform it through photosynthesis. This low productivity of the biomass does not meet the energy needs of the transportation sector. Besides, the production of ethanol is an energy-intensive process. The production of 1 thermal unit of ethanol requires the combustion of 0.76 unit of coal, petroleum or natural gas. Biodiesel can also be obtained from a variety of crops. The choice of biomass fuel will largely depend on the sustainability and energy efficiency of the production process.

Hydrogen is often mentioned as the energy source of the future. The steps in using hydrogen as a transportation fuel consist in: (1) producing hydrogen by electrolysis of water or by extracting it from hydrocarbons; (2) compressing or converting hydrogen into liquid form; (3) storing it on-board a vehicle; and (4) using fuel cells to generate electricity on demand from the hydrogen to propel a motor vehicle. Hydrogen fuel cells are two times more efficient than gasoline and generate near-zero pollutants. But hydrogen suffers from several problems. A lot of energy is wasted in the production, transfer and storage of hydrogen. Hydrogen manufacturing requires electricity production. Hydrogen-powered vehicles require 2–4 times more energy for operation than an electric car which does not make them cost-effective. Besides, hydrogen has a very low energy density and requires a very low temperature and very high pressure storage tank adding weight and volume to a vehicle. This suggests that liquid hydrogen fuel would be a better alternative for ship and aircraft propulsion.

Electricity is being considered as an alternative to petroleum fuels as an energy source. A pure battery electric vehicle is considered a more efficient alternative to a hydrogen-fuel propelled vehicle as there is no need to convert energy into electricity since the electricity stored in the battery can power the electric motor. Besides an electric car is easier and cheaper to produce than a comparable fuel-cell vehicle. The main barriers to the development of electric cars are the lack of storage systems capable of providing driving ranges and speeds comparable to those of conventional vehicles. The low energy capacity of batteries makes the electric car less competitive than internal combustion engines using gasoline. Yet, as technology improves, cost-effective batteries will become available.

Hybrid vehicles consisting of a propulsion system using an internal combustion engine supplemented by an electric motor and batteries, which provides opportunities combining the efficiency of electricity with the long driving range of an internal combustion engine. A hybrid vehicle still uses liquid fuel as the main source of energy but the engine provides the power to drive the vehicle or is used to charge the battery via a generator. Alternatively, the propulsion can be provided by the electricity generated by the battery. When the battery is discharged, the engine starts automatically without intervention from the driver. The generator can also be fed by using the braking energy to recharge the battery. Such a propulsion design greatly contributes to overall fuel efficiency. Given the inevitable oil depletion, the successful development and commercialization of hybrid vehicles appears on the medium term the most sustainable option to conventional gasoline-engine powered vehicles.

The diffusion of non-fossil fuels in the transportation sector has serious limitations. As a result, the price of oil will certainly continue to increase as more expensive fuel-recovery technologies will have to be utilized. But high oil prices are deflationary leading to recession in economic activity and the search for alternative sources of energy. Already, the potential peaking of conventional oil production is leading to the implementation of coal-derived oil projects. Coal liquefaction technology allows the transformation of coal into refined oil after a series of processes in an environment of high temperature and high pressure. While the cost-effectiveness of this technique is yet to be demonstrated, coal liquefaction is an important measure in the implementation of transportation fuel strategies in coal-rich countries, such as China and South Africa.

The costs of alternative energy sources to fossil fuels are higher in the transportation sector than in other types of economic activities. This suggests higher competitive advantages for the industrial, household, commercial, electricity and heat sectors to shift away from oil and to rely on solar, wind or hydro-power. Transportation fuels based on renewable energy sources might not be competitive with petroleum fuels unless future price increase is affected by different fuel taxes based on environmental impacts.

Transportation and peak oil

The extent to which conventional non-renewable fossil fuels will continue to be the primary resources for nearly all transportation fuels is subject to debate. But the gap between demand and supply, once considerable, is narrowing, an effect compounded by the peaking of global oil production. The steady surge in demand from developing economies, particularly China and India, requires additional outputs. This raises concern about the capacity of major oil producers to meet this rising world demand. The producers are not running out of oil, but the existing reservoirs may not be capable of producing on a daily basis the increasing volumes of oil that the world requires.

Reservoirs do not exist as underground lakes from which oil can easily be extracted. There are geological limits to the output of existing fields. This suggests that additional reserves need to be found to compensate for the declining production of existing fields.

Others argue that the history of the oil industry is marked by cycles of shortages and surplus. The rising price of oil will render cost effective oil recovery in difficult areas. Deep water drilling, extraction from tar sands and oil shale should increase the supply of oil that can be recovered and extracted from the surface. But there is a limit to the capacity of technological innovation to find and extract more oil around the world and the related risks can be very high. Technological development does not keep pace with surging demand. The construction of drilling rigs, power plants, refineries and pipelines designed to increase oil exploitation is a complex and slow process. The main concern is the amount of oil that can be pumped to the surface on a daily basis.

High fuel prices could stimulate the development of alternatives, but automotive fuel oil is relatively inelastic. Higher prices result in very marginal changes in demand for fuel. While $100 per barrel was for a long time considered a threshold that would limit demand for automotive fuel and lead to a decline in passenger and freight-km, evidence suggests that higher oil prices had limited impact on the average annual growth rate of world motorization. The analysis of the evolution of the use of fossil fuels suggests that in a market economy the introduction of alternative fuels is leading to an increase in the global consumption of both fossil and alternative fuels and not to the substitution of crude oil by bio-based alternative fuels. This suggests that in the initial phase of an energy transition cycle, the introduction of a new source of energy complements existing supply until the new source of energy becomes price competitive to be an alternative. The presence of both renewable and non-renewable types of fuels stimulates the energy market with the concomitant result of increasing greenhouse gas emissions. The production of alternative fuels adds up to the existing fossil fuels and does not replace it.

Under such circumstances, oil prices are bound to rise in a substantial way, sending significant price signals to the transport market. How the transport system will respond and adapt to higher energy prices is obviously subject to much debate and interpretation. The following potential consequences can be noted:

Road. As far as the automobile is concerned, higher oil prices could trigger changes in several phases. Initially, commuters would simply absorb the higher costs. Depending on their level of productivity, many economies could show a remarkable resilience. The next phase would see changes in commuting patterns (e.g. carpooling), attempts to use public transit, a rapid adoption of vehicles with high gasoline efficiency (in the United States, this could mark the downfall of the SUV) and a search for other transport alternatives. The existing spatial structure could also start to show signs of stress as the unsustainability of car-dependent areas becomes more apparent. There is already evidence that peak car mobility may have been reached in the United States. As high commuting costs and the inflationary effects of high oil prices on the economy become apparent many would no longer be able to afford to live in a suburban setting. Cities could start to implode. The trucking industry would behave in a similar way, first by lowering their profits and their operating expenses (e.g. scheduling, achieve FTL), but at some point, higher prices will be passed on to their customers.

Rail. This mode is set to benefit substantially from higher energy prices as it is the most energy-efficient land transportation mode. Rail is about three times more energy efficient than trucking. The level of substitution for passengers and freight remains uncertain and will depend on the current market share and level of service they offer. In North America, passenger rail has limited potential while in Europe and Pacific Asia passenger rail already assumes a significant market share. For rail freight, North American freight distribution has an advantage since rail accounts for a dominant share of tons-km while this figure is less significant for other regions of the world, mainly due to the distances involved and the fragmentation of the system. In many cases, there could be pressure towards the electrification of strategic long distance corridors and the development of more efficient cargo-handling facilities. Thus, growing energy prices are likely to affect long distance rail transportation differently depending on the geographical setting and the conditions of the existing system.

Air. This mode could be significantly impaired, both for passengers and freight. Air transportation is a highly competitive industry and the profit margins tend to be low. Fuels account for about 15 percent of the operating expenses of an air carrier, but because most of the other costs are fixed any variation in energy prices is reflected directly on air fares. A long term increase in energy prices, reflected in jet fuel, is likely to impact discretionary air travel (mainly tourism), but air freight, due to its high value, may be less impacted.

Maritime. This mode is likely to be relatively unaffected as it is the most energy efficient, but fuel is an important component of a ship's operating costs. The response of maritime shippers over higher energy prices tends to be lowering speed (slow steaming), which may have impacts on port call scheduling. In the long run, higher energy prices may however indirectly impact maritime transportation by lowering demand for long distance cargo movements and limiting port calls to ports having the most direct and efficient hinterland connections. In addition, this context may favor the development of short coastal and fluvial services where possible.

Higher energy prices will trigger notable changes in usage, modes, networks and supply chain management. From a macro-perspective, and since transportation is a very complex system, assessing the outcome of high energy prices remains hazardous. What appears very likely is a strong rationalization, a shift towards more energy-efficient modes as well as a higher level of integration between modes to create multiplying effects in energy efficiency. As higher transport costs play in, namely for containers, many manufacturing activities will reconsider the locations of production facilities to sites closer to markets (near-sourcing). While globalization was favored by cheap and efficient transport systems, the new relationships between transport and energy are likely to restructure the global network of production and distribution towards regionalization.

Concept 3 – Transport and sustainability

Sustainable development

An issue that has triggered concerns over the recent decades relates to the capacity of the global economy to accommodate an enduring demographic, economic and resource consumption growth. Since the 1970s, many statements have been made asserting that the world would be unable to sustain such growth without a possible socioeconomic and/or environmental breakdown. While these perspectives have been demonstrated to be inaccurate, since resource availability and the quality of life have increased, there are enduring concerns that at some point a threshold will be reached. Under such conditions, an emphasis on sustainable development has been advocated as a priority for future social and economic development.

Sustainable development is, however, a complex concept that is subject to numerous interpretations. It is not surprising that the subject is prone to much demagogy leading to confusion in terms of its nature, consequences and appropriate response. It is, however, generally agreed that a sustainable society favors conditions that benefit the environment, the economy and society without compromising the welfare of future generations. Still, as history clearly demonstrates, the conditions of future societies will largely depend upon the legacy of current societies on resources and the environment. All forms of assets (capital, real estate, infrastructures, resources) passed on to the next generation should be at least of equal value (utility) per capita. The basic definition of sustainability has been expanded to include three major points (often referred as the three Es):

Social equity. Relates to conditions favoring a distribution of resources among the current generation based upon comparative levels of productivity. This implies that individuals or institutions are free to pursue the ventures of their choice and reap the rewards for the risks they take and the efforts they make.

Economic efficiency. Concerns conditions permitting higher levels of economic efficiency in terms of resource and labor usage. It focuses on capabilities, competitiveness, flexibility in production and providing goods and services that supply a market demand. Under such circumstances, factors of production should be freely allocated and markets open to trade.

Environmental responsibility. Involves a "footprint" which is less than the capacity of the environment to accommodate. This includes the supply of resources (food, water, energy, etc.), but also the safe disposal of numerous forms of wastes. Its core tenets include the conservation and reuse of resources.

Since a growing share of the global population is urbanized, sustainability has increasingly become focused on urban areas. Major cities require a vast array of supporting infrastructures including energy, water, sewers and transport. A key to urban sustainability issues is linked with the provision and maintenance of a wide range of urban infrastructure. Every city has specific infrastructure and environmental problems. For instance, cities in developing countries have chronic deficiencies in the provision of the most basic infrastructure while their environmental conditions are deteriorating.

Figure 8.4 provides a typology of the sustainability level of urban passenger transportation systems from a sample of 64 cities across the world. The classification is mainly based on the assumption that a high level of reliance on the automobile coupled with low density levels is less sustainable than lower levels of automobile dependency and higher densities. Five major classes can be identified:

- **Class A**. Comparatively most sustainable transport. Very low automobile dependency, with public transport, walking and cycling more prominent than cars, which involves very low gasoline use per capita.
- **Class B**. Comparatively more sustainable transport. Low automobile dependency, public transport, walking and cycling equal with cars, low gasoline use.

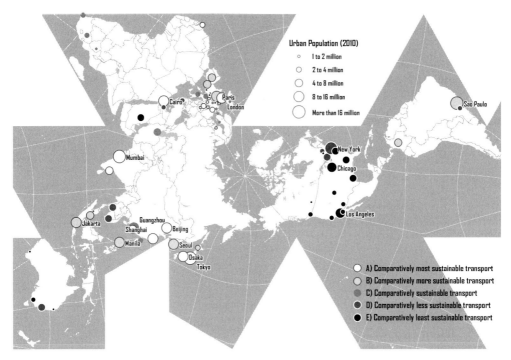

Figure 8.4   Sustainable urban passenger travel, selected cities

Source: Urban population data from United Nations, World Urbanization Prospects: The 2007 Revision Population Database. Kenworthy, J. and F. Laube (2001) The Millennium Cities Database for Sustainable Transport, International Union of Public Transport (UITP), Brussels, and Institute for Sustainability and Technology Policy (ISTP), Perth.

- **Class C**. Comparatively sustainable transport. Moderate automobile dependency, important role for public transport, walking and cycling, moderate gasoline use.
- **Class D**. Comparatively less sustainable transport. High automobile dependency, minor role for public transport (important support for peak-hour movements), walking and cycling, high gasoline use.
- **Class E**. Urban passenger transport systems that are comparatively the least sustainable. They involve a very high level of automobile dependency, a very limited role for public transit, walking and cycling, as well as very high gasoline use.

The typology has a distinct geography. North American and Australian cities are dominantly in class E with a few of the denser cities (e.g. New York, Toronto) in class D. This is mostly related to automobile dependency. European cities rank higher on the sustainable passenger transportation scale, mainly due to the higher role of public transit and denser urban settings. This characteristic is also shared with many Latin American cities. East Asian cities are generally on top of the urban passenger sustainability scale. It is worth underlining that even in Asia there are cities with a high level of automobile dependence such as Bangkok and Saigon.

Efficient and productive cities are not necessarily linked with high levels of energy consumption per capita. The first three classes of cities (A, B and C) are characterized by high densities and rely on public transit for mobility, while the last two classes

(D and E) have lower densities and rely on the automobile for most urban movements. A shortcoming of the above classification is that it considers only one dimension of urban transportation: the mobility of passengers. Cities are also subject to intense flows of goods since they are locations of production, consumption and distribution. Freight distribution should be considered an important component of urban sustainability.

Sustainable transportation

Transportation, as a core component supporting the interactions and the development of socioeconomic systems, has also been the object of much consideration of to what extent it is sustainable (Figure 8.5):

Sustainable transportation. The capacity to support the mobility needs of people, freight and information in a manner that is the least damaging to the environment.

The concept of sustainable transportation is intricately linked with the development of sustainable transport modes, infrastructures and logistics. Three major dimensions are considered for such a purpose:

Environment. A reduction of the environmental impacts of transportation is a likely strategy for sustainability. Transportation significantly contributes to harmful

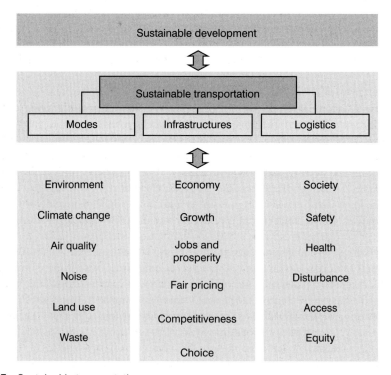

Figure 8.5 Sustainable transportation

Source: Adapted from UK Department of the Environment, Transport and the Regions (1999) Indicators of Sustainable Development.

emissions, noise and climate changes. Vehicles are becoming more environmentally efficient but there are more of them around. An improvement of the land use impacts of transportation, especially the impacts of infrastructure construction and maintenance, is also a strategic goal to achieve. The transportation system is also a generator of wastes (vehicles, parts, packaging, etc.) that must be reduced.

Economy. Transportation is a factor of economic growth and development. A sustainable strategy would aim to efficiently use transportation for the purpose of growth and for the creation of jobs. Transportation should also have a fair pricing strategy, meaning that uscrs are bearing the full costs (direct and indirect) of their usage of the transport system. A transport system where competition is fair and open is likely to promote modal choice and efficiency. In a system where transport is a public or private monopoly, price distortions and misallocations of capital are created which in the long run are likely to render the system unsustainable.

Society. Sustainable transportation should benefit society. It should be safe, should not impair human health and should minimize disturbance on communities. Access and equity are also two important principles as transportation should promote the access to goods and services for as many people as possible.

Sustainable development applied to transport systems requires the promotion of linkages between environmental protection, economic efficiency and social progress. Under the environmental dimension, the objective consists in understanding the reciprocal influences of the physical environment and the practices of the industry and that environmental issues are addressed by all aspects of the transport industry. Under the economic dimension, the objective consists of orienting progress in the sense of economic efficiency. Transport must be cost-effective and capable of adapting to changing demands. Under the social dimension, the objective consists in upgrading standards of living and quality of life.

Most agree that automobile dependence is related to an unsustainable urban environment. However, such an observation is at odds with the mobility choice and preferences of the global population where the automobile is rapidly adopted when income levels reach a certain threshold. Other transport alternatives commonly do not measure up to the convenience of the automobile. Private and flexible forms of transportation, such as the automobile, are thus fundamental to urban mobility and should not be discarded as options for the sake of sustainability. A bias is observed in the transport community towards an emphasis for public transit and non-motorized transportation as the dominant, if not sole, strategy towards sustainable transportation. Yet, almost all public transit systems are financially unsustainable, imposing burdens on the society. Freight transportation must also been considered in this process owing to the substantial growth of raw materials and goods being traded in a global economy. Freight transportation relies on much more environmentally sound modes such as rail and maritime transport.

Despite the apparent and projected success of measures to promote transport sustainability, they have their limits. Indeed, the built environment and transport infrastructure cannot change quickly enough to solve the bulk of problems related to unsustainable transport. Most of the investment that is already in place will remain in place for 50 years or more and new investment (in additional or improved infrastructure) will not represent much more than a few percentage points change in terms of reducing traffic congestion and its negative externalities. While policies, rules and regulations have a tendency to favor a misallocation of resources (such as compliance), users tend to instinctively react to price signals and discard modes that are becoming costly (unsustainable). Transportation and sustainability for both passengers and freight must also contend with mitigation versus adaptation issues:

- **Mitigation** concerns the improvement of productivity and efficiency of existing modes, terminals and managerial approaches so that environmental externalities are reduced. They tend to be short to medium term strategies.
- **Adaptation** is a change in the level of use and the market share of respective modes to better reflect a long term trend, such as higher energy prices and stricter environmental regulations.

There is a wide range of responses to environmental sustainability. The various outcomes for a sustainable environment involve three steps: (1) transport operations must conform to local, national and international regulations; (2) environmental costs of transport operations must be built into the price of providing transport facilities and services; (3) environmental performance must be introduced into the organization's management. Environmental sustainability represents a growing area of responsibility for transport companies, one that is forcing them to acquire expertise in environmental management. The most important challenge for the industry is to implement environmentally sustainable transport within competitive market structures while coping with changes in transport demand and improving transport supply. The issue of sustainable transport thus remains elusive.

CASE STUDY Green logistics

Authors: Jean-Paul Rodrigue, Brian Slack and Claude Comtois

Greenness and logistics

Most considerations in sustainable transportation focus on passengers, leaving freight issues somewhat marginalized. Logistics is at the heart of the operation of modern transport systems and implies a degree of organization and control over freight movements that only modern technology could have brought into being. It has become one of the most important developments in the transportation industry. Greenness has become a code word for a range of environmental concerns, and is usually considered positively. It is employed to suggest compatibility with the environment, and thus, like logistics, is something that is perceived as beneficial. When put together the two words suggest an environmentally friendly and efficient transport and distribution system.

> **Green logistics**. Supply chain management practices and strategies that reduce the environmental and energy footprint of freight distribution. It focuses on material handling, waste management, packaging and transport.

The loosely defined term covers several dimensions related to production planning, materials management and physical distribution, opening the door to a wide array of potential applications of environmentally friendly strategies along supply chains (Figure 8.6).

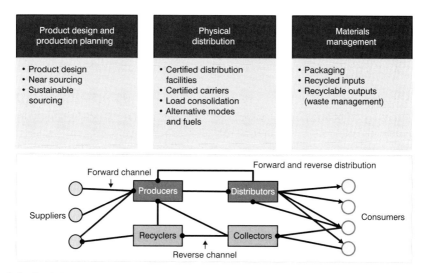

Figure 8.6 Logistic activities and their green dimensions

The green applications of logistics are numerous and cover three main dimensions:

- **Product design and production planning**. Developing products that have a lower environmental footprint, including their production process.
- **Physical distribution**. Ensuring that the mobility of freight related to logistics operations is performed in a sustainable and environmentally friendly manner.
- **Materials management**. Moving towards more efficient forms of materials use, including packaging and recycling so that what used to be an output can become an input.

All these dimensions can be individually or jointly applied. Since they involve different actors, concerted efforts are uncommon as each element of the supply chain pursues strategies that are judged to be the most effective along their respective channels. The conventional forward channel in freight distribution is well understood with raw materials, parts and finished goods flowing from suppliers to producers, distributors and, finally, to consumers. In many cases, there is also a reverse channel where wastes, packages and defective/obsolete products are "climbing back" up the supply chain. In some cases (such as a defective product), distributors will take back the merchandise, but in many others, a specialized segment of the distribution industry aims at collecting and then recycling goods and parts. Thus, reverse logistics (or reverse distribution) is concerned about the movements of previously shipped goods from customers back to manufacturers or distribution centers due to repairs, recycling or returns. There are several variants:

- An important segment is **customer-driven**, where domestic waste is set aside by home-dwellers for recycling. This has achieved wide popularity in many communities, notably because the public became involved in the process.
- A second type is where non-recyclable waste, including hazardous materials, is transported for disposal to designated sites. As landfills close to urban areas become scarce, waste has to be transported greater distances to disposal centers.

- A different approach is where reverse distribution is a continuous embedded process in which the organization (manufacturer or distributor) takes responsibility for the delivery of new products as well as their take-back. This means environmental consideration for the whole life cycle of a product (production, distribution, consumption and recycling/disposal).

Different stakeholders could be applying different strategies, all of which are labeled as green logistics. One corporation could be focusing on product packaging while another on alternative fuel vehicles; both are undertaking green logistics. However, a closer inspection of the concept and its applications raises a number of paradoxes and inconsistencies, which suggest that its application may be more difficult than initially expected. Although there has been much debate about what green logistics truly entails, the transportation industry has developed very narrow and specific interests about the issue. If transportation costs are reduced and assets such as vehicles, terminals and distribution centers better utilized, the assumption is that green logistics strategies are being implemented.

In common with many other areas of human endeavor, greenness became a catchword in the transportation industry in the late 1980s and early 1990s. It grew out of the emerging awareness of environmental problems, and in particular with well-publicized issues such as acid rain, CFCs, waste disposal and climate change. Environmental concepts, such as material flows or the carbon cycle, became readily applicable to supply chain management. The transportation industry was recognized as a major contributor to environmental issues through its modes, infrastructures and flows. The developing field of logistics was seen as an opportunity for the transportation industry to present a more environmentally friendly face. Yet, environmental perspectives and transportation sustainability issues remain predominantly focused on passenger transportation.

Interest in the environment by the logistics industry manifested itself most clearly in terms of exploiting new market opportunities. While traditional logistics seeks to organize forward distribution – that is the transport, warehousing, packaging and inventory management from the producer to the consumer – environmental considerations opened up markets for recycling and disposal, and led to an entire new subsector: reverse logistics. This reverse distribution involves the transport of waste and the movement of used materials. Even if the term reverse logistics is widely used, other names have been applied, such as reverse distribution, reverse-flow logistics, and even green logistics. Inserting logistics into recycling and the disposal of waste materials of all kinds, including toxic and hazardous goods, has become a major new market but it does not reflect the full extent of green logistics which is the greening of both the forward and reverse segments of supply chains.

Green logistics and its paradoxes

An overview of the standard characteristics of logistical systems reveals several inconsistencies with regards to the mitigation of environmental externalities. They take the form of five basic paradoxes:

Costs. The purpose of logistics is to reduce costs, notably transport costs. While these remain the most salient logistics costs, inventory carrying costs come second. In addition, economies of time and improvements in service reliability, including flexibility, are further objectives. Corporations involved in the physical distribution of freight are highly supportive of strategies that enable them to cut transport costs in a competitive setting. Economies of scale in transportation as well as higher load densities are common cost-saving strategies that concomitantly lead to environmental benefits in terms of lower fuel consumption per ton-km. On some occasions, the cost-saving strategies pursued by logistic operators can be at variance with environmental considerations that become externalized. This means that the benefits of logistics are realized by the users and eventually by the consumer if the benefits are shared along the supply chain. However, the environment assumes a wide variety of burdens and costs, which form a hierarchy ranging from costs internal to the supply chain to externalized costs. Society is becoming less willing to accept these costs, and pressure is increasingly being put on governments and corporations to include greater environmental considerations in their activities. A salient example concerns food supply chains that have been impacted by lower transport costs, enabling a diversification of the suppliers and longer transport chains. The concept of food-miles has been developed as an attempt to capture the full costs of food distribution by using the distance food is carried as a proxy.

Time. In logistics, time is often the essence. By reducing the time of flows, the velocity of the distribution system is increased and, consequently, its efficiency. This is achieved in the main by using the most polluting and least energy-efficient transportation modes. The significant increase of air freight and trucking is partially the result of time constraints imposed by logistical activities. The time constraints are themselves the result of an increasing flexibility of industrial production systems and of the retailing sector. Logistics offers door-to-door (DTD) services, mostly coupled with just-in-time (JIT) strategies. Other modes cannot satisfy the requirements such a situation creates as effectively. This leads to a vicious circle; the more DTD and JIT strategies are applied, the further the negative environmental consequences of the traffic it creates. The slow steaming strategy pursued by maritime shipping companies is further challenging time management within long distance supply chains.

Reliability. At the heart of logistics is the overriding importance of service reliability. Its success is based upon the ability to deliver freight on time with the least breakage or damage. Logistics providers often realize these objectives by utilizing the modes that are perceived as being most reliable. The least polluting modes are generally regarded as being the least reliable in terms of on-time delivery, lack of breakage and safety. Ships and railways have inherited a reputation for poor customer satisfaction. For instance, the schedule reliability of container shipping is around 50 percent, implying that about half the time a container ship will not arrive at a port terminal on the scheduled day. Lower reliability levels are linked with lower levels of asset utilization and higher inventory levels, which is wasteful and indirectly damaging to the environment. The reliability of the logistics industry is built around air and truck shipments which are the two least environmentally friendly modes.

Warehousing. Logistics is an important factor promoting globalization and international flows of commerce. Modern logistics systems economies are based on the reduction of inventories, as the speed and reliability of deliveries removes the need to store and stockpile. Consequently, a reduction in warehousing demands is one of the advantages of logistics. This means, however, that inventories have been transferred to a certain degree to the transport system, especially to roads but also to terminals. Inventories are actually in transit, contributing still further to congestion and pollution. The environment and society, not the logistical operators, are assuming the external costs. Not all sectors exhibit this trend, however. In some industrial sectors, computers for example, there is a growing trend for vertical disintegration of the manufacturing process, in which extra links are added to the supply chain. Intermediate plants where some assembly is undertaken have been added between the manufacturer and consumer. While facilitating the customizing of the product for the consumer, it adds an additional external movement of products in the production line.

Information technologies. Information technologies have led to new dimensions in retailing. One of the most dynamic markets is e-commerce. This is made possible by an integrated supply chain with data interchange between suppliers, assembly lines and freight forwarders. Even if for the online customers there is an appearance of a movement-free transaction, the distribution online transactions create may consume more energy than other retail activities. The distribution activities that have benefited the most from e-commerce are parcel-shipping companies such as UPS, FedEx or DHL that rely solely on trucking and air transportation. Information technologies related to e-commerce applied to logistics can obviously have positive impacts. So once again, the situation may be seen as paradoxical.

There is growing evidence that green logistics results in increased supply chain performance, particularly because it favors an integrated perspective of supply chains. The actors involved in logistical operations have a strong bias to perceive green logistics as a means to internalize cost savings, while avoiding the issue of external costs. The top environmental priority is commonly reducing packaging and waste. The rise in energy prices is conferring additional incentives for supply chain managers to improve upon logistics and will correspondingly push energy and emissions to the forefront. These observations support the paradoxical relationship between logistics and the environment that reducing costs does not necessarily reduce environmental impacts. By overlooking significant environmental issues, such as pollution, congestion, resource depletion, the logistics industry is still not very green. Green logistics remains an indirect outcome of policies and strategies aimed at improving the cost, efficiency and reliability of supply chains. A key aspect of more environmentally friendly freight distribution systems concerns city logistics where the "last mile" in freight distribution takes place as well as a large share of reverse logistics activities. Still, even in this context the driving force is not directly environmental issues, but factors linked with costs, time, reliability, warehousing and information technologies.

Bibliography

Banister, D. (2008) "The sustainable mobility paradigm", *Transport Policy*, 15(2): 73–80.

Banister, D. and K. Button (eds) (1993) *Transport, the Environment, and Sustainable Development*, London: Spon Press.

Black, W.R. (2010) *Sustainable Transportation: Problems and Solutions*, New York: Guilford Press.

Button, K. (1990) "Environmental externalities and transport policy", *Oxford Review of Economic Policy*, 6(2): 61–75.

Chapman, J.D. (1989) *Geography and Energy: Commercial Energy Systems and National Policies*, New York: Longman Scientific and Technical.

Davis, S., S. Diegel and R. Boundy (2009) *Transportation Energy Data Book*, Edition 28, US Department of Energy, ORNL-6984.

Gilbert, R and A. Perl (2008) *Transport Revolutions. Moving People and Freight without Oil*, London: Earthscan.

Goldman, T. and R. Gorham (2006) "Sustainable urban transport: four innovative directions", *Technology in Society*, 28: 261–73.

Hensher, D.A. and K.J. Button (eds) (2003) *Handbook of Transport and the Environment*, Handbooks in Transport no. 4, Amsterdam: Elsevier.

IEA/OECD (2009) *Transport, Energy and CO$_2$: Moving Toward Sustainability*, Paris: International Energy Agency.

Kenworthy, J.R. and F. Laube (eds) (2000) *An International Sourcebook of Automobile Dependence in Cities, 1960–90*, 2nd edition, Boulder: University Press of Colorado.

McKinnon, A., S. Cullinane, M. Browne and A. Whiteing (eds) (2010) *Green Logistics: Improving the Environmental Sustainability of Logistics*, London: Kogan Page.

Newman, P. and J.R. Kenworthy (1999) *Sustainability and Cities: Overcoming Automobile Dependence*, New York: Island Press.

Rodrigue, J-P., B. Slack and C. Comtois (2001) "Green logistics", in A.M. Brewer, K.J. Button and D.A. Hensher (eds) *The Handbook of Logistics and Supply-Chain Management*, Handbooks in Transport no. 2, London: Pergamon/Elsevier, pp. 339–51.

Ryley, T.J. and L. Chapman (eds) (2012) *Transport and Climate Change*, Bingley: Emerald.

Schiller, P.L., E.C. Bruun and J.R. Kenworthy (2010) *An Introduction to Sustainable Transportation*, London: Earthscan.

Tolley, R. (ed.) (2003) *Sustainable Transport: Planning for Walking and Cycling in Urban Environments*, New York: CRC Press.

UN-HABITAT (2009) *Planning Sustainable Cities, Global Report on Human Settlements 2009, United Nations Human Settlements Programme*, London: Earthscan.

US Environmental Protection Agency (2000) "The Lean and Green Supply Chain: A Practical Guide for Materials Managers and Supply Chain Managers to Reduce Costs and Improve Environmental Performance", Environmental Accounting Project, Washington, DC: EPA.

World Energy Council (2007) Transport Technologies and Policy Scenarios, World Energy Council.

9 Transport planning and policy

Since transportation is such an important component of contemporary society, capable of producing significant benefits, yet giving rise to many negative externalities, appropriate policies need to be devised to maximize the benefits and minimize the inconveniences. At the same time the allocation, design and construction of transport infrastructure and services must be subject to careful planning, both by public and private agencies. A distinction must be drawn between policy and planning, since the former usually relates to the strategies and goals while the latter refers to concrete actions. Because they both have to reflect the fundamental changes in society and contemporary issues and problems, policies and planning are constantly changing. For instance, the changing orientation of public policy led to deregulation in many transport sectors. Among the core policy issues, transport safety and security have come at the forefront. Natural and man-made disasters are also serious challenges for transport planning.

Concept 1 – The nature of transport policy

Authors: Brian Slack and Theo Notteboom

Policy and planning

The terms "policy" and "planning" are used very loosely and are frequently interchangeable in many transport studies. Mixing them together is misleading. Policy and planning represent separate parts of an overall process of intervention. There are circumstances where policy may be developed without any direct planning implications, and planning is frequently undertaken outside any of direct policy context. The following definitions are here used:

> **Transport policy** deals with the development of a set of constructs and propositions that are established to achieve particular objectives relating to social, economic and environmental development, and the functioning and performance of the transport system.
> **Transport planning** deals with the preparation and implementation of actions designed to address specific problems.

Thus, transport policy can be concomitantly a public and private endeavor, but governments are often the most involved in the policy process since they either own or manage many components of the transport system. Governments also often perceive that it is their role to manage transport systems due to the important public service they provide in addition to imposing a regulatory framework. Yet, many transport systems, such as maritime and air transportation, are privately owned. The private sector thus has much leverage into the policy process through its asset allocation decisions, which reflects in new public transport policy paradigms.

> **Public policy** is the means by which governments attempt to reconcile the social, political, economic and environmental goals and aspirations of society with reality. These goals and aspirations change as the society evolves, and thus a feature of policy is its changing form and character. Policy has to be dynamic and evolutionary.

A major distinction between planning and policy is that the latter has a much stronger relation with legislation. Policies are frequently, though not exclusively, incorporated into laws and other legal instruments that serve as a framework for developing planning interventions. Planning does not necessarily involve legislative action, and is more focused on the means of achieving a particular goal.

The relevance of transport policy

Transport policies arise because of the extreme importance of transport in virtually every aspect of economic, social and political activities of nation states. Transport is taken by governments of all types, from those that are interventionalist to the most liberal, as a vital factor in economic development. Transport is seen as a key mechanism in promoting, developing and shaping the national economy. Governments also seek to promote transportation infrastructure and services where private capital investment or services may not be forthcoming. Paradoxically, academics question the directness of the links between transport and economic development.

Transport frequently is an issue in national security. Policies are developed to establish sovereignty or to ensure control over national space and borders. The Interstate Highway Act of 1956, that provided the United States with its network of expressways, was formulated by President Eisenhower on the grounds of national security. Security was at the heart of the recent imposition of requirements on document clearance prior to the departure of freight from foreign countries to the USA.

Transport raises many questions about public safety and the environment. Issues of public safety have for a long time led to the development of policies requiring driving licenses, limiting the hours of work of drivers, imposing equipment standards, establishing speed limits, mandating highway codes, seat belts and other accident controls. More recently, environmental standards and control measures are being instituted, in response to the growing awareness of the environmental impacts of transport. Examples include banning leaded gasoline and mandating catalytic converters in automobiles.

Transport policy has been developed to prevent or control the inherent monopolistic tendency of many transport modes. Unrestrained competition leads to market dominance by a company thereby achieving monopoly power. Such dominance brings into question many issues affecting the public interest such as access (in a port would smaller shipping lines be excluded?), availability (would smaller markets continue to receive air service by a monopoly carrier?) and price (would the monopolist be in a position to charge high prices?). Other reasons for policy intervention include the desire to limit foreign ownership of such a vital industry over concerns that the system would be sidetracked to service more foreign than national interests.

In recent years, four trends have had significant consequences over the context in which transport policy takes place:

- Deregulation and privatization.
- A broader focus of policies, particularly in light of intermodalism and multimodalism.
- A move towards social and political issues behind transport projects as opposed to technical and engineering issues.
- Globalization increased interactions at the international level, both for freight and passengers.

Policy instruments

Governments have a large number of instruments at their disposal to carry out transport policy. Some are direct, such as public ownership, while others are indirect such as safety standards:

- An extremely important instrument is **public ownership**. The direct control by the state of transportation infrastructure, modes or terminals is very widespread. Most common is the provision by public agencies of transport infrastructure such as roads, ports, airports and canals. Public ownership also extends to include the operation of transport modes. In many countries airlines, railways, ferries and urban transit are owned and operated by public agencies.
- **Subsidies** represent an important instrument used to pursue policy goals. Many transport modes and services are capital intensive, and thus policies seeking to promote services or infrastructure that the private sector are unwilling or unable to provide may be made commercially viable with the aid of subsidies. Private railroad companies in the nineteenth century received large land grants and cash payments from governments anxious to promote rail services. In the USA, the Jones Act, that seeks to protect and sustain a US-flagged merchant fleet, subsidizes ship construction in US shipyards. Indirect subsidies were offered to the air carriers of many countries in the early years of commercial aviation through the awarding of mail contracts. Dredging of ship channels and the provision of other marine services such as pilotage and navigation aids are subsidies to facilitate shipping. Both public ownership and subsidies represent instruments that require the financial involvement of governments. Revenue generation is becoming an increasingly important instrument in transport policy.
- **Regulatory control** represents a means of influencing the shape of transportation that is very widely employed. By setting up public agencies to oversee particular sections of the transport industry, governments can influence the entire character and performance of the industry. The agencies may exert control on entry and exit, controlling which firms can offer transportation services, at what prices, to which markets. Thus while the actual services may be offered by private firms, the regulator in fact plays a determining role. Regulatory agencies in the USA such as the Civil Aeronautics Board played a critical role in shaping the US airline industry for decades.
- Many governments are major promoters of **research and development** in transportation. Government research laboratories are direct products of state investments in R&D, and much university and industry R&D is sustained by government contracts and programs. The fruits of this research are extremely important to the industry. It is a vital source for innovation and the development of new technologies such as intelligent vehicles and intelligent highway systems.
- **Labor regulations** pertaining to conditions of employment, training, and certification may not be directed purposefully at influencing transport, but as a policy they may exert significant effects over the industry.

- **Safety and operating standards**, such as speed limits, may have a similar effect. The restrictions on limiting the number of hours a truck driver may work may be instituted for safety reasons and for enhancing the working conditions of drivers, but they shape the economics of truck transport. In the same fashion speed limits help fix the distance of daily trips that one driver may undertake, thereby shaping the rate structure of the trucking industry.

Trends in policy development

Public policies reflect the interests of decision makers and their approaches to solving transport problems. These interests and approaches are both place specific (they apply to a particular area of jurisdiction) and time specific (they are established to reflect the conditions of transport and the intended solutions at a point in time). Policies change and evolve as conditions change and as new problems are recognized; they are dynamic. The dynamic nature of policy is reflected in the way policy instruments have been employed over the years. In the nineteenth century, when many of the modern transport systems were being developed, the prevailing political economy was one of laissez-faire, in which it was believed that the private sector should be the provider of transport services and infrastructure. Examples of private transport provision included:

- **Turnpikes**. The first British modern roads in the eighteenth century were the outcome of private trusts aiming at deriving income from tolls on roads they built and maintained. It was likely the first massive private involvement in transport infrastructure provision.
- **Canals**. Many of the earliest canals were built with private capital. One of the first canals that helped spark the industrial revolution in Britain was the Bridgewater Canal, built by the Duke of Bridgewater between 1761 and 1765 to haul coal from his mines to the growing industrial city of Manchester.
- **Urban transit**. In most North American cities public transit was operated by private firms. The earliest examples were horsecars that followed rail lines laid out on city streets. With electrification at the end of the nineteenth century, the horsecars were converted to streetcars and the network was greatly expanded. In the twentieth century buses were introduced by private companies operating on very extensive route systems.
- **Ships**. Most maritime shipping companies were private enterprises. Many were family businesses, some of which became large companies, such as the Cunard Line in the UK, MSC in Switzerland, or Maersk in Denmark. The main government involvement concerns military navies and ferries.
- **Railways**. Railways were developed by private companies during the nineteenth century, including such companies as Canadian Pacific and Union Pacific. In North America this has continued to the present day.

Many segments of the private transport sector were eventually captured by the government. In addition to the public ownership of transport modes, there emerged in the twentieth century a growing amount of regulatory control. The belief in liberal markets with little public interference was seriously reconsidered after the crash of 1929 and the economic downturn of the early 1930s. From that moment on governments were inclined to extend the scope of their responsibilities. The public sector was an important trigger for the reconstruction of Europe in the aftermath of the Second World War (e.g. the

Marshall plan), for the modernization of the industrial structure and for economic growth. Economic and social measures were directed towards the creation of the welfare state. The period from the 1940s to the 1970s were characterized by nationalization when socialist ideology was put into practice throughout the world. For example, the European transport industry saw the emergence of large national companies in public transport, freight rail, ferry services, deep sea shipping and the airline industry. These large nationalized companies could mobilize new sources and technologies, thereby contributing to the national objectives of economic growth and full employment.

While centrally planned economic systems (such as the Soviet Union, Eastern Europe and China) involved a complete control by the public sector, governments in Western Europe and North America were also major players in the market through market control systems up to the full nationalization of industries considered to be of strategic importance to economic development and external trade. The airline and the trucking industries saw entry limited by permits, and routes and rates were fixed by regulatory boards that had been set up to control the industries. At the same time greater safety regulations were being imposed and working conditions were increasingly being shaped by labor legislation.

By the 1960s, therefore, transportation had come under the sway of public policy initiatives that exerted an enormous influence on the industries and their spatial structures. At the same time, there was also a growing body of evidence that indicated that public ownerships and regulation were not always in the public interest. Transportation costs that were fixed by the regulatory authorities were maintained at higher levels than were necessary. Research demonstrated that many regulatory boards had been "captured" by those they were supposedly regulating, so that they were frequently acting to protect the industries rather than the public. At the same time there was a crisis of public finances in many countries, where the costs of operating the state-owned transportation industry were seen to be unsustainable. The theory of contestability repudiated traditional economic theory concerning monopoly power by arguing that the threat of entry of a new actor was sufficient to thwart a monopolist's ability to impose monopoly pricing. The key, therefore, is to relax entry thresholds, by allowing new firms to start up, a process that regulatory boards were impeding.

Deregulation and privatization policies spread, unequally, to many parts of the world. New Zealand has perhaps the most open transport policy, but many others, such as Canada and Australia, have made significant steps in this direction. In the EU, the pace of deregulation and privatization is proceeding unevenly. Subsidies to state-owned transport companies have been terminated and many airlines have been privatized. The government-owned railroads still exist in France, Germany, Italy and Spain, but the tracks have been separated from the traction and rail service operations, and have been opened up to new service providers. In Latin America, most of the state-owned transport sector has been deregulated. While the former centrally planned states have had to make the furthest adjustments to a more open market economy, several, such as China, have opened up large sections of the transport industry to joint ventures with foreign private enterprises. In China, many new highways and most of the major ports are being developed with private capital. Thus, at the beginning of the twenty-first century, transportation is under less direct government economic control worldwide than at any period over the last 100 years.

Changing nature of policy interventions

The recent trends in transport policy towards liberalization and privatization have not necessarily weakened government interventions. Government policy orientations have changed, however. Governments are beginning to exert greater control over environmental and security concerns, issues that are replacing former preoccupations with economic matters. For instance, because of biofuel policies aiming at ethanol production using corn, the unintended consequence was a surge on global food prices as more agricultural land was devoted to energy production instead of food production. The environment is becoming a significant issue for government intervention. Coastal zone legislation has made it increasingly difficult for ports to develop new sites in the USA. Air quality is a major factor influencing the allocation of US federal funds for urban transport infrastructure. In Europe, environmental issues are having an even greater influence on transport policy. The EU Commission is promoting rail and short sea shipping as alternatives to road freight transport. Projects are assessed on the basis of carbon dioxide reduction. All transportation projects are subject to extensive environmental assessments, which may lead to a rejection of proposals, despite strong economic justification. As a major source of atmospheric pollution and environmental degradation, the transportation industry can anticipate many further government environmental policy interventions.

Safety has always been a policy issue. Legislation imposing speed limits, mandating seat belts, and other measures have sought to make travel safer. These continue to proliferate. However, it is in the area of security that the most recent set of policy initiatives have been drawn. Screening of people and freight has become a major concern since 9/11. Both the US government and such international organizations as the International Maritime organization (IMO) and the International Civil Aviation Organization (ICAO) have instituted new measures that impact on operations, and represent additional costs to the transport industry. While there may have been some reduction of policy involvement involving economic regulations, the influence of public policy on transport overall is still powerful – but contentious at times.

Concept 2 – Transport planning

Author: Brian Slack

The traditional transport planning process

Transport planning is usually focused on specific problems or on broad transport concerns at a local level. It has been traditionally a preoccupation of lower tier governments, such as the state or municipality. Because of this fact, transport planning is most developed in the urban sphere, and it is there where most experience has been gathered. The planning process, however, has a number of similarities with the policy process. Identifying a problem, seeking options and implementing the chosen strategy are essential steps in planning too. Because it tends to deal with localized problems, the solutions adopted in transport planning tend to be much more exact and specific than policy directives.

For a long time planning was a field dominated by traffic engineers who gave it a distinctly mechanistic character, in which the planning process was seen as a series of rigorous steps undertaken to measure likely impacts and to propose engineering solutions. There were four major steps: trip generation, trip distribution, modal split and route selection. They involved the use of mathematical models, including regression analysis, entropy-maximizing models and critical path analysis. There are many reasons why the results of these models should be treated with caution:

- They are only **as good as the data they manipulate** and many times the data are inaccurate or incomplete.
- They are based on assumptions that the mathematical relationships between variables **remain constant**.
- They can be **manipulated** to produce the outcome that the analyst knows the client prefers.
- Because the predictions were **rarely subjected to subsequent evaluation**, their validity is largely questioned, and the modeler is happy to predict the future since projections rarely question the validity of the methodology.

The predictions of future traffic flows produced by the four-stage sequence are then used to identify planning options. Since the most common prediction of the modeling is that present capacities will be unable to cope with traffic growth, the tendency has been to produce planning solutions that call for an expansion of capacity. This has been referred to as predict and accommodate. It is the solution that has typified so much urban transport planning from the 1940s to the 1980s. It has given rise to the enormous expansion of highway construction that reinforces the dominance of the automobile. Rarely are there postmortems of the prediction models and, as has been learned by empirical observation, the issue of induced demand has distorted the actual traffic.

Contemporary transport planning

In cities traffic problems have increased significantly since the 1970s, despite a great deal of urban transport planning. There is a growing realization that perhaps planning has failed and that the wrong questions have been asked. Rather than estimate traffic increases and then provide capacity to meet the expected growth, it is now accepted that what is required is better management of the transport system through new approaches to planning. Just as urban planning requires the inputs of many specialists, transport planning is beginning to utilize multidisciplinary teams in order to broaden the scope of the planning process. Planning is still a multi-step process, but it has changed considerably:

- **Goals and objectives**. While the goal of traditional transport policy, improving accessibility, is still useful, it has to be considered in the context of other desirable goals. For instance improving safety and health, reducing emissions from vehicles, improving equity, enhancing economic opportunities, improving community livability, promoting mobility are all valid. But which goal(s) are pursued results in a very different planning process. Defining goals becomes a much more complicated stage in contemporary planning. Increasingly goals have turned to consider managing demand, rather than trying to build capacity.
- **Options**. Given the possible range of goals that transport planners have to consider, it becomes necessary to provide a set of possible options. Several objectives may be desirable, and thus it is important to consider what they imply. Several scenarios may have to be considered, and they must become important components of the planning process.
- **Identification of actors, institutions, stakeholders**. Given that transport planning has the potential to influence so many elements of society – economic wellbeing, environmental conditions, social integration – it is important that those affected by the transport problem and its potential resolution should be identified so that they

can be engaged. This would be a much wider list of affected parties than simply those involved in transportation activity itself, and requires recognizing a role for citizen participation.

- **Predicting outcomes, identifying benefits, and assessing costs**. The stage of predicting the outcomes for each of the options is a critical step in the process. Models continue to play an important role, but whereas the traditional models were based on the number of trips, increasingly modeling is becoming more activity based. Transport is seen in the context of scheduling household decisions in time and space. Demographic and social data are used extensively, and the mathematical models have become more sophisticated. Nevertheless there are roles for other types of analyses, including non-objective forecasts. The predicted outcomes must then be assessed as to their benefits and costs. These may be expressed in monetary terms, but many transport planning situations call for measurement in other terms, such as visual impacts, environmental dislocations and employment impacts.
- **Choosing course of action**. Evaluation of the scenarios has to consider the costs and benefits from the frequently conflicting perspectives of the stakeholders and actors. Extensive public consultation may be required. The information has to be disseminated and explained so that an informed public can participate in the debate. Ultimately it will be the politicians who decide, but they are swayed by the strength of the arguments presented by the transport professionals, and in publicly contentious cases by pressure brought to bear by citizens groups.

Transport demand management

In rejecting the former paradigm of building capacity, transport planners have turned increasingly to managing both demand and the transport system. Building roads has produced a car-oriented society in which the other modal alternatives have little opportunity to coexist. Car ownership is beyond the ability of the transport planner to control directly and the question remains if it should. But car use and ownership is affected by land use and density, both elements that planners can affect. High population densities, in particular, favor walking, bicycling and public transit use. It is for this reason that a great deal of attention in planning is being paid to densification and integration. This includes concentrating development along well-served transport corridors (transit-oriented development) and increasing densities in areas undergoing rehabilitation.

Managing the demand for transport is made up of a large number of small interventions that cumulatively can impact car use, but in particular improve the livability of cities. A sample of well-practiced and successful interventions includes:

- **Park and ride**. Parking spaces are provided, usually close to an expressway, where drivers can board buses that provide service to the city center. This has become a staple feature in the outer zones of many US and British cities. Its success is variable, however, and there is some evidence from the UK that park and ride may actually increase car use, as people who may have used regular bus services now use their cars to drive to the car parks.
- **Traffic calming**. Measures that seek to reduce the speed of vehicles in urban areas, such as speed bumps and street narrowing. For residential streets the goal is to make their use by car drivers unattractive because of the obstacles; for thoroughfares the objective is to reduce the average speeds. The measures indicate the need for much greater attention to street design and layout.

- **Priority lanes for buses and high occupancy vehicles**. Lanes on major thoroughfares and expressways that are reserved for buses, taxis and passenger vehicles with several occupants. This has become an important feature of transport planning in North America, where major highway expansion projects offer priority lanes. The goal is to encourage use of buses and high occupancy vehicles that can be seen to travel at higher speeds along the reserved lanes by other drivers who may be stuck in traffic jams.
- **Alternate work schedules**. Encouraging work hours other than the dominant 9 to 5 schedule. One of the great problems in transport planning is that demand is concentrated in two main peak periods. In the past, efforts were made to meet this demand by increasing road capacity, which was never sufficient, and resulted in an underuse of the capacity the other 20 hours each day. Promoting flexible schedules and encouraging telecommuting are policies that are seeking to spread out the demand for transport over more hours and even reducing the demand altogether.
- **Promoting bicycle use**. In some countries, particularly the Netherlands, the bicycle is an important mode of travel. It is a green and healthy mode, but in automobile-dependent cities, the bicycle does not share the roads easily with trucks and cars. Encouraging greater use of the bicycle requires significant planning adjustments, such as the provision of bicycle lanes and bike stands.
- **Car sharing**. Encouraging drivers to share car use with neighbors or co-workers.
- **Enhancing pedestrian areas**. In most cities vehicles dominate the streets. In many areas of high population density, the quality of life (enhanced safety, less pollution, etc.) and the visual attractiveness of streetscapes can be enhanced by excluding vehicles from streets altogether, or limiting access to public transport vehicles. In Europe this has become a distinctive feature of the historic cores of many cities.
- **Improving public transit**. For 50 years or more public transit use has declined in most cities. Yet it is the only major alternative to the car in these cities, and thus enhancing the use of transit has become a major planning objective. Improvements include making transit more attractive, by improving bus schedules and improving the appearance and comfort of transit vehicles and stations. At the same time efforts are underway to widen the range of transit alternatives. These include extending commuter rail services, and constructing new systems such as light and heavy rail modes.
- **Parking management**. Restricting on-street parking and charging higher rates for parking.

Pricing

While planning interventions may have a positive cumulative effect in shaping transport demand, some economists suggest that a more direct approach involving imposing more stringent cost measures on car users is necessary. It is widely accepted that car users pay only a small proportion of the actual costs of their vehicle use. Economists argue that the external costs should be borne by the users. As intuitively rational as this argument may be, there are several problems with its application:

- First, there are difficulties in **measuring externalities**, with considerable variations in estimates between different studies. Different types of use, speeds, engines, vehicle weight, driving conditions, etc. make it difficult to produce broadly accepted values. Decision makers have difficulty in agreeing to impose charges when there is a diversity of evidence about external costs.

- Second, there are **practical difficulties in collecting these costs**. One of the easiest (and most widely used) methods is a gasoline tax. It is a crude approach, however, because it imperfectly distinguishes between driving conditions and engine type – a fuel-efficient vehicle may have just as high consumption in heavy urban traffic as a gas-guzzler in a rural setting.
- Third, is the political **difficulty of imposing such additional costs on the public**. In North America, in particular, free access to roads is regarded as a birthright, and it is intensely unpopular to propose any new forms of revenue generation that hint at additional taxation.

The effectiveness of economic controls is evident by the experience of Hong Kong, where, despite high incomes, car ownership and use remains at a very low level. This is due in the main to the high cost of parking. An even more drastic example is Singapore, where extreme measures limiting car purchases, high vehicle licenses, electronic tolls on highways and cordon pricing in the downtown area have restrained car use. The use of pricing mechanisms may be less in other countries, but the trend towards greater application of some forms of tolling is accelerating. Cordon pricing has been applied in a number of jurisdictions where access to certain areas, usually the CBD, is tolled. The most famous application was the decision to charge private vehicles for entry into Central London in early 2003, a program that has proved to be successful, despite a great deal of opposition.

Another form of pricing is congestion or "fair" pricing. Here certain lanes of a highway are tolled, but at variable rates. When traffic is moving freely, the charges for the tolled lanes are nil. But as traffic builds up and speeds are reduced, the costs of using the reserved lanes increase. Collection of the tolls is electronic, and drivers are informed of the current charges by large signs. Drivers are given a choice, therefore, to stay in the slower lanes for free, or move to the tolled lanes at a cost that is proportionate to the speed on the congested lanes.

Intelligent vehicles and intelligent highways

Technology is seen by many transport planners as a solution to a wide range of transport problems. This is an approach that has achieved wide acceptance in the USA, where there has always been a strong emphasis on seeking engineering solutions to urban transport problems. It involves using information technologies (ITS) to provide better information and control over traffic flow and individual vehicle use. Many of the solutions involve the application of remote sensing techniques along with ITS.

One of the most promising approaches involves Interactive Highways. They are a means of communication between the road and driver that warn of approaching road conditions. Warnings include electronic message boards that suggest alternate routes to approaching motorists, and designated radio frequencies that give updated traffic reports. It is based on a closed-circuit TV system (CCTV) that records lane-by-lane occupancy, volume and speed. At the same time ramp meters record in real time the amount of traffic entering the highway. This information is analyzed and processed at a control center that can dispatch emergency equipment to accidents as they happen, and can inform other drivers of road conditions, accidents, construction and delays.

A further technology is Emergency Signal Priority. This is a means of giving emergency vehicles and public transport buses priority at traffic lights in congested areas. The system allows a vehicle equipped with a system emitter to send a coded infrared message to the system detector, installed at the traffic intersection. When activated, the detector receives the coded message and then either holds the existing green light until the vehicle passes through or changes the existing red light to a green light.

ITS is being applied in many further innovative ways to improve the efficiency of emergency vehicles. For example, in Montreal mathematical models are being used to predict where road accidents are likely to occur given the time of day, traffic volumes and weather conditions. Ambulances can be assigned to these zones. Once deployed and assigned to a specific event, optimal routing is determined and relayed to drivers. When the first responders have identified the extent and type of injuries, the information is relayed to a control center which determines availability of doctors and nurses at which hospital emergency room, and suggests a routing for the ambulance using a least-time model estimation.

ITS is providing many solutions to the problems of road pricing. Toll collection is increasingly using electronic means to collect tolls without requiring vehicles to stop at toll booths. In its simplest form, vehicles equipped with a transponder that emits details of the vehicle are allowed to pass through toll lanes without stopping to pay. Receptors at the booth record the passage and debit the account. This is at the heart of the cordon pricing and of most new toll systems in place.

This technology, however, is being wedded to global positioning systems (GPS), which is likely to produce radical changes in the way vehicular traffic is priced. This combination of technologies will permit a more effective means of applying road pricing than the road tax. Vehicles will be required to have an on-board unit that includes a GPS receiver, a set of digital maps showing jurisdictional boundaries, an odometer feed, a set of distance rate charges and a wireless communication system to report billing data. During each trip the GPS determines the jurisdictional zones, the odometer calculates the distance traveled in each zone, and the computer tabulates the running total of fees, and periodically signals the data to the billing agency.

Freight planning

The vast preponderance of transport planning, certainly at the urban level, has been devoted to passengers. The automobile and public transit issues have preoccupied planners because individual mobility can be a highly political issue (drivers are also voters). Yet freight traffic represents a significant part of many problems that planning seeks to address. The models and data inputs used in transportation planning are of little relevance when applied to freight movements. For example, demographic data, such as household size, the backbone of passenger analysis, are irrelevant for freight. The bi-polar daily peak of traffic movements applies only to passengers, freight movements being distributed in a different profile over a 24-hour period.

While trucks account for approximately 10 percent of vehicles on the road, their size, low maneuverability, noisiness and high pollution output make their presence particularly objectionable. Truck pickup and delivery in city centers is particularly problematic because of limited parking. At the same time trucks are vital to the economy and well-being of society since they provide critical supplies. Commerce is dominated by trucking, and the logistics industry in particular is dependent on road transport for pickup and delivery. Garbage pickup, snow removal and fire protection are among many essential services that are truck-oriented.

Planning for freight movements, such as city logistics, is still in its infancy. As a largely private sector activity it is difficult to control, and many of the decisions that affect trucking are made by the industry itself. The emergence of large logistics/ distribution centers on the outer fringes of metropolitan areas is taking place without public control or oversight. In Europe, some attempt to manage such development by establishing publicly promoted freight villages has only limited success.

Several cities are seeking to limit trucking as pressures keep mounting up. In many jurisdictions limits on heavy trucks in urban areas are in place, and there are restrictions on the times of delivery and pickup, which in some European cities extend to the exclusion of all trucks in the urban core during daytime hours. The question remains about to what extent constraining urban freight circulation impairs the economy.

All these steps are tackling the problem at the edges. In many cities there are no census data on freight traffic, so that planning in the few cases where it takes place is inevitably hit and miss. There needs to be a much greater focus on freight planning overall, since it is almost universally recognized that freight transport is important.

Concept 3 – Transport safety and security

Authors: Brian Slack and Jean-Paul Rodrigue

A new context in transport security

While issues of safety and security have been before transport planners and managers for many years, it is only recently that physical security has become an overriding issue. Concerns were already being raised before in the past, but the tragic events of 9/11 thrust the issue of physical security into the public domain as never before and set in motion responses that are reshaping transportation in unforeseen ways. In addition, threats to health, such as the spread of pandemics, present significant challenges to transport planning and operations. Because of the nature of transport systems, safety and security issues concern the modes and the terminals. Each involves a different set of issues.

As locations where passengers and freight are assembled and dispersed, terminals have particularly been a focus of concern about security and safety. Because railway stations and airports are some of the most densely populated sites anywhere, crowd control and safety have been issues that have preoccupied managers for a long time. Access is monitored and controlled, and movements are channeled along pathways that provide safe access to and from platforms and gates. In the freight industry security concerns have been directed in two areas: worker safety and theft. Traditionally, freight terminals have been dangerous work places. With heavy goods being moved around yards and loaded onto vehicles using large mobile machines or manually, accidents are systemic. Significant improvements have been made over the years, through worker education and better organization of operations, but freight terminals are still comparatively hazardous. The issue of thefts has been one of the most severe problems confronting all types of freight terminals, especially where high value goods are being handled. Docks, in particular, have been seen as places where organized crime has established control over local labor unions. Over the years access to freight terminals has been increasingly restricted, and the deployment of security personnel has helped control thefts somewhat.

In light of the emergence of global supply chains the emphasis in freight transport security is gradually shifting into a more comprehensive but complex approach including several dimensions and potential measures:

- **Dimensions**. Particularly concern the integrity of the cargo, the route and the information systems managing the supply chain.
- **Measures**. The set of procedures that can be implemented to maintain the integrity of the cargo, namely inspections, the security of facilities and personnel as well as of the data.

Still, in spite of the qualitative benefits, the setting and implementation of security measures come at a cost that must be assumed by the shippers and eventually by the consumers. It has been estimated that an increase of 1 percent in the costs of trading internationally would cause a decrease in trade flows in the range of 2 to 3 percent. Security-based measures could increase total costs between 1 and 3 percent. Additionally, the impacts are not uniformly assumed as developing countries, particularly export-oriented economies, tend to have higher transport costs. Security measures can affect them in a greater fashion.

Physical security of passengers

Airports have been the focus of security concerns for many decades. Hijacking aircraft came to the fore in the 1970s, when terrorist groups in the Middle East exploited the lack of security to commandeer planes for ransom and publicity. Refugees fleeing dictatorships also found taking over aircraft a possible route to freedom. In response, the airline industry and the international regulatory body, ICAO, established screening procedures for passengers and bags. This process seems to have worked in the short run, at least, with reductions in hijackings, although terrorists changed their tactics by placing bombs in unaccompanied luggage and packages, as for example in the Air India crash off Ireland in 1985 and the Lockerbie, Scotland, crash of Pan Am 103 in 1988. The growth in passenger traffic and the development of hub-and-spoke networks placed a great deal of strain on the security process. There were wide disparities in the effectiveness of passenger screening at different airports, and because passengers were being routed by hubs, the numbers of passengers in transit through the hub airports grew significantly. Concerns were being raised by some security experts, but the costs of improving screening and the need to process ever larger numbers of passengers and maintain flight schedules caused most carriers to oppose tighter security measures.

The situation was changed irrevocably by the events of September 11, 2001. The US government created the Department of Homeland Security which in turn established a Transportation Security Authority (TSA) to oversee the imposition of strict new security measures on the industry. Security involves many steps, from restricting access to airport facilities, fortifying cockpits, to the more extensive security screening of passengers. Screening now involves more rigorous inspections of passengers and their baggage at airports. For foreign nationals inspection employs biometric identification, which at present involves checking fingerprints, but in the future may include retinal scans and facial pattern recognition. A new system, the Computer Assisted Passenger Prescreening System (CAPPS II), requiring more personal information from travelers when they book their flights, is used to provide a risk assessment of each passenger. Passengers considered as high risk are further screened.

The imposition of these measures has come at a considerable cost. A significant factor has been the integration of screeners into the federal workforce, with important increases in salaries and training costs. The purchase of improved screening machines and the redesigning of airport security procedures have been important cost additions. These measures have also had a major influence on passenger throughputs. Clearing security has become the most important source of delays in the passenger boarding process. Passengers are now expected to arrive two hours before departure at the terminal in order to clear security.

Security issues have had a negative effect on the air transport industry as costs increased with delays and inconveniences to passengers increasing as well. Business travel, the most lucrative sub-market for the regular airlines, has suffered particularly sharp declines. Anecdotal evidence suggests that these passengers are switching to other modes for shorter trips, so as to avoid the time delays and aggravation caused by the security process, and to executive jets for longer trips.

Freight security

Security in the freight industry has always been a major problem. Illegal immigrants, drug smuggling, custom duty evasion, piracy and the deployment of substandard vessels have been some of the most important concerns. However, as in the air passenger business, the events of 9/11 highlighted a new set of security issues. The scale and scope of these problems in freight is of an even greater magnitude. The less regulated and greater international dimensions of the shipping industry in particular have made it vulnerable to security breaches. The number of ports, the vast fleet of global shipping and the range of products carried in vessels, and the difficulty of detection, has made the issue of security in shipping an extremely difficult one to address. The container, which has greatly facilitated globalization, makes it extremely difficult to identify illicit and/or dangerous cargoes. In the absence of scanners that can scan the entire box, manual inspection becomes a time-consuming and virtually impossible task considering the large volumes involved. Hubbing compounds the problem, as large numbers of containers are required to be handled with minimum delays and inconvenience.

In the USA the response was to enact the Maritime Transportation and Security Act in 2002. The basic elements of this legislation were adopted by the International Maritime Organization (IMO) in December 2002 as the International Ship and Port Security code (ISPS). There are three important features of these interventions. First, is the requirement of an Automated Identity System (AIS) for all vessels between 300 and 50,000 dwt. AIS requires vessels to have a permanently marked and visible identity number, and there must be a record maintained of its flag, port of registry and address of the registered owner. Second, each port must undertake a security assessment. This involves an assessment of its assets and facilities and an assessment of the effects of damages that might be caused. The port must then evaluate the risks and identify its weaknesses to its physical security, communication systems, utilities, etc. Third, is that all cargoes destined for the USA must receive customs clearance prior to the departure of the ship. In addition, it is proposed that biometric identification for seafarers is to be implemented and that national databases of sailors are to be maintained.

The ISPS code is being implemented in ports around the world. Without certification, a port would have difficulty in trading with the USA. Security is thus becoming a factor in a port's competitiveness. The need to comply with ISPS has become an urgent issue in ports large and small around the world. The costs of securing sites, of undertaking risk assessments and of monitoring ships all represent an additional cost of doing business, without any commercial return. US ports have been able to tap funding from the Department of Homeland Security, but foreign ports have to comply or risk the loss of business. In 2008 legislation in the USA required that all containers being shipped to the USA undergo screening. Foreign ports will be expected to purchase very expensive gamma-ray and x-ray scanners, and undertake screening of all US-bound containers, regardless of the degree of security threat. This is a further financial and operational headache foreign ports have to contend with. Security has become an additional element in determining competitive advantage.

Like its passengers counterpart, the airline freight industry is facing stringent security requirements. Since 2010 a new TSA regulation forces the screening of all cargo carried by air within the United States or internationally and this before being loaded. The Certified Cargo Screening Program (CCSP) forces airlines, freight forwarders and shippers to assume the costs of these security measures in an attempt to establish a secure air freight transport chain. This measure comes at difficult times for the air cargo industry which is already dealing with higher energy costs and weak demand. It will likely involve additional costs, delays and disruptions, undermining the competitiveness of air cargo. Already, several freight forwarders are considering maritime shipping instead of air cargo for long distance trade or the unusual combination of air and maritime shipping.

Concept 4 – Transportation and disasters

Transportation and national security

Transportation systems are designed to operate under normal conditions. Yet, disruptions such as those caused by an accident or by a storm are rather common and well mitigated. On occasion, a disruption at a much higher scale takes place to the extent that the security of a whole region or nation is compromised.

> A **disaster** involves extensive damage to people and physical infrastructure that is unforeseen in nature, scale and extent. It often implies that their risk of occurrence has not been properly assessed and a large share of the damage is the outcome of a lack of preparedness.

From an economic standpoint, the impacts of disasters are dependent on three factors: (1) the nature and level of incidence of disasters; (2) the level of exposure of populations and infrastructures; and (3) the level of vulnerability of populations and infrastructures. There are several drivers that have an impact on the threats and risk level of disasters on transportation systems:

- **Increased mobility**. The mobility of passengers (for tourism, business and migration) and freight has increased notably around the world, including the crossing of international boundaries. Air and maritime transportation are particularly illustrative. This trend has also been strengthened by trade agreements and reductions in tariffs promoted by organizations such as the World Trade Organization. There are more economic opportunities, but some risks, such as infectious diseases, can spread faster and more extensively.
- **Infrastructure and economic interdependency**. Infrastructures are increasingly interdependent, particularly transportation and energy infrastructures, so a disruption in one will have an effect on others. This interdependency is also economic as global or regional trade is based upon mutual specialization. Some parts and components are provided by a limited number of suppliers, which can be prone to risks in case of disruptions. The same applies to resources such as oil that require a continuous supply with limited margins to accommodate disruptions.

- **Centralization and concentration of distribution**. The principle of economies of scale often leads to a centralization of network structures and a concentration of economic activities. Most transportation systems are organized in hub-and-spoke networks, particularly for air transportation, but this characteristic is also prevalent in maritime shipping. Global trade is articulated by major gateways where a few control a large share of the commercial flows. At the more basic geographical level strategic passages impose bottlenecks for global maritime freight circulation.
- **Urbanization**. The emergence of large cities has led to acute concentrations of populations, a pattern significantly different than the more dispersed settlements that prevailed in rural societies. Concentration of population equates with a concentration of risk. Thus, any disaster affecting an urban area is compounding its impacts on par with the population and infrastructure density.

The transport industry has responded to these drivers with massive investments in infrastructure and facilities that have expanded the capacity and efficiency of transportation systems, both at the domestic and international levels. Added flows and capacities have in turn created increased demands on the management of physical distribution systems, which includes activities such as transportation, transshipment, warehousing, insurance and retailing. These are all of strategic importance to national economies. With the increasing reliance on distribution systems, any failure of transportation, due to intentional or non-intentional causes, can have very disruptive consequences and can compromise national security over four major issues:

- **Transportation supply**. Ensuring that transportation modes, routes, terminals and information systems are able to satisfy national security needs such as troop deployment and emergency relief.
- **Transportation readiness**. Maintaining the readiness of transportation to face time-sensitive national security needs.
- **Transportation vulnerability**. Reducing the vulnerability of the transportation modes, terminals and users to intentional harm or disruption from natural events.
- **Illegal use of transportation**. Reducing the trade of restricted or illegal goods (e.g. drugs, endangered species), and illegal immigration.

Potential threats and risks

The disruptions caused by disasters take place over complex transportation systems and are consequently difficult to evaluate. Particularly, the non-linearity character of complex systems implies that a disruption can have multiplying and feedback effects, many unforeseen. A disaster of high severity is likely to trigger a phase transition where the resulting transport conditions are very different to those ex-ante. A simple taxonomy of disasters reveals that they are natural or man-made in origin. Natural disasters fall into four main categories:

- **Extreme weather events**. Many climate events such as storms and blizzards occur regularly and tend to have minimal impacts on transport systems with delays, partial closures or diversions. Others, such as floods, cyclones (hurricanes), tornadoes and droughts can be of disastrous proportions. Tropical cyclones (hurricanes) are particularly harmful since they cover wide areas (a mid-sized cyclone can cover an area of 500 km in diameter), are moving slowly (25 km/hr) and are associated with high winds and rainfalls. Regional air transport and public transit systems are usually shut down and land transportation can be seriously impaired. For instance, Hurricane Sandy, which struck the New York/New Jersey coasts in November 2012, caused the preemptive shut down of all the airports and public transit systems of the region. Due to flooding and power outages, it took several days for the system to be brought back to normal operating conditions, which had substantial impacts on commuting, a problem compounded by gas shortages because of a lack of gas stations with power and difficulties to distribute fuel. The closing down of port facilities and terminals had substantial economic impacts. There are concerns that climate change may be linked with more recurrent extreme weather events, but so far the evidence remains tenuous. Beyond this debate the fact remains that extreme weather events will continue to regularly occur, but their frequency and scale is uncertain.

- **Geophysical**. Tectonic activity is the source of the most serious disasters. Earthquakes are salient forms of geophysical threats since they are difficult to predict. Tsunamis are also considered an emerging risk as a growing number of people live along coastal areas. The 2011 Tohoku earthquake in Japan is among the five largest in recorded history. While the damage by the earthquake was significant, it is the associated tsunamis that caused the most extensive damage to Japanese transport infrastructure. Although areas of high earthquake occurrence are readily identified, the specific location and scale of an event remains a probability that is often difficult to conceptualize in the planning of transport infrastructure. While volcanoes have always been localized and easily identifiable risks, the ash clouds they release have recently been a source of concern. For instance, the ash cloud released by the 2010 eruption of the Eyjafjallajökull volcano in Iceland forced the shutdown of most of the European and transatlantic air transport system for close to a week, stranding millions of passengers. Since in geological times air transportation is an extremely recent phenomenon, the probability and extent of ash cloud events remains uncertain. For instance, an event of the scale of the Krakatau 1883 eruption taking place today would have profound ramifications for the global maritime and air transport systems.
- **Geomagnetic storms**. They concern disturbances in the earth's magnetic structure, mostly the outcome of solar activity where the frequency of geomagnetic storms varies accordingly. Geomagnetic storms can impair power grids and have a higher probability of taking place around the north and south poles. Still, they are a less known and often underestimated risk. The largest geomagnetic storm in history took place in 1859, but since back then electrical systems were rudimentary, its impacts on human activities were marginal. Such an event taking place today would be heavy in consequences (e.g. hundreds of millions losing electric power) and would qualify as a disaster.
- **Sea level rise**. Although potential rises in sea levels are mainly attributed to anthropogenic causes (global warming), it qualifies as a natural disaster. There is a variety of scenarios about potential sea level rises, but evidence underlines a rise by one meter by 2100 (compared with a 2000 baseline) as almost a certainty. If the sea level rise accelerates, the one meter scenario could even be reached by 2050. Irrespective of the timing, sea level rise places critical transport infrastructure such as ports and airports at risk of damage and discontinuity in operations. For instance, a port terminal or an airport could not be directly impaired by sea level rise, but its access roads could be, compromising its commercial viability. Sea level rise would also amplify the impacts of extreme weather events, namely storm surges.

The second class of disasters concern those that are man-made and they can be intentional or unintentional:

- **Accidents**. The outcome of technical failures or human errors and where modes, infrastructure or terminals can be damaged, even destroyed, which includes injuries and the loss of life. Small-scale accidents occur very frequently, particularly over road transportation. However, transportation-related accidents are rarely considered as disasters because they are very punctual events not related to a massive loss of life and damage. The exception is an aircraft crash with a complete loss of life. With improved airline safety these events are, however, increasingly uncommon, particularly in relation to the growing amount of passenger-km flown each year around the world (Figure 9.1).

Like all modes of transportation, air transport is subject to accidents that can be due to human (67 percent) or technical (20 percent) causes, and rarely due to atmospheric conditions (6 percent). Since air accident statistics started to be systematically compiled, 50 percent and 21 percent of all accidents took place during landing and takeoff phases respectively. The evolution of the yearly number of fatalities is quite revealing. Up to the early 1970s the number of fatalities increased with some proportionality with the growth of air traffic. By the 1970s, in spite of substantial growth levels of air traffic, fatalities undertook a downward trend. This is jointly the outcome of better aircraft designs, improved pilot training, better navigation and control systems as well as comprehensive accident management aiming at identifying the causes and then possible mitigation strategies.

For instance, on August 2, 1985, Delta Flight 191 from Fort Lauderdale ran into a developing thunderstorm and wind shear conditions on its final approach to Dallas/Fort Worth airport (DFW). The strong and unstable winds forced the plane, a Lockheed L-1011, to hit the ground before the runway, resulting in 135 fatalities. An investigation of the cause of the accident suggested substantial revisions in airport emergency

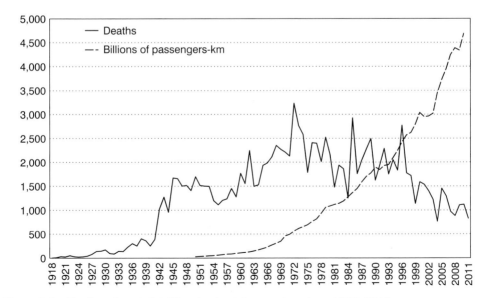

Figure 9.1 Number of yearly fatalities due to air transport crashes, 1918–2011

Source: Aircraft Crashes Record Office, Geneva.

procedures and upgrades in weather detection equipment. On June 1, 2009, Air France Flight 447 from Rio de Janeiro to Paris crashed in the Atlantic Ocean off the coast of Brazil. In this case temporarily defective air sensor speed readings and pilot error caused an aerodynamic stall that crashed the plane, resulting in 228 fatalities. After investigation, better training concerning stall recovery at high altitude was recommended.

- **Conflicts, terrorism and piracy**. Conflicts such as wars and civil unrest often result in the damaging of infrastructure with transportation commonly a voluntary or involuntary target. Due to the importance of global trade and the structure of maritime shipping networks, bottlenecks (strategic passages) are subject to risk of partial or complete closure. Terrorism has been a disruptive issue that has come to the forefront in the last two decades. For instance, the disruptions caused by the September 11, 2001, events can clearly be considered a disaster because of their scale and scope, such as the closing of the North American air transport system. The surge in global trade in the second half of the twentieth century created an environment where piracy is on the rise. Shipping lines are forced to pass through constrained areas, chokepoints, namely straits such as Malacca and Bab el Mandab along the heavily used Asia–Europe maritime routes, which make the interception of ships more feasible within a delimited area. The outcome of piracy on global supply chains has been small but not negligible as ships have changed their routing and insurance surcharges are being levied for cargo transiting through areas prone to piracy.
- **Economic and political shocks**. They are likely to play a growing role in the future, particularly financial issues as most developed nations have accumulated a staggering amount of debt that is likely to be defaulted on. Such an event would be associated with a lack of capital available for infrastructure construction, maintenance and oversight, rendering elements of the transport system more prone to risks, such as accidents.
- **Pandemics**. At the intersection of natural (biological) and man-made causes (people are vectors and a virus could be mutated by anthropogenic causes), a pandemic is an event of potential profound ramifications. Yet, the risk of the event itself is extremely difficult to assess. Although a pandemic would not directly damage transportation systems, transportation is intractably linked with such a disaster as it will act as a vector for its diffusion (particularly air transportation) and a shutting down of transportation services in the wake of a pandemic would compromise supply chains (food, energy, medical supplies).

As the freight transportation dimension is getting increasingly globalized and complex, supply chain risks are salient. Developing countries are particularly vulnerable to an array of disasters because infrastructure, including transportation, tends to be of lower quality and thus less resilient to disruptions. A fundamental element in addition to the risk is who bears the responsibility for it. International commercial transactions underline that the actor assuming the liability depends on the type of terms. Therefore, depending on the terms of the contract the same event may imply a different allocation of the liability.

Transportation disaster planning

Although a potential disaster can never be effectively planned, and even anticipated in some instances, there are a series of steps, known as Disasters Risk Management, which could reduce disruptions:

- **Risk Assessment**. The likelihood of an event and its potential impacts should be comprehensively assessed, such as a low to high probability over a defined time frame and over a specific area (e.g. a city). This should provide a prioritization of risks, but it remains a very uncertain process.
- **Preparedness**. In light of the potential risks a level of preparedness should be considered in terms of potential responses. This can involve the warehousing and positioning of relief material, such as fuel, parts and equipment, and the training of the labor force in emergency situations.
- **Mitigation**. Concerns the immediate reaction to the event and can involve the shutting down of transport systems (particularly public transit), the evacuation of populations and the mobilization of first response resources, namely distributing emergency relief (food, medical supplies). The goal is to control and attenuate the disruptions caused by the disaster.
- **Response**. Once the disaster has been mitigated, steps are implemented to bring back capacity with existing infrastructure. If a mode has been impaired, the usage of alternative modes and infrastructure has to be considered. The goal is to maintain operational as many elements of the transport system as possible.
- **Recovery**. Concerns all the steps necessary to recover the transport capacity that was lost during the disaster. It can involve repairs, restarting services that were discontinued as well as investments in new and improved infrastructures, modes and terminals. The goal is to bring back the capacity and level of service to pre-disaster conditions. With the lessons learned from the disaster, more resilient infrastructure and networks are a likely outcome.

The reconstruction time of transportation infrastructure tends to be slower than other infrastructure. Evidence from the 1995 Kobe earthquake underlines that electric power and telecommunications are first restored in a matter of weeks. Road and rail infrastructure can take several months while for port infrastructure it can be a matter of years. Transport infrastructure, particularly terminals, are much more capital intensive than utilities and require specialized and heavy equipment for their repair or construction. Highway and rail services can run at lower capacity and on alternative routes. If a port is shut down, other ports can generally be used (the same applies for air travel). While this is less efficient as it involves longer routes for imports or exports, it remains in most cases economically feasible.

CASE STUDY   Transportation and pandemics

Pandemics

There are approximately 1,500 microbes that are known to be a source of disease among the human population. Influenza can be one the most virulent among them because of its ability to mutate and be efficiently transmitted through the respiratory route. Under normal circumstances, influenza's impacts are relatively benign since populations have developed a level of immunity to its debilitating effects. Yet, it is estimated that between 1 to 1.5 million people per year die of influenza or related complications with a distinct seasonality that runs between October and March in the northern hemisphere and between May and September in the southern hemisphere. Influenza pandemics are thus considered to be among the most significant threats to the welfare of the global population.

Pandemic. An epidemic of infectious disease that spreads through human populations across a large area, even worldwide.

Over the last 300 years, ten major influenza pandemics have occurred. The 1918 pandemic (Spanish Flu) is considered to be the most severe yet: 30 percent of the world's population became ill and between 50 and 100 million died. One important factor why the Spanish Flu spread so quickly and so extensively was through modern transportation, which at the beginning of the twentieth century offered a global coverage. The virus was spread around the world by infected crews and passengers of ships and trains and severe epidemics occurred in shipyard and railway personnel.

Concerns about the emergence of a new pandemic are salient, particularly in light of recent outbreaks such as SARS (Severe Acute Respiratory Syndrome) in 2002–3 and the Swine Flu in 2009, which quickly spread because of the convenience and ubiquity of global air travel. The next influenza pandemic could be equally severe and widespread illness or absenteeism in freight transportation sectors can cause cascading disruptions of social and economic systems. The relationships between transportation and pandemics involve two major sequential dimensions:

- **Transportation as a vector**. With ubiquitous and fast transportation comes a quick and extensive diffusion of a communicable disease. From an epidemiological perspective, transportation can thus be considered as a vector, particularly for passenger transportation systems. The configuration of air transportation networks shapes the diffusion of pandemics. This issue concerns the early phases of a pandemic where transportation systems are likely to spread any outbreak at the global level.
- **Continuity of freight distribution**. Once a pandemic takes place or immediately thereafter, the major concerns shift to freight distribution. Modern economic activities cannot be sustained without continuous deliveries of food, fuel, electricity and other resources. However, few events can be more disruptive than a pandemic as critical supply chains can essentially shut down. Disruptions in the continuity of distribution are potentially much more damaging than the pandemic itself.

Vectors and velocities

The more efficient transportation, the more efficient is the vector that can transmit an infectious disease. International and long distance transport such as air and rail, modes and terminals alike, concentrates passengers and increases the risk of exposure. In the past, this could be an advantage as a ship could be quarantined, since there was ample time during the voyage for an infection to carry its course and the symptoms to become apparent. Today, it is a different matter as the velocity conferred by transportation systems for long distance travel is superior to incubation time of many flu variants (the period after the infection before symptoms are revealed). Since the incubation time for the average influenza virus is between 1 and 4 days, there is ample time for someone being infected to travel to the other side of the world before noticing symptoms. This represents the translocation phase and is the most crucial in a pandemic (Figure 9.2).

Once symptoms have developed, there is also a "denial phase" where an infected individual will continue traveling, particularly if going back to his/her place of origin. An infected individual beginning to show symptoms is likely to cancel outbound travel, but will do the utmost, even breaking quarantine (or warnings), to go back home. Thus, in a window of a few days before an outbreak could become apparent to global health

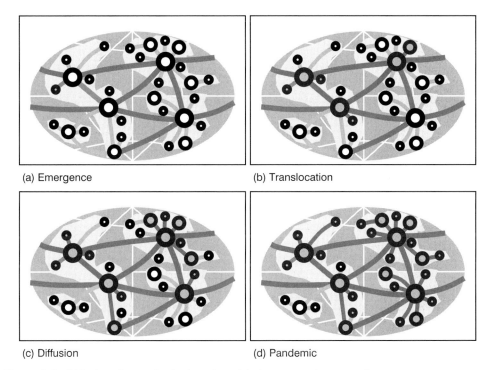

(a) Emergence

(b) Translocation

(c) Diffusion

(d) Pandemic

Figure 9.2 Diffusion of a pandemic through a global transportation network

authorities, a virus could have easily been translocated in many different locations around the world. At this point, the vector and velocity of modern transport system would ensure that an epidemic becomes a pandemic. In some cases, the velocity of global transportation systems is higher than at the regional level, which paradoxically implies that a virus can spread faster at the global level – between major gateways – than at the regional level.

Once an outbreak becomes apparent, the global passenger transportation system, such as air travel and passenger rail, can quickly be shut down in whole or in part, either voluntarily (more likely if the outbreak is judged to be serious) or by the unwillingness of passengers to be exposed to risks. The latter is what happened during the SARS outbreak in 2003. For instance, while the public transportation systems of several large Chinese cities were still operated, the number of users precipitously dropped because of risk avoidance. The SARS outbreak also had a substantial impact on the global airline industry. After the disease hit, flights in Pacific Asia decreased by 45 percent from the year before. During the outbreak, the number of flights between Hong Kong and the United States fell 69 percent. It is quite clear that this impact would pale in comparison to that of a 12–36-month worldwide influenza pandemic.

Continuity of freight distribution

However dramatic the impacts of modern transportation as a high velocity vector for a pandemic, a potential greater risk resides in the geographical and functional structure of supply chains because the continuity of freight distribution could be compromised. Up to the mid-twentieth century, the scale of production, transport and retail was dominantly local (food) or regional (durable goods such as cars). Since then, globalization expanded substantially the scale at which a wide array of goods is distributed. Thus, the interconnectedness of the global economy, while being a net advantage from a supply chain standpoint, could make the next influenza pandemic more devastating than the ones before it. Even the slightest disruption in the availability of parts, finished goods, workers, electricity, water and petroleum could bring many aspects of contemporary life to a halt. The global economy has been favored by the exploitation of comparative advantages and a more tight management of supply chains. Inventories are kept to a minimum. Virtually no production surge capacity exists. As a consequence, most markets depend on the timely delivery of many critical products (such as pharmaceuticals, medical supplies, food and equipment parts) and services (such as communications support).

The transportation industry has also consolidated into a small number of global and national mega-players to achieve massive economies of scale. This is the case for the two most important global freight transportation modes: maritime shipping and air cargo. Since the frequency, speed and reliability of shipments are high under normal circumstances, manufacturers have relocated their facilities to lower cost locations. Because transportation costs are lower than inventory management costs, retailers and secondary manufacturers employ "just-in-time" inventory systems – their "stockpile" is flowing in the transportation stream as inventory in transit. Most supply chains are restocked on a continuous basis, on par with the demand, which is labeled pull logistics. The typical efficiency, and potential non-resiliency, of critical supply chains as a function of transportation would be placed under stress during a pandemic. The most important include:

- **Food**. Modern food production and distribution relies on low levels of inventory, particularly to avoid wastes of perishable products on store shelves. On average, supermarkets have between 2 to 5 days of inventory of perishable goods (dairy, produce, meat) and about 1 to 2 weeks for other goods (pasta, canned goods, etc.). It is worth underlining that these figures are for a normal and stable demand. In the case of a pandemic, available food supplies could quickly be exhausted through hoarding behavior. Food security is therefore defined by the ability of the transportation workers to move food from producers, to the bulk-storage facilities, to the processor and lastly to the grocer.
- **Energy**. The provision and distribution of energy is critical to the functioning of a modern economy and society. For instance, about 40 percent of the world's supply of electricity is generated by burning coal (50 percent for the United States). Coal power plants maintain a fairly low stockpile, about 30 days, and rely on a constant supply from major coal mining regions, which tend to be far away. While a pandemic would not directly damage energy systems, many energy distribution systems could be threatened through the removal of essential personnel from the workplace for weeks or months and impaired transportation capabilities to supply power plants.
- **Medical supplies**. A pandemic is obviously associated with a surge in the use of medical facilities, equipment and pharmaceutical products. Global drug production is controlled by a few large conglomerates that maintain a limited number of facilities at selected locations. Commonly, a single drug is produced at a single plant. If global distribution systems were impaired during a pandemic, it would be difficult for many essential drugs to reach patients while limited stockpiles maintained at medical facilities would quickly run out. For instance, over 95 percent of all generic drugs used in the United States are made offshore, primarily in China and India. A

similar pattern applies to critical medical equipment such as ventilators. Even simple respiratory masks could quickly run out. All these shortages are likely to result in additional deaths.

It is very likely that a pandemic would quickly exhaust available food, energy and medical resources, replacements will not be forthcoming. Thus, supply chain issues are expected to seriously compound the impacts of an influenza pandemic.

Bibliography

Baumol, W.J. (1984) "Deregulation and the theory of contestable markets", *Yale Journal on Regulation*, 1: 111–37.

Flyvbjerg, B. (2009) "Survival of the unfittest: why the worst infrastructure gets built – and what we can do about it", *Oxford Review of Economic Policy*, 25(3): 344–67.

Goetz, A.R. (2002) "Deregulation, competition, and antitrust implications in the US airline industry", *Journal of Transport Geography*, 10: 1–19.

Hogwood, B. and Gunn, L.A (1984) *Policy Analysis for the Real World*, Oxford: Oxford University Press.

Kappenman, J. (2012) "A perfect storm of planetary proportions", *IEEE Spectrum*, February.

Kenworthy, J.R. and P.W.G. Newman (1990) "Cities and transport energy: lessons from a global survey", *Ekistics*, 344/345: 258–68.

Litman, T. (2005) *Rail Transit in America: A Comprehensive Evaluation of Benefits*, Victoria, BC: Victoria Transport Policy Institue.

Luke, T.C. and J-P. Rodrigue (2008) "Protecting public health and global freight transportation systems during an influenza pandemic", *American Journal of Disaster Medicine*, 3(2): 99–107.

OECD (2011) *Future Global Shock – Improving Risk Governance*, Paris: OECD Publishing.

Osterholm, M.T. (2005) "Preparing for the next pandemic", *Foreign Affairs*, July/August, pp. 24–37.

Sorenson, P. and Taylor, B. (2005) "Paying for roads: new technology for an old dilemma", *Access*, 26: 2–9.

Transportation Research Board (2006) *Critical Issues in Transportation*, Washington, DC: The National Academies, http://onlinepubs.trb.org/onlinepubs/general/CriticalIssues06.pdf.

Victoria Transport Policy Institute (2012) *Transportation Demand Encyclopedia*, Victoria, BC: Victoria Transport Policy Institue.

World Economic Forum (2012) *New Models for Addressing Supply Chain and Transport Risk*, World Economic Forum.

⑩ Methods in transport geography

Transportation is not a science, but a field of inquiry and application. As such, it tends to rely on a set of specific methodologies since transportation is a performance-driven activity and this performance can be measured and compared. Transportation planning and analysis are interdisciplinary by nature, involving among others, civil engineers, economists, urban planners and geographers. Each discipline has developed methodologies dealing with their respective array of problems. Still, transportation is an infrastructure-intensive activity, implying that engineering has been the dominant methodological paradigm for transportation studies.

Method 1 – Methods in transport geography

Transportation and methodologies

Two common traits of transportation studies, regardless of disciplinary affiliation, are a heavy reliance on empirical data and the intensive use of data analytic techniques, ranging from simple descriptive measures to more complex modeling structures. In some respects, transport geography stands out from many other fields of human geography by the nature and function of its quantitative analysis (Figure 10.1).

There is a growing complexity and data requirements for the four common models in transport geography. Each is building upon the other, implying for instance that the estimation of accessibility cannot be assessed without information about distance and that spatial interactions are derived from accessibility assessments:

Distance. The most fundamental element of geography in general and transport geography in particular. Distance can be represented in different manners, from a simple Euclidean distance calculation to a complex estimation of a logistical distance that considers all the tasks necessary for the realization of a movement.

Accessibility. Defined as the measure of the capacity of a location to be reached by, or to reach, different locations. Therefore, the capacity and the arrangement of transport infrastructure are key elements in the determination of accessibility. It is thus based upon the concept of location and distance.

Spatial interaction. A realized movement of people, freight or information between an origin and a destination. It is a transport demand/supply relationship expressed over a geographical space. Routing is a specific category of spatial interaction that considers a given set of origins and destinations for which specific (often optimal) routes are found.

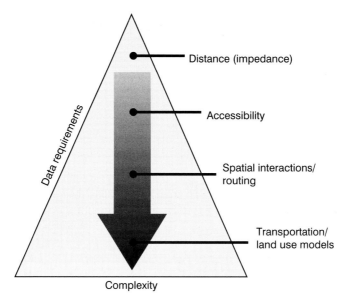

Figure 10.1 Models in transport geography

Transportation/land use models. A complex framework trying to assess the numerous relations and feedback effects between transportation and the spatial structure.

In fact, transport geography was one of the main forces in the quantitative revolution that helped redefine geography in the 1960s with the use of inferential statistics, abstract models and new theories. Although this perspective provided much needed rigor, it also favored a disconnection between empirical and theoretical approaches. Like in economics, the quantitative revolution led to a mechanistic perspective where concordance to reality became somewhat secondary; realities were made to fit into models. Even if contemporary transport geography has a more diversified approach, the quantitative dimension still plays an important part in the discipline.

Thus, in addition to providing a conceptual background to the analysis of movements of freight, people and information, transport geography is very much an applied science. The main goal of methods aims to improve the efficiency of movements by identifying their spatial constraints. These constraints can be capacity, cost, time, or environmental impacts, but more often in combination. It is consequently possible to identify relevant strategies and policies and provide some scenarios about their possible consequences. There are various ways of classifying the methods that are used by transport geographers:

- Whether they are qualitative or quantitative.
- Whether they deal with infrastructures (e.g. terminals) or flows.
- Whether they provide interpolation or extrapolation.
- Whether the technique provides description, explanation or optimization.
- According to the level of data aggregation, the nature of the assumptions or the complexity of the calculations.

Like in geography, spatial and temporal process cannot be considered separately. A basic taxonomy can divide them into transport-related methods and multidisciplinary methods.

Transport-related methods

A first group of methods concern those directly related to the study of transportation since most draw their origins from transport planning. The methods mainly used in transport geography include:

- **Network analysis** (also referred to as graph theory), which is used to study transport network forms and structures, particularly how they change in time. For example, one could use network analysis to study the evolution of the hub-and-spoke configuration of airline services.
- Transport geographers also play a key role in studying **land use–transport interactions**. Numerical models have been developed, which, over time, have become increasingly complex.
- Transport geographers are also interested in **flow and location allocation models** that can be used to define such things as school district boundaries or the location for a new retail outlet. These techniques are optimization procedures rather than methods for describing or understanding current transport systems.

In addition, there are various methods of general use in transportation studies that are readily applicable in transport geography:

- First, a diverse set of techniques is used in **urban transportation modeling** exercises, the purpose of which is to understand and predict urban spatial patterns.
- Second, **traffic surveys** that are used to gather empirical information about movements such as their routing and frequency.

Multidisciplinary methods

Multidisciplinary methods include the whole range of methods that were not specifically developed for transportation studies, but are readily applicable to its analysis. They are labeled as multidisciplinary since they can be applied to a wide range of issues irrespective of the discipline. First, there are methods that are central to geography, but are not restricted to the study of transportation systems:

Cartography is the most obvious example of a geographic technique. Indeed, various types of maps are used in the analysis of transport systems, including land use maps, depictions of transport infrastructure, isoline maps of transportation costs or schematics of transportation activity patterns.

Geographic information systems (GIS), which are an outgrowth of digital cartography, provide a set of tools for storing, retrieving, analyzing and displaying spatial data from the real world. GIS technology has been applied to some large-scale transportation planning and engineering applications. More often, however, GIS are applied in a prescriptive way to small-scale problems, for example to plot optimal routes for buses, delivery trucks or emergency vehicles.

There are also various **statistics** that have been developed or modified by geographers to describe urban economic systems. Examples include the Gini coefficient and indexes of concentration and specialization.

Second, there are various methods that are used in many different applications, including transportation analysis. They underline that transportation analysts are not restricted to those methods that have been developed with transportation in mind, but to whatever is relevant to a specific problem. In fact, many methods that were initially developed for other problems have widespread use in transportation studies:

Some methods are used to **collect primary data**, e.g. questionnaires and interviews, while others are used to analyze data. Some of the analytic techniques are straightforward to implement and interpret; graphs (e.g. scattergrams, distance-decay curves) and tables (e.g. origin-destination matrices) are two examples. Others are more complex, such as inferential statistics like the t-test, analysis of variance, regression and chi-square.

Increasingly, transportation studies are concerned with **impacts and public policy issues**. They rely more on qualitative information such as policy statements, rules and regulations. Various types of impacts are considered, including economic (e.g. community development), social (e.g. access to services), environmental (e.g. air or water pollution) and health (e.g. road accidents). The broad fields of environmental impact assessment, risk assessment and policy analysis are relevant to these issues.

The development and application of methods to transport studies in general and transport geography in particular has been increasingly complex, particularly as improvements in information technologies made available more powerful analytical tools. For instance, a commercial geographic information system package has analytical and modeling capabilities well beyond what is undertaken by most researchers, analysts or policy makers. Future developments are therefore more likely to focus on empirical data analysis using known methods but with more extensive datasets. This will lead to more detailed and consequential analysis of real world transport phenomena and help better connect theoretical knowledge and real world applications.

Method 2 – Graph theory: definition and properties

Authors: Jean-Paul Rodrigue and César Ducruet

Basic graph definition

A graph is a symbolic representation of a network and of its connectivity. It implies an abstraction of the reality so it can be simplified as a set of linked nodes.

Graph theory is a branch of mathematics concerned with how networks can be encoded and their properties measured. It has been enriched in the last decades by growing influences from studies of social and complex networks.

The goal of a graph is representing the structure, not the appearance of a network. The conversion of a real network into a planar graph is a straightforward process which follows some basic rules:

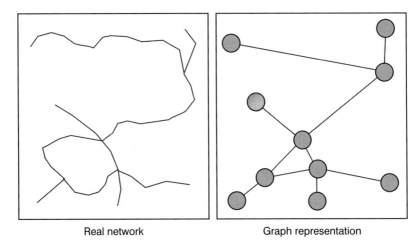

Real network   Graph representation

Figure 10.2   Graph representation of a real network

- The most important rule is that **every terminal and intersection point becomes a node**.
- Each connected node is then linked by a **straight segment**.

The outcome of this abstraction is the actual structure of the network (Figure 10.2). The real network, depending on its complexity, may be confusing in terms of revealing its connectivity (what is linked with what). A graph representation reveals the connectivity of a network in the best possible way. Other rules can also be applied, depending on the circumstances:

- A node that is not a terminal or an intersection point can be added to the graph if along that segment **an attribute is changing**. For instance, it would be recommended to represent as a node the shift from two lanes to four lanes along a continuous road segment, even if that shift does not occur at an intersection or terminal point.
- A "**dummy node**" can be added for aesthetical purposes, especially when it is required that the graph representation remains comparable to the real network.
- Although the relative location of each node can remain similar to their real world counterpart, this is not required.

In transport geography most networks have an obvious spatial foundation, namely road, transit and rail networks which tend to be defined more by their links than by their nodes. This it is not necessarily the case for all transportation networks. For instance, maritime and air networks tend to be defined more by their nodes than by their links since links are often not clearly defined. A telecommunication system can also be represented as a network, while its spatial expression can have limited importance and would actually be difficult to represent. Mobile telephone networks or the Internet, possibly the most complex graphs to be considered, are relevant cases of networks having a structure that can be difficult to symbolize. However, cellular phones and antennas can be represented as nodes while the links could be individual phone calls.

Servers, the core of the Internet, can also be represented as nodes within a graph while the physical infrastructure between them, namely fiber optic cables, can act as links. Consequently, all transport networks can be represented by graph theory in one way or the other.

The following elements are fundamental at understanding graph theory:

Graph. A graph G is a set of vertices (nodes) v connected by edges (links) e. Thus $G = (v, e)$.

Vertex (Node). A node v is a terminal point or an intersection point of a graph. It is the abstraction of a location such as a city, an administrative division, a road intersection or a transport terminal (stations, terminuses, harbors and airports).

Edge (Link). An edge e is a link between two nodes. The link (i, j) is of initial extremity i and of terminal extremity j. A link is the abstraction of a transport infrastructure supporting movements between nodes. It has a direction that is commonly represented as an arrow. When an arrow is not used, it is assumed the link is bi-directional.

Sub-graph. A sub-graph is a subset of a graph G where p is the number of sub-graphs. For instance $G' = (v', e')$ can be a distinct sub-graph of G. Unless the global transport system is considered in its whole, every transport network is in theory a sub-graph of another. For instance, the road transportation network of a city is a sub-graph of a regional transportation network, which is itself a sub-graph of a national transportation network.

Buckle (loop). A link that makes a node correspond to itself is a buckle.

Planar graph. A graph where all the intersections of two edges are a vertex. Since this graph is located within a plane, its topology is two-dimensional. This is typically the case for power grids, road and railway networks, although great care must be inferred to the definition of nodes (terminals, warehouses, cities).

Non-planar graph. A graph where there are no vertices at the intersection of at least two edges. This implies a third dimension in the topology of the graph since there is the possibility of having a movement "passing over" another movement such as for air and maritime transport. A non-planar graph has potentially many more links than a planar graph.

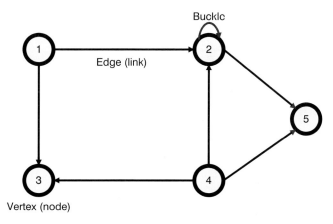

Figure 10.3 Basic graph representation of a transport network

Links and their structures

A transportation network enables flows of people, freight or information, which are occurring along its links. Graph theory must thus offer the possibility of representing movements as linkages, which can be considered over several aspects:

Connection. A set of two nodes as every node is linked to the other. Considers if a movement between two nodes is possible, whatever its direction. Knowing connections makes it possible to find if it is possible to reach a node from another node within a graph.

Path. A sequence of links that are traveled in the same direction. For a path to exist between two nodes, it must be possible to travel an uninterrupted sequence of links. Finding all the possible paths in a graph is a fundamental attribute in measuring accessibility and traffic flows.

Chain. A sequence of links having a connection in common with the other. Direction does not matter.

Length of a link, connection or path. Refers to the label associated with a link, a connection or a path. This label can be distance, the amount of traffic, the capacity or any attribute of that link. The length of a path is the number of links (or connections) in this path.

Cycle. Refers to a chain where the initial and terminal node is the same and that does not use the same link more than once in a cycle.

Circuit. A path where the initial and terminal node corresponds. It is a cycle where all the links are traveled in the same direction. Circuits are very important in transportation because several distribution systems are using circuits to cover as much territory as possible in one direction (delivery route).

Clique. A clique is a maximal complete sub-graph where all vertices are connected.

Cluster. Also called community, it refers to a group of nodes having denser relations with each other than with the rest of the network. A wide range of methods are used to reveal clusters in a network, notably they are based on modularity measures (intra- versus inter-cluster variance).

Ego network. For a given node, the ego network corresponds to a sub-graph where only its adjacent neighbors and their mutual links are included.

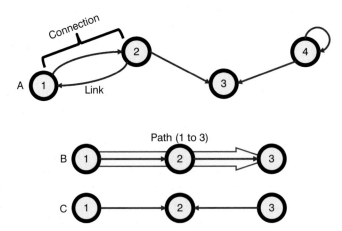

Figure 10.4 Connections and paths

Loan Receipt
Liverpool John Moores University
Library Services

Borrower Name: Wan,Chengpeng
Borrower ID: ********

The geography of transport systems /
31111014308942
Due Date: 15/12/2016 23:59

Total Items: 1
08/12/2016 13:52

Please keep your receipt in case of
dispute.

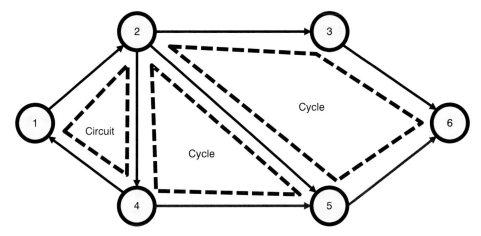

Figure 10.5 Cycles and circuits

Nodal region. A nodal region refers to a subgroup (tree) of nodes polarized by an independent node (which largest flow link connects a smaller node) and a number of subordinate nodes (which largest flow link connects a larger node). Single or multiple linkage analysis methods are used to reveal such regions by removing secondary links between nodes while keeping only the heaviest links.

Dual graph. A method in space syntax that considers edges as nodes and nodes as edges. In urban street networks, large avenues made of several segments become single nodes while intersections with other avenues or streets become links (edges). This method is particularly useful to reveal hierarchical structures in a planar network.

Basic structural properties

The organization of nodes and links in a graph conveys a structure that can be described and labeled. The basic structural properties of a graph are:

Symmetry and asymmetry. A graph is symmetrical if each pair of nodes linked in one direction is also linked in the other. By convention, a line without an arrow represents a link where it is possible to move in both directions. However, both directions have to be defined in the graph. Most transport systems are symmetrical but asymmetry can often occur as it is the case for maritime (pendulum) and air services. Asymmetry is rare on road transportation networks, unless one-way streets are considered.

Assortativity and disassortativity. Assortative networks are those characterized by relations among similar nodes, while disassortative networks are found when structurally different nodes are often connected. Transport (or technological) networks are often disassortative when they are non-planar, due to the higher probability for the network to be centralized into a few large hubs.

Completeness. A graph is complete if two nodes are linked in at least one direction. A complete graph has no sub-graph and all its nodes are interconnected.

Connectivity. A complete graph is described as connected if for all its distinct pairs of nodes there is a linking chain. Direction does not have importance for a graph to be connected, but may be a factor for the level of connectivity. If p > 1 the graph is not connected because it has more than one sub-graph (or component). There are various levels of connectivity, depending on the degree at which each pair of nodes is connected.

Complementarity. Two sub-graphs are complementary if their union results in a complete graph. Multimodal transportation networks are complementary as each sub-graph (modal network) benefits from the connectivity of other sub-graphs.

Root. A node r where every other node is the extremity of a path coming from r is a root. Direction has an importance. A root is generally the starting point of a distribution system, such as a factory or a warehouse.

Trees. A connected graph without a cycle is a tree. A tree has the same number of links as nodes plus one. ($e = v - 1$). If a link is removed, the graph ceases to be connected. If a new link between two nodes is provided, a cycle is created. A branch of root r is a tree where no links are connecting any node more than once. River basins are typical examples of tree-like networks based on multiple sources connecting towards a single estuary. This structure strongly influences river transport systems.

Articulation node. In a connected graph, a node is an articulation node if the sub-graph obtained by removing this node is no longer connected. It therefore contains more than one sub-graph ($p > 1$). An articulation node is generally a port or an airport, or an important hub of a transportation network, which serves as a bottleneck. It is also called a bridge node.

Isthmus. In a connected graph, an isthmus is a link that is creating, when removed, two sub-graphs having at least one connection. Most central links in a complex network are often isthmuses, the removal of which by reiteration helps reveal dense communities (clusters).

Method 3 – Graph theory: measures and indices

Authors: César Ducruet and Jean-Paul Rodrigue

Measures at the network level

Several measures and indices can be used to analyze the network efficiency. Many of them were initially developed by Kansky and can be used for:

- Expressing the relationship between values and the network structures they represent.
- Comparing different transportation networks at a specific point in time.
- Comparing the evolution of a transport network at different points in time.

Outside the description of the network size by the number of nodes and edges, and its total length and traffic, several measures are used to define the structural attributes of a graph: the diameter, the number of cycles and the order of a node.

Diameter (d). The length of the shortest path between the most distanced nodes of a graph is the diameter. d measures the extent of a graph and the topological length between two nodes.

The diameter enables the development of a network in time to be measured. The higher diameter, the less linked a network tends to be. In the case of a complex graph, the diameter can be found with a topological distance matrix (Shimbel distance), which computes for each node pair its minimal topological distance. Graphs whose extent remains constant, but with a higher connectivity, have lower diameter values. Planar networks often have a large diameter due to the presence of many intermediate stops between two distant nodes.

Number of cycles (u). The maximum number of independent cycles in a graph. This number (u) is estimated through the number of nodes (v), links (e) and sub-graphs (p). Trees and simple networks have a value of 0 since they have no cycles. The more complex a network is, the higher the value of u, so it can be used as an indicator of the level of development and complexity of a transport system.

$$u = e - v + p$$

Indices at the network level

Indices are more complex methods to represent the structural properties of a graph since they involve the comparison of a measure over another. Some indices take into account spatial features (distance, surface) as well as the level of activity (traffic), while others solely rest on the topological dimension of the network.

Cost (C). Represents the total length of the network measured in real transport distances where aij is the presence (1) or absence (0) of a link between i and j and lij the length of the link. This measure can also be calculated based on two other dimensions of the network: the Minimum Spanning Tree (MST) and the Greedy Triangulation (GT). The MST represents the shortest and/or lowest cost subtree of the network; it can be obtained by applying, among other shortest path algorithms, the Kruskal algorithm, which allows finding the lowest cost route connecting all nodes in the network. The GT refers to the maximal connected planar graph keeping the same number of nodes as in the original network but adding all possible links without breaking its planarity. Such operations take into account both the topology and the geography of the network, while comparing the latter with its optimal configurations. More efficient networks have relative costs near to 1, while less efficient networks are closer to 0.

$$Cost = \sum_{i,j} a_{ij} l_{ij}$$

$$Cost_{rel} = \frac{Cost - Cost^{MST}}{Cost^{GT} - Cost^{MST}}$$

Detour index. A measure of the efficiency of a transport network in terms of how well it overcomes distance or the friction of distance. The closer the detour index gets to 1, the more the network is spatially efficient. Networks having a detour index of 1 are rarely, if ever, seen and most networks would fit on an asymptotic curve getting close to 1, but never reaching it. For instance, the straight distance, $D(S)$, between two nodes may be 40 km but the transport distance, $D(T)$, real distance, is 50 km. The detour index is thus 0.8 (40 / 50). The complexity of the topography is often a good indicator of the level of detour.

$$DI = \frac{D(S)}{D(T)}$$

In order to derive a measure of relative efficiency, the **detour index relative efficiency** (*Erel*) is thus the ratio between the detour index calculated from the original network and the detour index calculated either from the *MST* (minimum spanning tree) or the *GT* (greedy triangulation).

$$E_{rel} = \frac{DI - DI^{MST}}{DI^{GT} - DI^{MST}}$$

Network density. Measures the territorial occupation of a transport network in terms of km of links (*L*) per square kilometers of surface (*S*). The higher it is, the more a network is developed.

$$ND = \frac{L}{S}$$

Pi index. The relationship between the total length of the graph *L(G)* and the distance along its diameter *D(d)*. It is labeled as Pi (π) because of its similarity with the real Pi value (3.14), which expresses the ratio between the circumference and the diameter of a circle. A high index shows a developed network. It is a measure of distance per units of diameter and an indicator of the shape of a network.

$$\Pi = \frac{L(G)}{D(d)}$$

Eta index. Average length per link. Adding new nodes will cause a decrease of Eta as the average length per link declines.

$$\eta = \frac{L(G)}{e}$$

Theta index. Measures the function of a node, which is the average amount of traffic per intersection. The higher theta is, the greater the load of the network. The measure can also be applied to the number of links (edges).

$$\Theta = \frac{Q(G)}{v}$$

Beta index. Measures the level of connectivity in a graph and is expressed by the relationship between the number of links (*e*) over the number of nodes (*v*). Trees and simple networks have Beta value of less than one. A connected network with one cycle has a value of 1. More complex networks have a value greater than 1. In a network with a fixed number of nodes, the higher the number of links, the higher the number of paths possible in the network. Complex networks have a high value of Beta. The rich-club coefficient is the Beta index applied to relations among larger order (degree) nodes; it verifies whether the connectivity is higher among larger degree nodes than for the whole network.

In Figure 10.6 four graphs are of growing connectivity. Graphs A and B are not fully connected and their Beta value is lower than 1. Graph C is connected and has a Beta value of 1. Graph D is even more connected with a Beta value of 1.25.

Alpha index. A measure of connectivity which evaluates the number of cycles in a graph in comparison with the maximum number of cycles. The higher the alpha index, the more a network is connected. Trees and simple networks will

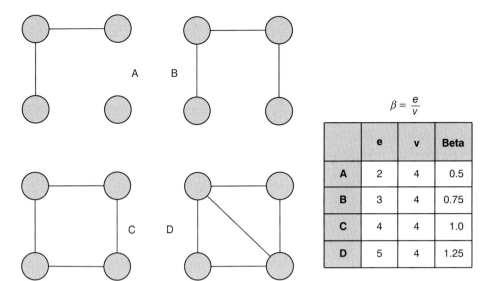

$$\beta = \frac{e}{v}$$

	e	v	Beta
A	2	4	0.5
B	3	4	0.75
C	4	4	1.0
D	5	4	1.25

Figure 10.6   Beta index

have a value of 0. A value of 1 indicates a completely connected network. Measures the level of connectivity independently of the number of nodes. It is very rare that a network will have an alpha value of 1, because this would imply very serious redundancies. This index is also called Meshedness Coefficient in the literature on planar networks.

The graphs in Figure 10.7 have a growing level of connectivity. While graph A has no cycles, graph D has the maximum possible number of cycles for a planar graph.

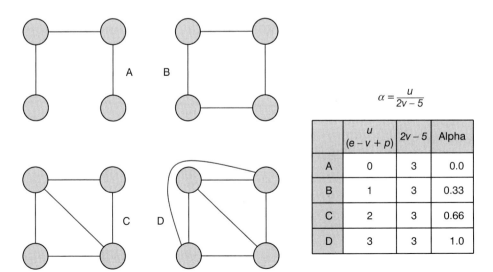

$$\alpha = \frac{u}{2v - 5}$$

	$\dfrac{u}{(e - v + p)}$	$2v - 5$	Alpha
A	0	3	0.0
B	1	3	0.33
C	2	3	0.66
D	3	3	1.0

Figure 10.7   Alpha index

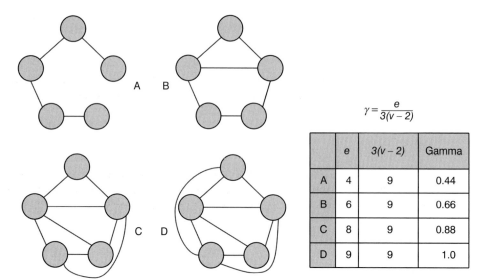

$$\gamma = \frac{e}{3(v-2)}$$

	e	3(v − 2)	Gamma
A	4	9	0.44
B	6	9	0.66
C	8	9	0.88
D	9	9	1.0

Figure 10.8 Gamma index

Gamma index (*g*). A measure of connectivity that considers the relationship between the number of observed links and the number of possible links. The value of gamma is between 0 and 1 where a value of 1 indicates a completely connected network and would be extremely unlikely in reality. Gamma is an efficient value to measure the progression of a network in time.

Being solely based on the number of nodes and links, Alpha, Beta, and Gamma indices remain limited in revealing structural differences between two networks of equal size. More robust measures have thus been proposed by physics, which take into account the internal complexity of the graph.

Measures and indices at the node level

Numerous measures exist for highlighting the situation of a node in a network. Some of them are made at the "local level" based on links with adjacent nodes, while others on the "global level" consider the node's situation in the whole network.

Order (degree) of a node (*o*). The number of its attached links and is a simple but effective measure of nodal importance. The higher its value, the more a node is important in a graph as many links converge to it. Hub nodes have a high order, while terminal points have an order that can be as low as 1. A perfect hub would have its order equal to the summation of all the orders of the other nodes in the graph and a perfect spoke would have an order of 1. The percentage of nodes directly connected in the entire graph is thus a measure of reachability. An isolate is a node without connections (degree equals to 0). The difference between in-degree and out-degree in a directed graph (digraph) may underline interesting functions of some nodes as attractors or senders. The order may be calculated at different depths: adjacent nodes (depth 1), adjacent nodes of adjacent nodes (depth 2), etc. The weighted degree is simply the total of values associated with links.

$$k_i = C_D(i) = \sum_j^N x_{ij}$$

Koenig number (or associated number, eccentricity). A measure of farness based on the number of links needed to reach the most distant node in the graph.

$$e(x) = \max_{y \in X} d(x, y)$$

Shimbel index (or Shimbel distance, nodal accessibility, nodality). A measure of accessibility representing the sum of the length of all shortest paths connecting all other nodes in the graph. The inverse measure is also called closeness centrality or distance centrality.

$$A_i = \sum_{j=1}^N d_{ij}$$

Hub dependence (*hd*). A measure of node vulnerability that is the share of the highest traffic link in total traffic (weighted degree). Weak nodes depending on few links will have a high hub dependence, especially if they locate in the neighborhood of a large node, while hubs will have a more even traffic distribution among their connections. It indicates to what extent removing the largest traffic link would affect the node's overall activity. The measure can be extended to more links (2, 3 . . . 10 largest flow links).

Several critiques have been addressed to such indexes as they do not always take into account the real length, quality and weight of the links; networks of equal size may exhibit contrasted topological forms. However, they remain useful for describing the changing structure of one given network.

Method 4 – Geographic Information Systems for Transportation (GIS-T)

Authors: Shih-Lung Shaw and Jean-Paul Rodrigue

GIS in transportation

In a broad sense a geographic information system (GIS) is an information system specializing in the input, management, analysis and reporting of geographical (spatially related) information. Among the wide range of potential applications GIS can be used for, transportation issues have received a lot of attention. A specific branch of GIS applied to transportation issues, commonly labeled as GIS-T, is one of the pioneer GIS application areas.

Geographic Information Systems for Transportation (GIS-T) refers to the principles and applications of applying geographic information technologies to transportation problems.

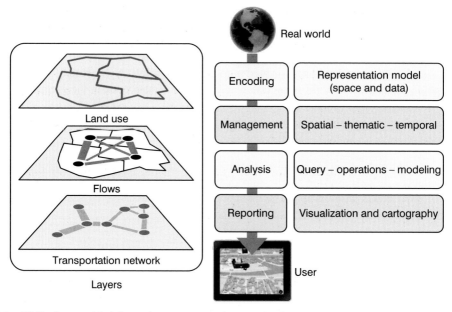

Figure 10.9 Geographic information systems and transportation

The four major components of a GIS – encoding, management, analysis and reporting – have specific considerations for transportation (Figure 10.9):

Encoding. Deals with issues concerning the representation of a transport system and its spatial components. To be of use in a GIS, a transport network must be correctly encoded, implying a functional topology composed of nodes and links. Other elements relevant to transportation, namely qualitative and quantitative data, must also be encoded and associated with their respective spatial elements. For instance, an encoded road segment can have data related to its width, number of lanes, direction, peak-hour traffic, etc.

Management. The encoded information is often stored in a database and can be organized along spatial (by region, country, census units, etc.), thematic (for highway, transit, railway, terminals, etc.) or temporal (by year, month, week, etc.) considerations. It is important to design a GIS database that organizes a large amount of heterogeneous data in an integrated and seamless environment such that the data can be easily accessed to support various transportation application needs.

Analysis. Considers the wide array of methodologies and tools available for transport issues. They can range from a simple query over an element of a transport system (e.g. what is the peak-hour traffic of a road segment?) to a complex model investigating the relationships between its elements (e.g. if a new road segment was added, what would be the impacts on traffic and future land use developments?).

Reporting. A GIS would not be complete without its visualization and data reporting capabilities for both spatial and non-spatial data. This component is particularly important as it offers interactive tools to convey complex information in a visual format (displayed or printed). A GIS-T thus becomes a useful tool to inform people who otherwise may not be able to visualize the hidden patterns and relationships embedded in the datasets (e.g. potential relationships among traffic accidents, highway geometry, pavement condition and terrain).

Information in a GIS is often stored and represented as layers, which are a set of geographical features linked with their attributes. On Figure 10.9 a transport system is represented as three layers related to land use, flows (spatial interactions) and the network. Each has its own features and related data.

GIS-T research can be approached from two different, but complementary, directions. While some GIS-T research focuses on issues of how GIS can be further developed and enhanced in order to meet the needs of transportation applications, other GIS-T research investigates the questions of how GIS can be used to facilitate and improve transportation studies. In general, topics related to GIS-T studies can be grouped into three categories:

- **Data representations**. How can various components of transport systems be represented in a GIS-T?
- **Analysis and modeling**. How can transport methodologies be used in a GIS-T?
- **Applications**. What types of applications are particularly suitable for GIS-T?

GIS-T data representations

Data representation is a core research topic of GIS. Before a GIS can be used to tackle real world problems, data must be properly represented in a digital computing environment. One unique characteristic of GIS is the capability of integrating spatial and non-spatial data in order to support both display and analysis needs. There have been various data models developed for GIS. The two basic approaches are object-based data models and field-based data models:

- An **object-based data model** treats geographic space as populated by discrete and identifiable objects. Features are often represented as points, lines, and/or polygons.
- On the other hand, a **field-based data model** treats geographic space as populated by real world features that vary continuously over space. Features can be represented as regular tessellations (e.g. a raster grid) or irregular tessellations (e.g. triangulated irregular network – TIN).

GIS-T studies have employed both object-based and field-based data models to represent the relevant geographic data. Some transportation problems tend to fit better with one type of GIS data model than the other. For example, network analysis based on the graph theory typically represents a network as a set of nodes interconnected with a set of links. The object-based GIS data model therefore is a better candidate for such transportation applications. Other types of transportation data exist which require extensions to the general GIS data models. One well-known example is linear referencing data (e.g. highway mileposts). Transportation agencies often measure locations of features or events along transportation network links (e.g. a traffic accident occurred at the 52.3 milepost on a specific highway). Such a one-dimensional linear referencing system (i.e. linear measurements along a highway segment with respect to a pre-specified starting point of the highway segment) cannot be properly handled by the two-dimensional Cartesian coordinate system used in most GIS data models. Consequently, the dynamic segmentation data model was developed to address the specific need of the GIS-T community.

Origin-destination (O-D) flow data are another type of data that are frequently used in transportation studies. Such data have been traditionally represented in matrix forms (i.e. as a two-dimensional array in a digital computer) for analysis. Unfortunately, the relational data model widely adopted in most commercial GIS software does not provide adequate support for handling matrix data. Some GIS-T software vendors therefore have developed additional file formats and functions for users to work with matrix data in a GIS environment. The above examples illustrate how the conventional GIS approaches can be further extended and enhanced to meet the needs of transportation applications.

Developments of enterprise and multidimensional GIS-T data models also have received significant attention. Successful GIS deployments at the enterprise level (e.g. within a state department of transportation) demand additional considerations to embrace the diversity of application and data requirements. An enterprise GIS-T data model is designed to allow "each application group to meet the established needs while enabling the enterprise to integrate and share data". The needs of integrating 1-D, 2-D, 3-D and temporal data in support of various transportation applications also have called for the implementation of multidimensional (including spatio-temporal) data representations.

Modern information and communication technologies (ICT) such as the Internet and cellular phones have changed the ways that people and businesses conduct their activities. These changing activity and interaction patterns in turn lead to changing spatio-temporal traffic patterns. Our world has become more mobile and dynamic than ever before due to modern ICT. With the advancements of location-aware technologies (e.g. Global Positioning System, cellular phone tracking system and Wi-Fi positioning system), it is now feasible and affordable to collect large volumes of tracking data at the individual level. Consequently, how to best represent and manage dynamic data of moving objects (people, vehicles or shipments) in a GIS environment presents new research challenges to GIS-T, especially when we have to deal with the Big Data issues.

In short, one critical component of GIS-T is how transportation-related data in a GIS environment can be best represented in order to facilitate and integrate the needs of various transportation applications. Existing GIS data models provide a good foundation of supporting many GIS-T applications. However, due to some unique characteristics of transportation data and application needs, many challenges still exist to develop better GIS data models that will improve rather than limit what we can do with different types of transportation studies.

GIS-T analysis and modeling

GIS-T applications have benefited from many of the standard GIS functions (query, geocoding, buffer, overlay, etc.) to support data management, analysis and visualization needs. Like many other fields, transportation has developed its own unique analysis methods and models. Examples include shortest path and routing algorithms (e.g. traveling salesman problems, vehicle routing problem), spatial interaction models (e.g. gravity model), network flow problems (e.g. minimum cost flow problem, maximum flow problem, network flow equilibrium models), facility location problems (e.g. p-median problem, set covering problem, maximal covering problem, p-centers problem), travel demand models (e.g. the four-step trip generation, trip distribution, modal split, traffic assignment models, and more recent activity-based travel demand models) and land use-transportation interaction models.

While the basic transportation analysis procedures (e.g. shortest path finding) can be found in most commercial GIS software, other transportation analysis procedures and models (e.g. travel demand models) are available only selectively in some commercial software packages. Fortunately, the component GIS design approach adopted by GIS software companies provides a better environment for experienced GIS-T users to develop their own custom analysis procedures and models.

It is essential for both GIS-T practitioners and researchers to have a thorough understanding of transportation analysis methods and models. For GIS-T practitioners, such knowledge can help them evaluate different GIS software products and choose the one that best meets their needs. It also can help them select appropriate analysis functions available in a GIS package and properly interpret the analysis results. GIS-T researchers, on the other hand, can apply their knowledge to help improve the design and analysis capabilities of GIS-T. Due to the increasing availability of tracking data that include both spatial and temporal elements, development of spatio-temporal GIS analysis functions to help better understand the dynamic movement patterns in today's mobile world has attracted significant research attention in recent years.

GIS-T applications

GIS-T is one of the leading GIS application fields. Many GIS-T applications have been implemented at various transportation agencies and private firms. They cover much of the broad scope of transportation and logistics, such as infrastructure planning and management, transportation safety analysis, travel demand analysis, traffic monitoring and control, public transit planning and operations, environmental impacts assessment, intelligent transportation systems (ITS), routing and scheduling, vehicle tracking and dispatching, fleet management, site selection and service area analysis, and supply chain management. Each of these applications tends to have its specific data and analysis requirements. For example, representing a street network as centerlines may be sufficient for transportation planning and vehicle routing applications. A traffic engineering application, on the other hand, may require a detailed representation of individual traffic lanes. Turn movements at intersections also could be critical to a traffic engineering study, but not to a region-wide travel demand study.

These different application needs are directly relevant to the GIS-T data representation and the GIS-T analysis and modeling issues discussed above. When a need arises to represent transportation networks of a study area at different scales, what would be an appropriate GIS-T design that could support the analysis and modeling needs of various applications? In this case, it is desirable to have a GIS-T data model that allows multiple geometric representations of the same transportation network. Research on enterprise and multidimensional GIS-T data models discussed above aims at addressing these important issues of better data representations in support of various transportation applications.

With the rapid growth of the Internet and wireless communications in recent years, a growing number of Internet-based and wireless GIS-T applications can be found. Websites such as Google Maps and Mapquest are frequently used by people to get driving directions. Global positioning system (GPS) navigation systems also are available as a built-in device in vehicles or as a portable device. Coupled with wireless communications, these devices can offer real-time traffic information and provide helpful location-based services (LBS) (e.g. finding the closest ATM location and reporting the estimated travel time to reach the ATM location based on the current traffic conditions). Another trend observed in recent years is the growing number of GIS-T applications in the private sector, particularly for logistics applications. Since many businesses involve operations at geographically dispersed locations (e.g. supplier sites, distribution centers/warehouses, retail stores and customer location), GIS-T can be a useful tool for a variety of logistics applications. Many of these logistics applications are based on the GIS-T analysis and modeling procedures such as the routing and facility location problems.

GIS-T is interdisciplinary in nature and has many possible applications. Transportation geographers, who have appropriate backgrounds in both geography and transportation, are well positioned to pursue GIS-T studies.

Method 5 – The notion of accessibility

Defining accessibility

Accessibility is a key element to transport geography and to geography in general since it is a direct expression of mobility either in terms of people, freight or information. Well-developed and efficient transportation systems offer high levels of accessibility (if the impacts of congestion are excluded), while less-developed ones have lower levels of accessibility. Thus accessibility is linked with an array of economic and social opportunities.

Accessibility is defined as the measure of the capacity of a location to be reached by, or to reach different, locations. Therefore, the capacity and the arrangement of transport infrastructure are key elements in the determination of accessibility.

All locations are not equal because some are more accessible than others, which imply inequalities. The notion of accessibility consequently relies on two core concepts:

The first is **location** where the relativity of space is estimated in relation to transport infrastructures, since they offer the means to support movements.
The second is **distance**, which is derived from the connectivity between locations. Connectivity can only exist when there is a possibility to link two locations through transportation. It expresses the friction of distance and the location which has the least friction relatively to others is likely to be the most accessible. Commonly, distance is expressed in units such as in kilometers or in time, but variables such as cost or fuel consumption can also be used.

There are two spatial categories applicable to accessibility problems, which are interdependent:

The first type is known as **topological accessibility** and is related to measuring accessibility in a system of nodes and paths (a transportation network). It is assumed that accessibility is a measurable attribute significant only to specific elements of a transportation system, such as terminals (airports, ports or subway stations).
The second type is known as **contiguous accessibility** and involves measuring accessibility over a surface. Under such conditions, accessibility is a measurable attribute of every location, as space is considered in a contiguous manner (Figure 10.10).

Space/time convergence is far from being a uniform process as differences in transport infrastructures and basic landscape constraints have a discriminatory effect on accessibility. Figure 10.10 represents travel time, from less than 1 hour to 10 days, to the nearest city of more than 50,000 people. It is the outcome of an overlay of several friction of distance factors, including the road and rail networks, navigable rivers, shipping lanes and land cover. It can be considered as a proxy for global accessibility with only 10 percent of the world's population being more than 48 hours away from a large city. While it depicts the general ease of accessing urban markets, it does not depict well the

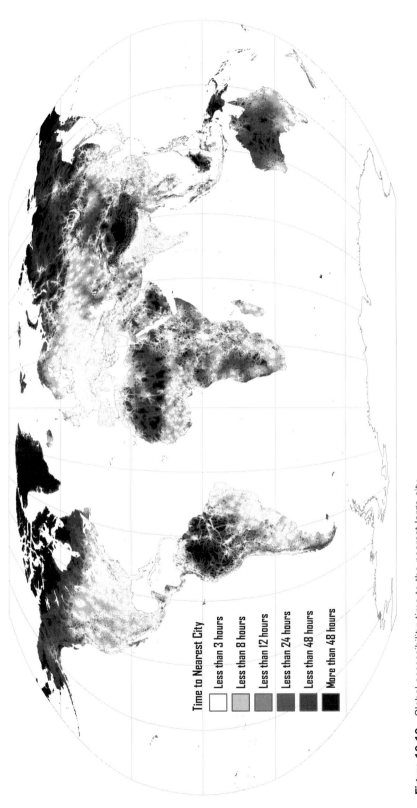

Figure 10.10 Global accessibility: time to the nearest large city

Source: Nelson, A. (2008) Estimated travel time to the nearest city of 50,000 or more people in year 2000. Global Environment Monitoring Unit – Joint Research Centre of the European Commission, Ispra, Italy.

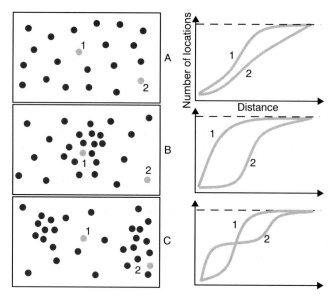

Figure 10.11 Accessibility and spatial structure

effectiveness of global freight flows. For instance, South Asia appears highly accessible because of the density of large cities, but the quality and capacity of inland transport infrastructures are generally poor.

Accessibility is a good indicator of the underlying spatial structure since it takes into consideration location as well as the inequality conferred by distance to other locations.

Due to different spatial structures, two locations of the same importance will have different accessibilities (Figure 10.11). For a spatial structure where locations are uniformly distributed (A), locations 1 and 2 have different accessibilities, with location 1 being the most accessible. As distance (Euclidean) increases, location 1 has access to a larger number of locations than location 2. To access all locations, location 2 would require about double the traveled distance than location 1. This is particularly the case as the spatial structure changes to one having a concentration of activities around location 1 (B). In this case, the number of locations that can be reached by location 1 increase rapidly and then eventually peaks. The third example (C) has a spatial structure having roughly two foci. Although the number of locations that can be reached from location 2 initially increases faster than for location 1, it catches up and is actually the most accessible, but by a lesser margin.

Connectivity and total accessibility

The most basic measure of accessibility involves network connectivity where a network is represented as a connectivity matrix ($C1$), which expresses the connectivity of each node with its adjacent nodes. The number of columns and rows in this matrix is equal to the number of nodes in the network and a value of 1 is given for each cell where this is a connected pair and a value of 0 for each cell where there is an unconnected pair. The summation of this matrix provides a very basic measure of accessibility, also known as the degree of a node:

$$C1 = \sum_{j}^{n} C_{ij}$$

- $C1$ = degree of a node.
- cij = connectivity between node i and node j (either 1 or 0).
- n = number of nodes.

The network on Figure 10.12 can be represented as a connectivity matrix, which is rather simple to construct:

- **Size of the connectivity matrix**: involves a number of rows and cells equivalent to the number of nodes in the network. Since the above network has 5 nodes, its connectivity matrix is a five by five grid.
- **Connection**: Each cell representing a connection between two nodes receives a value of 1 (e.g. Cell B – A).
- **Non-connection**: Each cell that does not represent a direct connection gets a value of 0 (e.g. Cell D – E).
- If all connections in the network are bi-directional (a movement is possible from node C to node D and vice versa), the connectivity matrix is **transposable**.

Adding up a row or a column gives the degree of a node. Node C is obviously the most connected since it has the highest summation of connectivity comparatively to all other nodes. However, this assumption may not hold true on a more complex network because of a larger number of indirect paths which are not considered in the connectivity matrix.

The connectivity matrix does not take into account all the possible indirect paths between nodes. Under such circumstances, two nodes could have the same degree, but may have different accessibilities. To consider this attribute, the total accessibility matrix (T) is used to calculate the total number of paths in a network, which includes direct as well as indirect paths. Its calculation involves the following procedure:

$$T = \sum_{k=1}^{D} Ck$$

$$C1 = \sum_{j}^{n} C_{ij}$$

$$Ck = \sum_{i}^{n} \sum_{j}^{n} c_{ij}^{1} \times c_{ji}^{k-1} \left(\forall k \neq 1 \right)$$

- D = the diameter of the network.

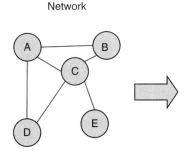

	Network		Connectivity matrix				

	A	B	C	D	E
A	0	1	1	1	0
B	1	0	1	0	0
C	1	1	0	1	1
D	1	0	1	0	0
E	0	0	1	0	0

Figure 10.12   Connectivity matrix

Thus, total accessibility would be a more comprehensive accessibility measure than network connectivity.

Geographic and potential accessibility

From the accessibility measure developed so far, it is possible to derive two simple and highly practical measures, defined as geographic and potential accessibility. Geographic accessibility considers that the accessibility of a location is the summation of all distances between other locations divided by the number of locations. The lower its value, the more a location is accessible.

$$A(G) = \sum_{i}^{n} \sum_{j}^{n} d_{ij} \bigg/ n$$

Although geographic accessibility can be solved using a spreadsheet (or manually for simpler problems), geographic information systems have proven to be a very useful and flexible tool to measure accessibility, notably over a surface simplified as a matrix (raster representation). This can be done by generating a distance grid for each place and then summing all the grids to form the total summation of distances grid. The cell having the lowest value is thus the most accessible place.

The construction of a geographic accessibility matrix, A(G), is a rather simple undertaking:

- Build the **valued graph matrix** (L). In Figure 10.13 the L-matrix shows the shortest distance in kilometers between five nodes (node A to node E).
- Build the **geographic accessibility matrix** A(G). The A(G) matrix is similar to the L-matrix except that the summation of rows and columns is divided by the number of locations in the network. The summation values are the same for columns and rows since this is a transposable matrix. The most accessible place is node C, since it has the lowest summation of distances.

$$A(P) = \sum_{i}^{n} P_{i} + \sum_{j}^{n} P_{j} \big/ d_{ij}$$

- $A(P)$ = potential accessibility matrix.
- d_{ij} = distance between place i and j (derived from valued graph matrix).
- P_j = attributes of place j, such as its population, retailing surface, parking space, etc.
- n = number of locations.

The potential accessibility matrix is not transposable since locations do not have the same attributes, which brings the underlying notions of emissiveness and attractiveness:

- **Emissiveness** is the capacity to leave a location, the sum of the values of a row in the A(P) matrix.
- **Attractiveness** is the capacity to reach a location, the sum of the values of a column in the A(P) matrix.

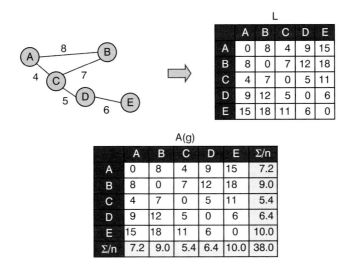

Figure 10.13 Geographic accessibility

By considering the same valued graph matrix (L) as the previous example and the population matrix P, the potential accessibility matrix, P(G), can be calculated:

- The value of all corresponding cells (A – A, B – B, etc.) equals the value of their respective attributes (P).
- The value of all non-corresponding cells equals their attribute divided by the corresponding cell in the L-matrix.

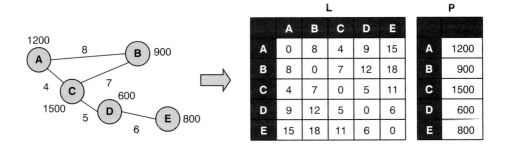

P(G)

i\j	A	B	C	D	E	Σi
A	1200.0	150.0	300.0	133.3	80.0	1863.3
B	112.5	900.0	128.6	75.0	50.0	1266.1
C	375.0	214.3	1500.0	300.0	136.4	2525.7
D	66.6	50.0	120.0	600.0	100.0	936.6
E	53.3	44.4	72.7	133.3	800.0	1103.7
Σj	1807.4	1358.7	2121.3	1241.6	1166.4	7695.4

Figure 10.14 Potential accessibility

The higher the value, the more a location is accessible, node C being the most accessible. The matrix being non-transposable, the summation of rows is different from the summation of columns, bringing forward the issue of attractiveness and emissiveness. Node C has more emissiveness than attractiveness (2525.7 versus 2121.3), while node B has more attractiveness than emissiveness (1358.7 versus 1266.1).

Method 6 – Network data models

Nature and utility

Graph theory developed a topological and mathematical representation of the nature and structure of transportation networks. However, graph theory can be expanded for the analysis of real-world transport networks by encoding them in an information system. In the process, a digital representation of the network is created, which can then be used for a variety of purposes such as managing deliveries or planning the construction of transport infrastructure. This digital representation is highly complex, since transportation data are often multi-modal, can span several local, national and international jurisdictions and have different logical views depending on the particular user. In addition, while transport infrastructures are relatively stable components, vehicles are very dynamic elements.

It is thus becoming increasingly relevant to use a data model where a transportation network can be encoded, stored, retrieved, modified, analyzed and displayed. Obviously, geographic information systems have received a lot of attention over this issue since they are among the best tools to store and use network data models. Network data models are an implicit part of many GIS, if not an entire GIS package of its own.

The core purpose of a network data model is to provide an accurate representation of a network as a set of links and nodes. **Topology** is the arrangement of nodes and links in a network. Of particular relevance are the representations of location, direction and connectivity. Even if graph theory aims at the abstraction of transportation networks, the topology of a network data model should be as close as possible to the real world structure it represents. This is especially true for the usage of network data models in a GIS.

Figure 10.15 represents the basic topology of an urban transport network composed of linked nodes. It has been encoded into a network data model to represent the reality

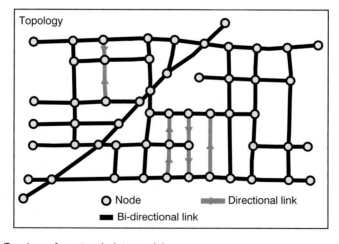

Figure 10.15 Topology of a network data model

as closely as possible, both topologically and geographically. Topologically, each node has been encoded with the connectivity it permits, such as if a left turn is possible or not (although this attribute is not displayed here). Further, a direction has been encoded in each link (directional or bi-directional) to represent one-ways. Geographically, each node is located at a coordinate that matches to the actual intersection it represents. In addition, the links between each node have been decomposed into several segments (not implicitly shown) to respect the positional accuracy of the road they represent.

Cartography allows the visualization of a transport network for the purpose of reckoning and simple navigation and serves to indicate the existence of a network. Different elements of the network can have a symbolism defined by some of their attributes. For instance, a highway link may be symbolized as a thick line with a label such as its number, while a street may be symbolized as an unlabeled simple line. The symbolized network can also be combined with other features such as landmarks to provide a better level of orientation to the user. This is commonly the case for road maps used by the general public.

By using attributes encoded in the network data model, such as road type, each segment can be displayed to reflect its importance. For instance, Figure 10.16 displays three road classes (highway, Main Street and street) differently. Descriptive labels for the most important elements and directional signs for one-ways have also been added. To enrich the cartographic message, additional layers of information have been added, namely landmarks. Nodal attributes can also have a cartographic utility, such as displaying if an intersection has traffic lights.

Transportation network models can be used to derive a precise location, notably through a linear referencing system. For instance, the great majority of addresses are defined according to a number and a street. If address information in embedded in the attributes of a network data model, it becomes possible to use this network for **geocoding** and to pinpoint the location of an address, or any location along the network, with reasonable accuracy.

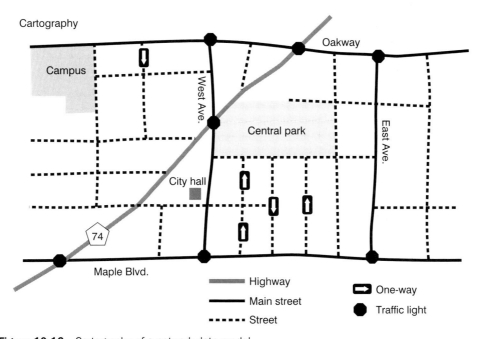

Figure 10.16   Cartography of a network data model

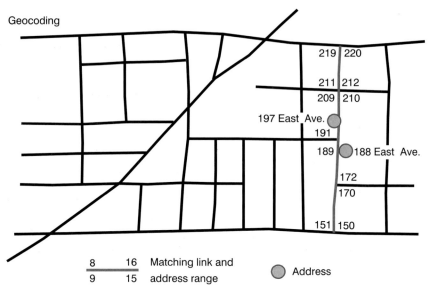

Figure 10.17 Geocoding in a network data model

Geocoding is possible if a linear referencing system is embedded in a network data model. One of the most common linear referencing systems is the address system, where each link has a corresponding street name and address range. The address range on Figure 10.17 illustrates even (right side) and odd (left side) addresses, very common attributes in most network data models such as TIGER (developed by the US Census Bureau). For instance, finding the approximate location of the address "197 East Ave." would first imply querying the network data model to find all the links that have "East Ave." as a name attribute. Then, the appropriate address range is found and the location interpolated. "197" corresponds to the 191–209 address range, located on the left side of East Ave. Its approximate location would be at 1/3 [1 – (209–197) / (209–191)] of the length of the link that has the 191–209 address range. The same procedure can be applied to the "188 East Ave." address, which in this case would be located at 1/4 of the length of the link that has the 172–210 address range.

Network data models may be used to find optimal paths and assign flows with capacity constraints in a network. While **routing** is concerned by the specific behavior of a limited number of vehicles, **traffic assignment** is mainly concerned by the system-wide behavior of traffic in a transport network. This requires a topology in which the relationship of each link with other intersecting segments is explicitly specified. Impedance measures (e.g. distance) are also attributed to each link and will have an impact on the chosen path or on how flows are assigned in the network. Routing and traffic assignment at the continental level is generally simple since small variations in impedance are of limited consequences. Routing and traffic assignment in an urban area is much more complex as it must consider stop signs, traffic lights and congestion, in determining the impedance of a route.

Routing in a network data model can be simulated if impedance is available to links and nodes. For links, impedance is often characterized by travel time, while turn penalties are often used to characterize impedance at nodes, which is how difficult (if possible) it is to turn in one direction, as opposed to another. Figure 10.18 represents a typical routing "traveling salesperson" type of problem. Starting and ending at a warehouse, a

Routing

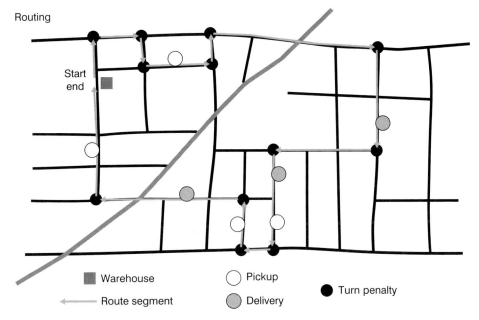

Figure 10.18 Routing in a network data model

delivery truck has a set of deliveries and pickups to perform. The locations of those pickup and delivery points could have been derived from address matching (geocoding). Considering link and node (turn penalties) impedance attributes that are encoded in the network data model, it is possible to plot an optimal route minimizing travel time that would satisfy basic constraints related to the start and end points, pickup and delivery points, as well as link and turn penalty impedances.

Layer-based approach

Most conventional GIS data models separate information in layers, each representing a different class of geographical elements symbolized as points, lines and polygons in the majority of cases. As such, a network data model must be constructed with the limitation of having points and lines in two separate layers; thus the layer-based approach. Further, an important requirement is that the geometry of the network matches the reality as closely as possible since these networks are often part of a geographic information system where an accurate location and visualization is a requisite. This has commonly resulted in the fragmentation of each logical link into a multitude of segments, with most of the nodes of these segments mere intermediate cosmetic elements. The topology of such network data models is not well defined, and has to be inferred. However, these network data models benefit from the attribute linking capabilities of the spatial database models they are derived from. Among the most significant attributes that can be attached to network layers are:

- **Classification and labeling**. Each segment can be classified into categories such as its function (street, highway, railway, etc.), importance (number of lanes) and type (paved, non-paved). Also, a complex labeling structure can be established with prefixes, proper names and suffixes.

- **Linear referencing system**. Several systems to locate elements along a segment have been established. One of the most common is the address system where each segment is provided with an address range. Through linear interpolation, a specific location can be derived (geocoding).
- **Segment travel costs**. Can consider a vast array of impedance measures. Among the most common is the length of the segment, a typical travel time or a speed limit. Congestion can also be assessed, either as a specific value of impedance or as a mathematical function.
- **Direction**. To avoid unnecessary and often unrealistic duplication of links, especially at the street level, a directional attribute can be included in the attribute table.
- **Overcrossing and undercrossing**. Since the great majority of layer-based network models are planar, they are ill-designed to deal with non-planar representations. A provision must be made in the attribute table to identify segments that are over-crossing or undercrossing a segment they are intersecting with.
- **Turn penalties**. An important attribute to ensure accurate routing within a network. Each intersection has different turn constraints and possibilities. Conventionally in road transportation, a right turn is assumed to have a lesser penalty than a left turn. The opposite applies for countries where driving is on the left (e.g. the UK).

The layer-based approach is consequently good to solve the cartography and geocoding issues. However, it is ill-suited to comprehensively address routing and assignment transport problems.

Object-oriented approach

The object-oriented approach represents the latest development in spatial data models. It assumes that each geographical feature is an object having a set of properties and a set of relationships with other objects. As such, a transportation network is an object composed of other objects, namely nodes and links. Since topology is one of the core concepts defining transportation networks, relationships expressing it are embedded in object-oriented representations. The basic elements of an object-oriented transportation network data model are:

- **Classes**. They categorize objects in a specific taxonomy, which has a proper set of properties and relationships. The two basic classes of a network are obviously nodes and links, but each of these classes can be subdivided into subclasses. For instance, a link can be subdivided as a road link, a rail link, a walkway, etc.
- **Properties**. They refer to a set of measurable characteristics that are associated with a specific class. For instance, the properties of a road class could be its length, number of lanes, name, surface, speed limit, etc.
- **Relationships**. They describe the type of logical relations objects have with one another. Instance (is-a) and membership (is-in) are among the most common relations. For example, a street is an instance of the road class, which itself is an instance of a transport infrastructure. A specific road segment can be considered part of a specific transport system through a membership relation. From these relations inheritance can be derived, where the characteristics of one object can be passed to another. Using the previous example, it is logical to derive that a street is a transport infrastructure, thus the object street inherits the properties of the object transport infrastructure.

By their structure, especially with their embedded topology, an object-oriented transport network data model would be effective to solve the routing issue in transport. However, object-oriented data models are still in the design phase with proposals such as UNETRANS (Unified NEtwork-TRANSportation data model) hoping to become accepted standards. The potential of the object-oriented approach for GIS remains to be seen as well as the amount of effort required to convert or adapt existing transport network databases, which are mainly layer-based, into the new representational structure.

Method 7 – Transport technical and economic performance indicators

Authors: Jean-Paul Rodrigue and Claude Comtois

Network and operational indicators

Multimodal transportation networks rest upon the combinatory costs and performance of transport modes, or what is referred to as economies of scope. For instance, a single container shipped overseas at the lowest cost from its origin can go from road, to maritime, to railway and to road again before reaching its destination. For passengers, a commuter may also undertake a journey involving a sequence of modes such as walking, riding a bus and then a subway. Freight shippers and carriers therefore require quantitative tools for decision making in order to compare performances of various transport modes and transport networks. Time-efficiency becomes a set imperative for both freight and passenger transit in private as well as in public sector activities.

Performance indicators are widely used by geographers and economists to empirically assess the technical performance of differing transport modes, in other words their capacity to move goods or passengers around. They are ratios since a value such as traffic or capacity does not express a performance but the ratio of traffic over capacity is a performance indicator. Technical performance should not be confused with economic performance, which mostly relates to how much transport output (e.g. traffic) can be supported with specific inputs (e.g. capital or labor). Performance measures are relative, implying that they mean little by themselves and must be interpreted within a comparative framework, which can be across space (e.g. systems or jurisdictions) or across time (e.g. monthly variations). Basic technical performance calculations can be particularly useful for the analysis of global network performance as well as for modal comparison, analysis and evaluation by bridging both physical attributes (length, distance, configuration, etc.) and time-based attributes (punctuality, reliability, etc.) of networks. Some indicators are currently used to measure the technical performance of freight and passenger transport:

- **Passenger or freight density**. A standard measure of transport efficiency.
- **Mean distance traveled**. A measure of the ground covering capacity of networks and different transport modes. Used to measure the relative performance of transport modes.
- **Mean per capita ton output** (freight) or **mean number of trips per capita** (passenger).
- **Mean utilization coefficient**. Especially useful with increasing complexity of logistics associated with containerization of freight (i.e. the problem of empty returns). Can also be used to measure transit ridership.

More specifically, such indicators are of great utility by allowing cross-temporal analysis of a transport nexus or given transport modes. Another fundamental dimension of technical performance concerns operations and relate to specific parts of the transport network such as a segment or a terminal. The most salient indicators include:

- **Transport time/speed/turnover**. An expression of the velocity of passengers or freight along segments (speed) and at terminals or distribution centers (turnover).
- **Reliability**. The consistence of operations within defined parameters such as capacity, safety, duration and punctuality.
- **Punctuality**. The on-time performance of transport services. Particularly important for scheduled services such as flights, public transit, railways and containerized maritime shipping.
- **Load factor**. The level of transport asset utilization of modes and terminals in relation to their capacity. High load factors may be indicative of congestion.

On-time arrivals are a commonly used performance measure in air transportation since passengers plan their trips according to the arrival schedule. In 2011, 79.62 percent of all flights in the United States were running on time, implying that 20.38 percent of all flights arrived more than 15 minutes late. The main sources of delays as reported by airlines are:

- **Air carrier delay** (5.15 percent). The cause of the cancellation or delay was due to circumstances within the airline's control (e.g. maintenance or crew problems, aircraft cleaning, baggage loading, fueling, etc.).
- **Extreme weather** (0.51 percent). Significant meteorological conditions (actual or forecasted) that, in the judgment of the carrier, delays or prevents the operation of a flight, such as a tornado, thunderstorm, blizzard or hurricane.

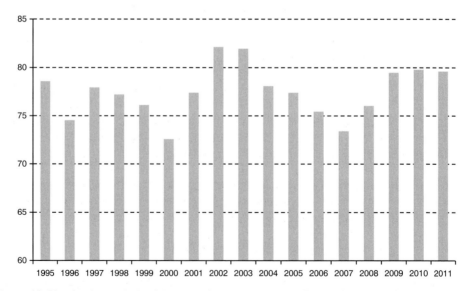

Figure 10.19 On-time arrivals of domestic flights in the United States, 1995–2011 (in %)

Source: Research and Innovative Technology Administration (RITA), US Department of Transportation (US DOT).

- **National Aviation System** (5.80 percent). Delays and cancellations attributable to the national aviation system that refer to a broad set of conditions, such as non-extreme weather conditions, airport operations, heavy traffic volume, and air traffic control.
- **Late-arriving aircraft** (6.75 percent). A previous flight with the same aircraft arrived late, causing the present flight to depart late. This is the outcome of propagation effects on schedule integrity.
- **Security** (0.04 percent). Delays or cancellations caused by evacuation of a terminal or concourse, re-boarding of aircraft because of security breach, inoperative screening equipment and/or long lines in excess of 29 minutes at screening areas.

Fluctuations in the flight delay pattern are mainly attributed to the growth and decline in air traffic. If air transport grows rapidly, the outcome tends to be a decreasing performance of on-time arrivals as the system is trying to cope with additional demands with a similar level of capacity (e.g. 1995–2000 or 2003–7). On the other hand, if air transportation stops growing or even declines, the performance improves since some flights are removed (e.g. 2000–2002 or 2007–10).

Road traffic performance

Technical performance indicators have dominantly been applied to road transportation, although other modes such as air and maritime transport are also increasingly monitored. There are two major operational types of traffic influencing the capacity of modern roads, which are continuous and discontinuous traffic. The capacity of a road is the maximal hourly flow of people or vehicles that can be supported by any link. This value is influenced by three major concepts.

- **Road conditions**. Physical attributes of the road such as its type (paved, non-paved), number of lanes, width of lanes, design speed and the vertical and horizontal alignment.
- **Traffic conditions**. Attributes of the traffic using the road such as its temporal distribution and its direction.
- **Control conditions**. Attributes of the control structures and existing traffic laws such as speed limit, one-ways and priority.

Considering the above conditions, the capacity of a road is about 1,000 vehicles per lane per hour for continuous traffic roads and about 500 vehicles per lane per hour for discontinuous traffic roads. The operational goal of traffic planning is thus to make it so that road, traffic and control conditions ensure an adequate, if not optimal, service. Several guidelines will favor such a goal such as wide enough lanes for a safe maximum speed in both directions and limited grades to limit speed differentials. The capacity of a road is also linked to the level of service, which is a qualitative measure of operational conditions of roads and its perception by users. The spatial distribution of bottlenecks, notably within urban areas, also has a strong impact on capacity as they are the choking points of the whole road transport system. Traffic can be valued according to three primary measures, which are speed, volume or density:

- **Speed** is a rate of distance covered per unit of time. The average speed is the most commonly used measure to characterize traffic on a road.
- **Volume** is the number of vehicles observed at a point or a section over a period of time.

- **Density** is the number of vehicles that occupies a section at any point in time. For example, a road section having a volume of 1,000 vehicles per hour with an average speed of 50 km/hr will have a density of 20 vehicles/km.

The critical density is the density at which the volume is maximal and the critical speed is the speed at which the volume is maximal.

Economic performance indicators

Undoubtedly, transportation plays a considerable role in the economy with its omnipresence throughout the production chain, at all geographic scales. It is an integral constituent of the production–consumption cycle. Economic impact indicators help to show the relationship between transport systems and the economy as well as information on the economic weight of this type of activity. Geographers should be familiar with basic econometric impact indexes. Maritime transport is still the most cost-efficient way to transport bulk merchandise over long distances. On the other hand, while air transport is recognized for its unsurpassed time-efficiency versus other modes over long distances, it remains an expensive option. Thus, vertical integration, or the absorption of transportation activities by producers, illustrates the search for these two efficiency attributes by gaining direct control over inputs.

The relationship between transport systems and their larger economic frame becomes clear when looking at the restructuring patterns that carriers and firms are currently making. Structural mutations, best illustrated by the popularity of just-in-time practices, are fueled by two opposing yet effective forces: transporters seek to achieve economies of scale while having to conform to an increasingly "customized" demand.

Factor substitution is a commonly adopted path in order to reduce costs of production and attain greater efficiency. Containerization of freight by substituting labor for capital and technology is a good illustration of the phenomenon. Measures of capital productivity for such capital-intensive transport means are of central importance; an output/capital ratio is then commonly used. While the output/labor ratio performs the same productivity measurement but for the labor input (this form of indicator can be used for each factor of production in the system), a capital/labor ratio aims at measuring which factor predominates within the relationship between capital and labor productivity. The above set of indicators therefore provides insights on the relative weight of factors within the production process.

More scale-specific indicators can also be used to appreciate the role of transport within the economy. Knowing freight transport both contributes to and is fueled by a larger economic context, freight output can be confronted against macro-economic indicators: an output/GDP ratio measures the relationship between economic activity and traffic freight, in other words the traffic intensity. At the local level, the status of the transport industry within the local economy is given by a transport sector income/local income ratio. Still at a micro-scale, finally, a measure of the relative production value of freight output is provided by an output/local income ratio.

Underlying objectives of the application of such indicators are as varied as they are numerous. Efficiency indicators constitute valuable tools to tackle project viability questions as well as to measure investment returns and cost/subsidy recovery of transport systems. Input–output analyses making use of some of the above indicators are also instrumental to the development of global economic impact indexes and productivity assessment concepts such as the Total Factor Productivity (TFP) and to identify sources of productivity gains.

Method 8 – The Gini coefficient

The Lorenz curve

The Gini coefficient was developed to measure the degree of concentration (inequality) of a variable in a distribution of its elements. It compares the Lorenz curve of a ranked empirical distribution with the line of perfect equality (Figure 10.20). This line assumes that each element has the same contribution to the total summation of the values of a variable. The Gini coefficient ranges between 0, where there is no concentration (perfect equality), and 1 where there is total concentration (perfect inequality).

The Lorenz curve is a graphical representation of the proportionality of a distribution (the cumulative percentage of the values). To build the Lorenz curve, all the elements of a distribution must be ordered from the most important to the least important. Then, each element is plotted according to their cumulative percentage of X and Y, X being the cumulative percentage of elements and Y being their cumulative importance. For instance, out of a distribution of 10 elements (N), the first element would represent 10 percent of X and whatever percentage of Y it represents (this percentage must be the highest in the distribution). The second element would cumulatively represent 20 percent of X (its 10 percent plus the 10 percent of the first element) and its percentage of Y plus the percentage of Y of the first element.

The Lorenz curve is compared with the perfect equality line, which is a linear relationship that plots a distribution where each element has an equal value in its shares of X and Y. For instance, in a distribution of 10 elements, if there is perfect equality, the 5th element would have a cumulative percentage of 50 percent for X and Y. The perfect inequality line represents a distribution where one element has the total cumulative percentage of Y while the others have none. The Gini coefficient is defined graphically as a ratio of two surfaces involving the summation of all vertical deviations between the Lorenz curve and the perfect equality line (A) divided by the difference between the perfect equality and perfect inequality lines (A+B).

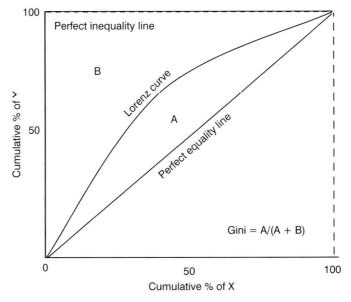

Figure 10.20 The Lorenz curve

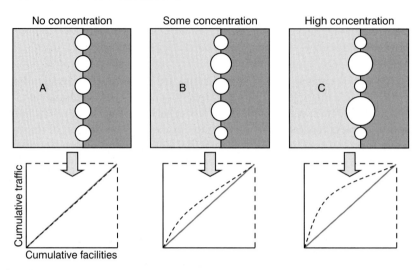

Figure 10.21 Traffic concentration and Lorenz curves

Figure 10.21 represents a simple system of 5 ports along a coast. In case A, the traffic for each port is the same, so there is no concentration and thus no inequality. The Lorenz curve of this distribution is the same as the perfect equality line; they overlap. In case B, there is some concentration of the traffic in two ports and this concentration is reflected in the Lorenz curve as it is different from the perfect equality line. Case C represents a high level of concentration in two ports and the Lorenz curve is significantly different from the perfect equality line.

The world's 50 largest container ports and airport terminals show a level of concentration in their distribution, with container ports having the highest level as depicted on its Lorenz curve (Figure 10.22). The 10 largest terminals account for about 49.7 percent of the traffic handled by world's largest 50 container ports. Air passenger traffic has a much lower concentration level (the 10 largest airports handling 30.8 percent of the traffic of the 50 largest airports), which is reflective of its widespread use as a mode of long distance transportation. Geographical concentration is also significant with most of the activity taking place around East Asia, Western Europe and North America.

Calculating the Gini coefficient (G)

The Gini coefficient represents the area of concentration between the Lorenz curve and the line of perfect equality as it expresses a proportion of the area enclosed by the triangle defined by the line of perfect equality and the line of perfect inequality. The closer the coefficient is to 1, the more unequal the distribution. Figure 10.23 shows a hypothetical set of terminals with varying amounts of traffic. X refers to the traffic proportion if the traffic was distributed evenly throughout all the terminals. Y refers to the actual proportion of traffic of each terminal. sX and sY are cumulative percentages of Xs and Ys (in fractions) and N is the number of elements (observations).

Geographers and many others have used the Gini coefficient in numerous instances, such as assessing income distribution among a set of contiguous regions (or countries) or to measure other spatial phenomena such as industrial location. Its major purpose as a method in transport geography has been related to measuring the concentration of traffic, mainly at terminals, such as assessing changes in port system concentration. Economies

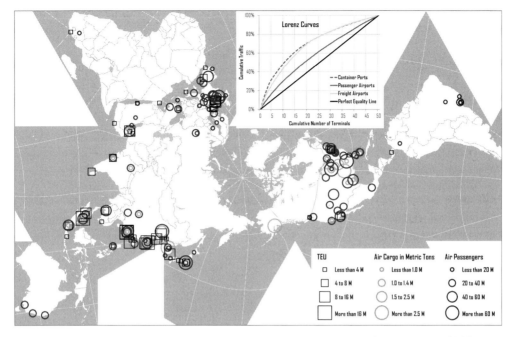

Figure 10.22 World's 50 largest container ports, passenger airports and freight airports, 2010

Sources: Airport Council International and Containerization International.

$$G = \left| 1 - \sum_{i=1}^{N} (\sigma Y_{i-1} + \sigma Y_i)(\sigma X_{i-1} - \sigma X_i) \right| = 0.372$$

Y	σX (Cumulative)	σY (Cumulative)	σY$_{i-1}$ + σY$_i$ (A)	σX$_{i-1}$ − σX$_i$ (B)	A*B
0.25	0.10	0.25	0.25	0.10	0.025
0.20	0.20	0 45	0.70	0.10	0.070
0.15	0.30	0.60	1.05	0.10	0.105
0.10	0.40	0.70	1.30	0.10	0.130
0.08	0.50	0.78	1.48	0.10	0.148
0.07	0.60	0.85	1.63	0.10	0.163
0.05	0.70	0.90	1.75	0.10	0.175
0.05	0.80	0.95	1.85	0.10	0.185
0.03	0.90	0.98	1.93	0.10	0.193
0.02	1.00	1.00	1.98	0.10	0.198
1.00					1.392

Figure 10.23 Calculation of the Gini coefficient

of scale in transportation can favor the concentration of traffic at transport terminals, while other considerations such as accessibility to regional markets can be perceived as a countervailing force to concentration. So, the temporal variations of the Gini coefficient reflect changes in the comparative advantages of a location within the transport system.

Method 9 – Spatial interactions and the gravity model

Overview

One methodology of particular importance to transport geography relates to how to estimate flows between locations, since these flows, known as spatial interactions, enable the demand (existing or potential) for transport services to be evaluated.

A **spatial interaction** is a realized movement of people, freight or information between an origin and a destination. It is a transport demand/supply relationship expressed over a geographical space.

Spatial interactions cover a wide variety of movements such as journeys to work, migrations, tourism, the usage of public facilities, the transmission of information or capital, the market areas of retailing activities, international trade and freight distribution. Economic activities are generating (supply) and attracting (demand) flows. The simple fact that a movement occurs between an origin and a destination underlines that the costs incurred by a spatial interaction are lower than the benefits derived from such an interaction. As such, a commuter is willing to drive one hour because this interaction is linked to an income, while international trade concepts, such as comparative advantages, underline the benefits of specialization and the ensuing generation of trade flows between distant locations. Three interdependent conditions are necessary for a spatial interaction to occur (Figure 10.24):

Complementarity. There must be a supply and a demand between the interacting locations. A residential zone is complementary to an industrial zone because the first is supplying workers while the second is supplying jobs. The same can be said concerning the complementarity between a store and its customers and between an industry and its

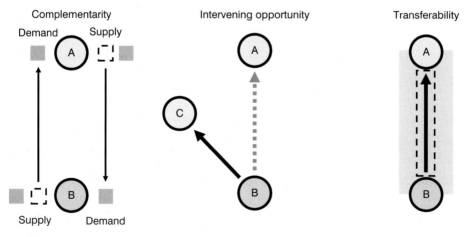

Figure 10.24 Conditions for the realization of a spatial interaction

suppliers (movements of freight). If location B produces/generates something that location A requires, then an interaction is possible because a supply/demand relationship has been established between those two locations; they have become complementary to one another. The same applies in the other direction (A to B), which creates a situation of reciprocity common in commuting or international trade.

Intervening opportunity. There must not be another location that may offer a better alternative as a point of origin or as a point of destination. For instance, in order to have an interaction of a customer to a store, there must not be a closer store that offers a similar array of goods. If location C offers the same characteristics (namely complementarity) as location A and is also closer to location B, an interaction between B and A will not occur and will be replaced by an interaction between B and C.

Transferability. Freight, persons or information being transferred must be supported by transport infrastructures, implying that the origin and the destination must be linked. Costs to overcome distance must not be higher than the benefits of related interaction, even if there is complementarity and no alternative opportunity. Transport infrastructures (modes and terminals) must be present to support an interaction between B and A. Also, these infrastructures must have a capacity and availability which are compatible with the requirements of such an interaction.

Spatial interaction models seek to explain spatial flows. As such it is possible to measure flows and predict the consequences of changes in the conditions generating them. When such attributes are known, it is possible for example to better allocate transport resources such as highways, buses, airplanes or ships since they would reflect the transport demand more closely.

Origin/destination matrices

Each spatial interaction, as an analogy for a set of movements, is composed of an origin/destination (O/D) pair. Each pair can itself be represented as a cell in a matrix where rows are related to the locations (centroids) of origin, while columns are related to locations (centroids) of destination. Such a matrix is commonly known as an origin/destination matrix, or a spatial interaction matrix.

Figure 10.25 represents movements (O/D pairs) between five locations (A, B, C, D and E). From this graph, an O/D matrix can be built where each O/D pair becomes a cell. A value of 0 is assigned for each O/D pair that does not have an observed flow. In the O/D matrix the sum of a row (T_i) represents the total outputs of a location (flows originating from), while the sum of a column (T_j) represents the total inputs of a location (flows bound to). The summation of inputs is always equal to the summation of outputs. Otherwise, there are movements that are coming from or going to outside the considered system. The sum of inputs or outputs gives the total flows taking place within the system (T). It is also possible to have O/D matrices according to age group, income, gender, etc. Under such circumstances they are labeled sub-matrices since they account for only a share of the total flows.

In many cases where spatial interaction information is relied on for planning and allocation purposes, origin/destination matrices are not available or are incomplete. Palliating this lack of data commonly requires surveys. With economic development, the addition of new activities and transport infrastructures, spatial interactions have a tendency to change very rapidly as flows adapt to a new spatial structure. The problem is that an origin/destination survey is very expensive in terms of effort, time and costs. In a complex spatial system such as a region, O/D matrices tend to be quite large. For instance, the

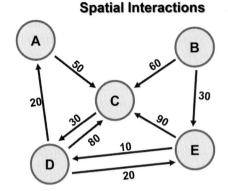

Spatial Interactions

O/D Matrix

	A	B	C	D	E	Ti
A	0	0	50	0	0	50
B	0	0	60	0	30	90
C	0	0	0	30	0	30
D	20	0	80	0	20	120
E	0	0	90	10	0	100
Tj	20	0	280	40	50	390

Figure 10.25 Constructing an origin/destination (O/D) matrix

consideration of 100 origins and 100 destinations would imply 10,000 separate O/D pairs for which information has to be provided. In addition, the data gathered by spatial interaction surveys is likely to rapidly become obsolete as economic and spatial conditions change. It is therefore important to find a way to estimate as precisely as possible spatial interactions, particularly when empirical data are lacking or incomplete. A possible solution relies on using a spatial interaction model to complement and even replace empirical observations.

Spatial interaction models

The basic assumption concerning many spatial interaction models is that flows are a function of the attributes of the locations of origin, the attributes of the locations of destination and the friction of distance between the concerned origins and the destinations. Figure 10.26 shows the general formulation of the spatial interaction model.

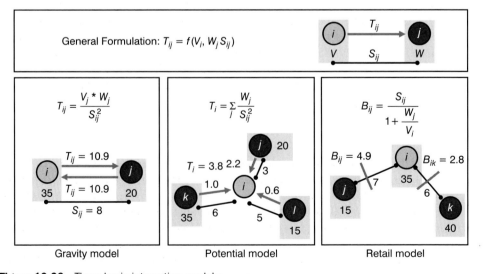

General Formulation: $T_{ij} = f(V_i, W_j S_{ij})$

$$T_{ij} = \frac{V_i * W_j}{S_{ij}^2}$$

$$T_i = \sum_j \frac{W_j}{S_{ij}^2}$$

$$B_{ij} = \frac{S_{ij}}{1 + \frac{W_j}{V_i}}$$

$T_{ij} = 10.9$

$T_{ij} = 10.9$

$S_{ij} = 8$

Gravity model

$T_i = 3.8$

Potential model

$B_{ij} = 4.9$

$B_{ik} = 2.8$

Retail model

Figure 10.26 Three basic interaction models

- T_{ij}: Interaction between location i (origin) and location j (destination). Its units of measurement are varied and can involve people, tons of freight, traffic volume, etc. It also relates to a time period such as interactions by the hour, day, month or year.
- V_i: Attributes of the location of origin i. Variables often used to express these attributes are socioeconomic in nature, such as population, number of jobs available, industrial output or gross domestic product.
- W_j: Attributes of the location of destination j. It uses similar socioeconomic variables to the previous attribute.
- S_{ij}: Attributes of separation between the location of origin i and the location of destination j. Also known as transport friction. Variables often used to express these attributes are distance, transport costs or travel time.

The attributes of V and W tend to be paired to express complementarity in the best possible way. For instance, measuring commuting flows (work-related movements) between different locations would likely consider a variable such as working age population as V and total employment as W. From this general formulation, three basic types of interaction models can be constructed:

- **Gravity model**. Measures interactions between all the possible location pairs by multiplying their attributes, which is then pondered by their level of separation. Separation is often squared to reflect the growing friction of distance. On Figure 10.26, two locations (i and j) have a respective "weight" (importance) of 35 and 20 and are at a distance (degree of separation) of 8. The resulting interaction is 10.9, which is reciprocal.
- **Potential model**. Measures interactions between one location and every other location by the summation of the attributes of each other location pondered by their level of separation (again squared to reflect the friction of distance). On Figure 10.26, the potential interaction of location i (T_i) is measured by adding the ratio "weight"/squared distance for each other locations (j, k and l). The potential interaction is 3.8, which is not reciprocal.
- **Retail model**. Measures the boundary of the market areas between two locations competing over the same market. It assumes that the market boundary between two locations is a function of their separation pondered by the ratio of their respective weights. If two locations have the same importance, their market boundary would be halfway between. On Figure 10.26, the market boundary between locations i and j (B_{ij}) is at a distance of 4.9 from i (and consequently at a distance of 2.1 from j).

The gravity model

The gravity model is the most common formulation of the spatial interaction method. It is named as such because it uses a similar formulation to Newton's law of gravity. Gravity-like representations have been applied in a wide variety of contexts, such as migration, commodity flows, traffic flows, commuting and evaluating boundaries between market areas. Accordingly, the attraction between two objects is proportional to their mass and inversely proportional to their respective distance. Consequently, the general formulation of spatial interactions can be adapted to reflect this basic assumption to form the elementary formulation of the gravity model:

$$T_{ij} = k \frac{P_i P_j}{D_{ij}}$$

- P_i and P_j: Importance of the location of origin and the location of destination.
- D_{ij}: Distance between the location of origin and then location of destination.
- k is a proportionality constant related to the rate of the event. For instance, if the same system of spatial interactions is considered, the value of k will be higher if interactions were considered for a year comparatively to the value of k for one week.

Thus, spatial interactions between locations i and j are proportional to their respective importance divided by their distance. The gravity model can be extended to include several calibration parameters:

$$T_{ij} = k \frac{P_i^\alpha P_j^\lambda}{D_{ij}^\beta}$$

P, d and k refer to the variables previously discussed.

- β (beta): A parameter of transport friction related to the efficiency of the transport system between two locations. This friction is rarely linear as the further the movement the greater the friction of distance. For instance, two locations serviced by a highway will have a lower beta index than if they were serviced by a road.
- λ (lambda): Potential to generate movements (emissiveness). For movements of people, lambda is often related to an overall level of welfare. For instance, it is logical to infer that for retailing flows, a location having higher income levels will generate more movements.
- α (alpha): Potential to attract movements (attractiveness). Related to the nature of economic activities at the destination. For instance, a center having important commercial activities will attract more movements.

A significant challenge related to the usage of spatial interaction models, notably the gravity model, is related to their calibration. Calibration consists in finding the value of each parameter of the model (constant and exponents) to ensure that the estimated results are similar to the observed flows. If it is not the case, the model is almost useless as it predicts or explains little. It is impossible to know if the process of calibration is accurate without comparing estimated results with empirical evidence.

In the two formulations of the gravity model that have been introduced, the simple formulation offers a good flexibility for calibration since four parameters can be modified. Altering the value of beta, alpha and lambda will influence the estimated spatial interactions (Figure 10.27). Furthermore, the value of the parameters can change in time due to factors such as technological innovations, new transport infrastructure and economic development. For instance, improvements in transport efficiency generally have the consequence of reducing the value of the beta exponent (friction of distance). Economic development is likely to influence the values of alpha and lambda, reflecting a growth in mobility.

Variations of the beta, alpha and lambda exponents have different impacts on the level of spatial interactions. For instance, the relationship between distance and spatial interactions will change according to the beta exponent. If the value of beta is high

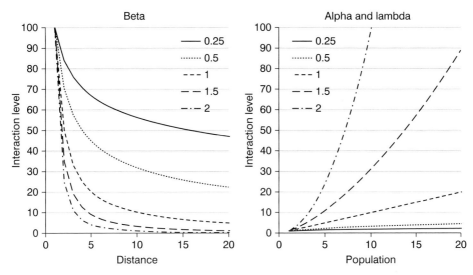

Figure 10.27   Effects of beta, alpha and lambda on spatial interactions

(higher than 0.5), the friction of distance will be much more important (steep decline of spatial interactions) than with a low value of beta (e.g. 0.25). A beta of 0 means that distance has no effects and that interactions remain the same whatever the concerned distance. Alpha and lambda exponents have the same effect on the interaction level. For a value of 1, there is a linear relationship between population (or any attribute of weight) and the level of interactions. Any value higher than 1 implies an exponential growth of the interaction level as population grows.

Calibration can also be considered for different O/D matrices according to age, income, gender, type of merchandise and modal choice. A part of the scientific research in transport and regional planning aims at finding accurate parameters for spatial interaction models. This is generally a costly and time-consuming process, but a very useful one. Once a spatial interaction model has been validated for a city or a region, it can then be used for simulation and prediction purposes, such as how many additional flows would be generated if the population increased or if better transport infrastructures (lower friction of distance) were provided.

Method 10 – Market area analysis

Market size and shape

Each economic activity has a location, but the various demands (raw materials, labor, parts, services, etc.) and flows each location generates also have a spatial dimension called a market area.

A **market area** is the surface over which a demand or supply offered at a specific location is expressed. For a factory it includes the areas to where its products are shipped; for a retail store it is the tributary area from which it draws its customers.

Transportation is particularly important in market area analysis because it impacts on the location of the activities as well as their accessibility. The size of a market area is a function of its threshold and range:

- **Market threshold.** Minimum demand necessary to support an economic activity such as a service. Since each demand has a distinct location, a threshold has a direct spatial dimension. The size of a market has a direct relationship with its threshold.
- **Market range.** The maximum distance each unit of demand is willing to travel to reach a service or the maximum distance a product can be shipped to a customer. The range is a function of transport costs, time or convenience in view of intervening opportunities. To be profitable, a market must have a range higher than its threshold.

Figure 10.28 considers a fairly uniform distribution of customers on an isotropic plain and a single market where goods and services may be purchased. If each customer is willing to purchase one unit per day and the market needs to sell 11 units per day to cover its costs (production or acquisition), then the threshold of the market would be the yellow circle of distance $D(T)$ from the market. However, 29 customers per day, including customers 1 and 2, patronize the market, of which an extra 18 are beyond the threshold distance $D(T)$. They contribute directly to the profitability of the market. The market range of all these customers is below distance $D(R)$. Beyond this range, customers are unwilling to go to the market, such as customer 3. There are different thresholds according to the variety of products or services that can be offered on a market. A threshold may be as low as 250 people for a convenience store or as high as 150,000 people for a theater. If the demand falls below the threshold level, the activity will run at a loss and will eventually fail. If the demand increases above the minimum, the activity will increase its profits, which may also lead to increased competition from new service activities. The frequency of use of goods or services is important in assessing the

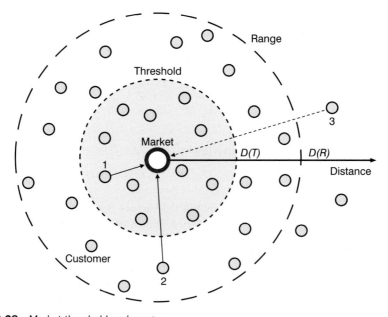

Figure 10.28 Market threshold and range

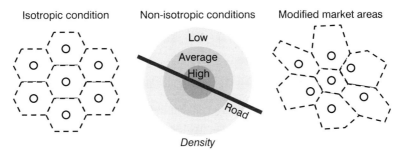

Figure 10.29 Non-isotropic conditions and the shape of market areas

extent of the market threshold, which is often linked to the level of income. A movie theater needing 500 visitors per night will require a threshold population of around 150,000 if the average number of visits is one per year. But, if the average number of visits is three per year, the population threshold drops to 50,000. Three movie theaters instead of one can be supported by the same population.

In the case of a single market area its shape in an isotropic plain is a simple concentric circle having the market range as radius. Since the purpose of commercial activities is to service all the available demand, when possible, and that the range of many activities is limited, more than one location is required to service an area. For such a purpose, a hexagonal-shaped structure of market areas represents the optimal market shape under a condition of isotropy. This shape can be modified by non-isotropic conditions mainly related to variations in density and accessibility.

Under isotropic conditions each market has the same polygonal evenly spaced area (Figure 10.29). This theoretical condition is obviously rarely found in reality. The two most important non-isotropic conditions impacting on the shape of market areas are differences in density (depicted here as concentric circles) and accessibility (depicted here as a road). The modified market areas are the possible outcome of non-isotropic conditions.

Economic definition of a market area

A market depends on the relationships between supply and demand. It acts as a price fixing mechanism for goods and services. Demand is the quantity of a good or service that consumers are willing to buy at a given price. It is high if the price of a commodity is low, while in the opposite situation – a high price – demand would be low. Outside market price, demand can generally be influenced by the following factors:

- **Utility**. While goods and services that are necessities (such as food) do not see much fluctuation in demand, the demand for items deemed of lesser utility (even frivolous) would vary according to income and economic cycles.
- **Income level**. Income, especially disposable income, is directly proportional with consumption. A population with a high income level has much more purchasing power than a population with a low income.
- **Inflation**. Involves an increase in the money supply in relation to the availability of assets, commodities, goods and services. Although it directly influences prices, inflation is outside the supply–demand relationship and decreases purchasing power, if wages are not increased accordingly.

- **Taxation**. Sale and value-added taxes can have an inhibiting effect on sales of goods and services as they add to the production costs and claim a share of consumer income.
- **Savings**. The quantity of capital available in savings can provide a potential to acquire consumption goods. Also, people may restrain from consuming if saving is a priority, namely in periods of economic hardship.

Supply is the amount of goods or services that firms or individuals are able to produce taking account of a selling price. Outside price, supply can generally be influenced by the following factors:

- **Profits**. Even if the sales of a product are limited, if profits are high an activity providing goods or services may be satisfied with this situation. This is particularly the case for luxury goods. If profits are low, an activity can cease, thus lowering the supply.
- **Competition**. Competition is one of the most important mechanisms for establishing prices. Where competition is absent (an oligopoly), or where there is too much (over-competition), prices artificially influence supply and demand.

According to the market principle, supply and demand are determined by the price, which is an equilibrium between both. It is often called equilibrium price or market price. This price is a compromise between the desire of firms to sell their goods and services at the highest price possible and the desire of consumers to buy goods and services at the lowest possible price.

For many economists, the market is a point where goods and services are exchanged and does not have a specific location, since it is simply an abstraction of the relationships between supply and demand. It is important to nuance this reasoning since most of the time consumers must move in order to acquire a good or service. The producer must also ship a commodity to a place where the consumer can buy it, whether at the store or at his/her residence (in the case of online shopping). The concept of distance thus must be considered concomitantly with the concept of market. In those conditions, the real price includes the market price plus the transport price from the market to the location of final consumption.

Competition over market areas

Competition involves similar activities trying to attract customers. Although the core foundation of competition for a comparable good or service is price, there are several spatial strategies that impact the price element. The two most common are:

Market coverage. Activities offering the same service will occupy locations in order to offer goods or services to the whole area. This aspect is well explained by the central place theory and applies well for sectors where spatial market saturation is a growth strategy (fast food, coffee shops, etc.). The range of each location will be a function of customer density, transport costs and the location of other competitors.

Range expansion. Existing locations try to expand their ranges in order to attract more customers. Economies of scale resulting in larger retail activities are a trend in that direction, namely the emergence of shopping malls. Taken individually, each store would have a limited range. However, as a group they tend to attract additional customers from wider ranges for many reasons. First, a complementarity of goods or services is offered. A customer would thus find it convenient to be able to buy clothes, shoes and personal care products at the same location. Second, a diversity of the same goods or services is offered (more choice) even if they compete between one another. Third, other related amenities are provided such as safety, food, indoor walking space, entertainment and also parking space.

Making market area competition models operational has been the object of numerous approaches. Initial work undertaken in the first half of the twentieth century focused on simple market competition (Hotelling's law), which was the foundation of market area analysis by considering factors such as retail location and distance decay. It is assumed that the product was uniform so customers would buy from the most convenient location (nearest seller) and that the friction of distance was linear and isotropic (Figure 10.30). The total price for the customer is thus the market price plus the transport price (time or effort spent to go to the market). Under such circumstances, two competitors will select locations A and B for optimal market coverage. With P1 being the market price, the market boundary would be F1 (point of cost indifference) since right of F1, customers would get a lower price at location B instead of at location A and left of F1, customers would get a lower price at location A. If, for any reasons, location A is able to lower the market price from P1 to P2, then its market area would expand at the expense of location B, from F1 to F2.

Later, factors such as market size were taken into consideration (Reilly's law) enabling complex market areas to be built. Since market areas are often non-monopolistic, this factor was included with market areas becoming ranges of probabilities that customers will attend specific locations (Huff's law). The purpose of Reilly's law of retail gravitation (1931) is to find a point of indifference between two locations, so the trading area of each can be determined. This point is assumed to be a function of the distance between two locations pondered by their respective size (population often used for this purpose). A location can thus be more attractive than the other. For instance, on Figure 10.31 two locations are 75 km apart. According to the Hotelling principle, the point of indifference should be halfway in between (35 km). However, since location A has a larger population (more weight), it is assumed that it will draw more customers. Under such circumstances, the point of indifference is 45.9 km away from location A.

Huff's retail model (1963) assumes that customers have a choice to patronize a location in view of other alternatives. Thus, a market area is expressed as a continuous line of probabilities, unless there are no other alternative locations. The point of indifference becomes

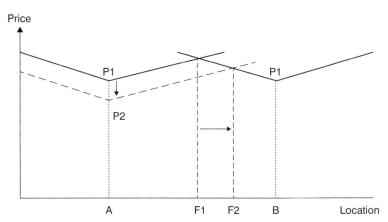

Figure 10.30 Hotelling's principle of market competition

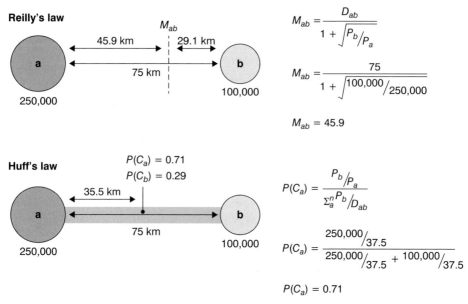

Figure 10.31 Reilly's and Huff's laws

the point of equal probability that a customer will patronize one location or another. On Figure 10.31, a customer has a greater chance (0.71) to patronize location A at the midpoint than patronize location B (0.29). The advantage of Huff's retail model is that it leaves room for customer choice. Although market areas are particularly relevant for retail analysis, the methodology also applies to time-dependent activities, such as freight distribution since distribution centers are located to service specific national or regional markets.

Geographic information systems and market areas analysis

Geographic information systems (GIS) have become fundamental tools to evaluate market areas, especially in retailing. With basic data, such as a list of customers and their addresses (or ZIP codes), it is relatively simple to evaluate market areas with a reasonable level of accuracy, a task that would have been much more complex beforehand. With GIS, market area analysis left the realm of abstraction to become a practical tool used by retailers and service providers in complex real world situations. In the spatial representation of a GIS, the market area is a polygon which can be measured and used to perform operations such as intersection (zones of spatial competition) or union (area serviced). Among the major methods a GIS can use to evaluate market areas are (Figure 10.32):

> **Buffer creation**, a common GIS procedure, associates each concentric circle with a distance or a time value. They can include the threshold and the range of a store. Another dimension of this method concerns Thiessen polygons where the market area is calculated as halfway distances from the location of other competitors.
> **Share by polygon** can be estimated as an aggregation of individual customers within a geographical unit of reference (ZIP code, census bloc, etc.) or a statistical calculation based on a set of representative variables, such as distance, population, income and age. A **star map** is a vector creation where segments have different origins (one for each customer) but the destination is the same (the store). It depicts a market area as a set

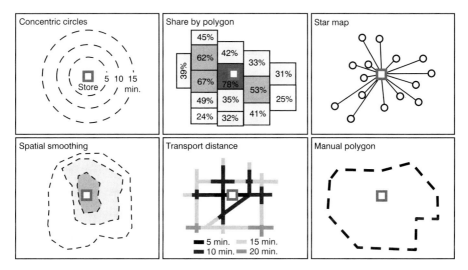

Figure 10.32 GIS Methods to estimate market areas

of customers connected to a store. Qualitative and quantitative attributes can be attached to each vector, such as frequency or sales.

Spatial smoothing is the outcome of statistical modeling that interpolates data from a known set of points (customers) to a continuous surface. The density of customers thus becomes a statistical surface expressing the market area.

Transport distance measures the accessibility, expressed in distance or time, of road segments to the store. It takes into consideration the different capacities of road segments (number of lanes, driving speed, turn penalties, etc.), which may be quite different from Euclidean distances such as those calculated by buffer creation. A new layer is created where each former road vector is segmented according to distance/time decay through a routing procedure that originates from the store.

Manual polygons are created with tracing where the analyst evaluates a market area from a set of assumptions, often based on specific expertise and empirical knowledge about that market. For instance, the analyst may empirically know that for various reasons few customers may be coming from a nearby neighborhood, excluding it from the market area. It may also been known that few customers are coming from further away than a specific street, making that street a boundary for the market area.

Method 11 – The policy process

Author: Brian Slack

Problem definition

Policies are developed in response to the existence of a perceived problem or an opportunity; they never exist in a vacuum. The context is extremely important because it will shape the actions being considered. For example:

Who has **identified the problem**? Is it widely recognized by society as a whole or is limited in scope to a local pressure group for example? In the case of the former there may be a greater willingness to intervene than in the latter, depending on the political power exerted by the pressure group.

Do the public authorities **have the interest or will to respond**? There are usually many more problems than the policy makers are willing to address. Many issues remain unaddressed.

Do the public authorities wish to wield the **instruments necessary to carry out a policy response**? The problem may be recognized, but public authorities may have little ability to effect change. Such is the problem of many environmental problems that require global solutions.

What is the **timescale**? How pressing is the problem, and how long would a response take? Policy makers are notoriously prone to attempt only short-term interventions, since their mandates are usually of relatively short duration. Long-term issues may not attract policy makers because the results of any policy intervention may be decades away.

These questions lie at the heart of the need to correctly identify the problem or opportunity. No policy response is likely to be effective without a clear definition of the issue. The following elements need to be considered in defining a problem:

Who has **identified the problem**, and why should it be seen to be a problem? Many problems exist, but few are taken up because they are not brought before a wide audience.

Is there **agreement on the problem**? If there is no agreement that a problem exists, it is unlikely that a strong policy response will be forthcoming. Effective policies are more likely to be formulated if there is widespread recognition of a problem and its causes. A problem for the Kyoto Accord on global warming is that decision makers in the USA have not been convinced that the problem is due to human-induced carbon dioxide emissions.

Is it an issue that can be **addressed by public policy**? The price of oil is regarded by many as a problem, but individual countries have no power to affect the price of this commodity.

Is it **too soon to develop a policy**? This argument was used by the lobby in California that opposed stricter emission controls on vehicles in the early 1990s, based on the argument that the technology of alternative energy for vehicles was not sufficiently advanced.

Is the problem **seen differently by groups with different values**? Environmentalists see many transport issues differently than many other interest groups. Divergence of opinions may affect how the problem is addressed.

Is the problem **fully understood**? Do we know the causal relationships that may be necessary to provide a solution? Transport and development, the role of transport in climate change are issues around which there is a debate.

Can the **relationships** between the factors that make up the problem be **quantified**? Problem definition is better when it is possible to measure the scale and scope of the issues involved.

In defining the problem or opportunity and to help address the questions above, background studies are required. The state of affairs needs to be assessed which will identify the actors, the issues and the possible means that are available. It is also important to forecast trends in order to identify whether the issue is likely to change.

Policy objectives and options

The eventual success of a policy depends upon establishing clear goals. If there are multiple objectives they must be consistent. They must be flexible enough to change over time as the circumstances evolve. In simple terms the objectives must:

- Identify the present conditions and situation.
- Indicate what the goals are.
- Identify the barriers to achieving the goals.
- Identify what is needed from other agencies and the private sector.
- Determine how success will be judged and measured.
- Identify what steps are required to achieve success.

Having defined the problem and objectives, policy options must be formulated and evaluated. In many cases more than one solution has to be considered for policy adoption. The objectives may be realized in many different ways. Best practices from other jurisdictions may be considered, and all other possible solutions need to be considered. By evaluating the options it may be possible to identify the one that best meets the goals that have been established and at the same time is the best fit for local circumstances. These types of evaluations are referred to as ex ante, because the outcomes are being assessed even before the policy is put into practice. Although one can never completely anticipate the outcome of different prospective policy options, ex ante evaluations are capable of bringing to light what problems may develop when the preferred option is implemented. Thus, when the future policy is to be evaluated (ex post), problems of data, reporting and identification of success criteria may have been already anticipated and resolved through an earlier ex ante assessment.

Many types of evaluation methods are employed in both ex ante and ex post assessments. These include cost–benefit analysis, multi-criteria analysis, economic impact and Delphi forecasting. Because evaluation takes place at several of the steps in the policy process, it is now regarded as a critically important issue. New ideas involving managing the policy process include performance-based management, where evaluation is built into the entire process. It means in the policy process, a great deal of attention has to be paid to how the goals, results and benefits are to be measured. The selection of indicators has to be agreed upon by policy managers from the inception.

Policy implementation

The implementation of the selected option represents a critical aspect of the policy process. The most carefully crafted policy that is widely accepted by those it affects can flounder because of improper implementation. It is impossible to define an optimal implementation procedure because of the wide range of socioeconomic circumstances in which policies are applied, and also because of the diversity of policies themselves. However, a 10-step model of policy implementation can be considered:

1 Policies must **not face insurmountable external constraints**. By this is meant that the policy must not exceed the jurisdictional or constitutional limits of the agency. This is a common issue in federal states, where different transport modes may be under different jurisdictions. Other examples include cases where the transport issue cannot be resolved because of international borders. However, transnational agreements, especially within the European Union, have considerably reduced external constraints in transport policy implementation.

2 In implementing the policy there must be an **adequate time frame and resources**. The policy may be appropriate, but may fail because its implementation took longer or was more expensive than budgeted.

3 The implementing agency must have **adequate staff and resources** to carry out the policy. A growing problem with environmental legislation is that the agencies do not have the means to ensure guidelines and standards are enforced.

4 The premises of **policy and theory must be compatible**. At one time public ownership was seen as a valid policy alternative. Today it may be a valid option in theory in some circumstances, but is not politically acceptable.

5 **Cause-and-effect relationships** in the policy must be direct and uncluttered. A successful policy must be seen to be based on clear and unambiguous relationships. Complex policies are more likely to be misunderstood.

6 **Dependency relationships** should be kept to a minimum. If the agency in charge of implementing the policy has to rely on others to it carry out, the more fragmented will become the authority. The implementing agency will become more dependent on others with not necessarily the same interests.

7 The basic objectives of the policy need to be **agreed upon and understood**. All actors in the policy process must possess a clear understanding of the policy and what is required to carry it out. It goes without saying that all those involved must understand the policy and have knowledge about their roles in carrying it out. Information and training are essential elements in the policy process.

8 Tasks must be **specified in an appropriate sequence**. Implementation is a process with connected steps from conception to the end. If the steps are not carried out in the correct sequence the policy may fail. Difficulties may arise, for example, if evaluation is completed without the indicators of success being agreed upon beforehand, or if another agency is involved before necessary pre-conditions for its participation have been completed.

9 **Communication and coordination** need to be on the same wavelength. Those implementing the policy have to possess the same information base, have to interpret it in the same way and communicate well with each other.

10 There must be **compliance**. Those agencies involved in implementing the policy must work towards total compliance. Many times policies are formulated but their compliance is lacking (see 3 and 7 above).

Policy evaluation and maintenance

The implementation stage is not the final step in the policy process. The effectiveness of the policy needs to be assessed after a certain period of time, and steps must be taken to ensure that there are resources and means to maintain a successful policy. In the past, this tended to be overlooked, and after a while policies would be sidetracked by other newer initiatives. The long-term effect was the presence of many different policy initiatives frequently with conflicting goals. The result was that policies in place frequently conflicted with each other in terms of goals or implementation measures.

Ongoing program evaluation is thus central to the maintenance of policy. This has tended to be a difficult issue for managers who today find their programs being assessed by methods and data requirements that were never built into the policy initially. Performance-based management has become an essential tool in the policy process as a result. Under this system evaluation is built into all stages of the policy process, and indicators are agreed upon by the managers who carry out the programs as well as the units that undertake evaluation. If performance criteria are not met, then policy goals and objectives need to be revised, if not abandoned altogether. This step can be controversial since it is an indication that the policy has failed.

Bibliography

Arlinghaus, S. L., W.C. Arlinghaus, and F. Harary (2001) *Graph Theory and Geography: An Interactive View*, New York: John Wiley.

BTS (2001) Special Issue on Methodological Issues in Accessibility, *Journal of Transportation and Statistics*, 4(2/3), Bureau of Transportation Statistics, Sept./Dec.

Butler, J.A. (2008) *Designing Geodatabases for Transportation*, Redlands, CA: ESRI Press.

Colwell, P.F. (1982) "Central place theory and the simple economic foundations of the gravity model", *Journal of Regional Science*, 22(4): 541–6.

Environmental Protection Agency (1997) *Evaluation of Modeling Tools for Assessing Land Use Policies and Strategies*, EPA420-R-97-007, Ann Arbor, MI: EPA.

FHWA (2001) *Transportation Performance Measures Toolbox*, Operations, Federal Highway Administration, Washington, DC: US DOT.

Fotheringham, A.S. and M.E. O'Kelly (1989) *Spatial Interaction Models: Formulations and Applications*, London: Kluwer Academic.

Guimera, R. and L.A.N. Amaral (2005) "Cartography of complex networks: modules and universal roles", *Journal of Statistical Mechanics*, 2: 1–12.

Holmes, T.J. (2006) "The diffusion of Wal-Mart and economies of density", University of Minnesota, Department of Economics.

Huff, D.L. and G.F. Jenks (1968) "A graphic interpretation of the friction of distance in gravity models", *Annals of the Association of American Geographers*, 58(4): 814–24.

Isard, W. (1956) *Location and Space-Economy: A General Theory Relating to Industrial Location, Market Areas, Land Use, Trade, and Urban Structure*, Cambridge, MA: MIT Press.

Jiang, B. and Claramunt, C. (2004) "Topological analysis of urban street networks", *Environment and Planning B*, 31: 151–62.

Kansky, K. (1963) "Structure of transportation networks: relationships between network geography and regional characteristics", Research Papers 84, University of Chicago, Department of Geography.

Kuby, M. and N. Reid (1992) "Technological change and the concentration of the U.S. general cargo Ports System: 1970–88", *Economic Geography*, 68(3): 272–88.

Lo, C.P. and A.K.W. Yeung (2002) *Concepts and Techniques of Geographic Information Systems*, Upper Saddle River, NJ: Prentice Hall.

Miller, H.J. and S-L. Shaw (2001) *Geographic Information Systems for Transportation: Principles and Applications*, New York: Oxford University Press.

Nystuen, J.D. and M.F. Dacey (1961) "A graph theory interpretation of nodal regions", Regional Science Association, Papers and Proceedings 7: 29–42.

O'Kelly, M.E. (1998) "A geographer's analysis of hub-and-spoke networks", *Journal of Transport Geography*, 6(3): 171–86.

Oum, T.H., M.W. Tretheway and W.G. Waters (1992), "Concepts, methods and purposes of productivity measurement in transportation", *Transportation Research A*, 26A(6): 493–505.

Shaw, S-L. (2010) "Geographic information systems for transportation: from a static past to a dynamic future", *Annals of GIS*, 16(3): 129–40.

—— (2011) "Guest Editorial – Geographic information systems for transportation: an introduction", *Journal of Transport Geography*, 19(3): 377–8.

Taaffe, E., H.L. Gauthier and M.E. O'Kelly (1998) *Geography of Transportation*, 2nd edition, Upper Saddle River, NJ: Prentice Hall.

Thill, J.C. (ed.) (2000) *Geographic Information Systems in Transportation Research*, Oxford: Elsevier Science.

Ullman, E.L. (1956) "The role of transportation and the bases for interaction", in W.L. Thomas Jr. et al. (eds) *Man's Role in Changing the Face of the Earth*, Chicago, IL: University of Chicago Press.

Victoria Transport Policy Institute (2011) "Defining, evaluating and improving accessibility", *Transport Demand Management Encyclopedia*, http://www.vtpi.org/tdm/tdm84.htm.

White, R. and G. Engelen (1993) "Cellular automata and fractal urban form: a cellular modelling approach to the evolution of urban land-use patterns", *Environment and Planning A*, 25(8): 1175–99.

Conclusion

Issues and challenges in transport geography

Authors: Jean-Paul Rodrigue, Claude Comtois and
Brian Slack

Transport geography seeks to understand the spatial organization of movements. It has emerged as a full-fledged field within geography with a strong propensity to include concepts and methods from other disciplines such as economics, engineering, environmental sciences and sociology. Because transportation systems are involved in a wide variety of scales and modes, from local public transit to global maritime shipping, the approach has a tendency to be partitioned. It is indeed difficult to reconcile perspectives such as pedestrian mobility issues related to land use or the selection of air cargo hubs. Irrespective of the scale and the mode, transport geography shares several common issues and challenges.

Congestion

The issue of congestion is likely to remain as one of great ongoing issues in transport geography because there are unprecedented demands for transportation being generated by a global economy that is ever more dependent upon the transport industry. The causes of congestion are well understood, even if the solutions are not. Congestion occurs across modes and locations and arises from two causes. Most important is when demand for mobility exceeds the capacity to support it. It can also occur when random events bring about a temporary disruption to service, such as an accident or a natural hazard such as flooding. In the case of the second set of causes, it is possible to mitigate their effects if the occurrence is frequent, such as accidents, or if the risks are great, as for example flooding in a flood plain. A common and attractive solution is to increase capacity. However, as has been shown, increasing capacity engenders a hidden demand, so that adding lanes to an expressway tends to attract even more circulation. Furthermore, demand is increasing ceaselessly, so that the practicality of this solution may be questioned.

The expected growth of the demand is likely to have major impacts on the nature and form of the future transport industry. In the short term at least, road transport is likely to continue its dominance of the transport industry. There are two basic reasons for this assertion. In the developed world automobiles and trucks already dominate the market, and the spatial patterns of people, industries and services have adjusted themselves somewhat to the demands of these modes. Such low-density, space-extensive patterns are pushing the traffic congestion ever further out, and make it very difficult for other higher capacity modes to compete. At the same time the demand for mobility is growing as a result of the rapid industrialization in developing countries such as China and India. There too a modal shift is occurring in favor of road transport. Increasing prosperity in these countries represents a great potential for growth in road transport.

Congestion is not limited to internal urban-generated traffic. International trade is likely to continue to be dominated by maritime transport (in terms of weight) and air transport (in terms of value). This has already led to concentration of traffic at a relatively small number of gateways and hubs, which are capable of extracting scale economies. For example, the 20 largest container ports handled more than 49 percent of global traffic in 2010. The traffic concentration, however, is already producing capacity problems in many of these hubs. International trade has grown at a rate faster than economic growth as measured by GDP in recent decades and there are expectations that hub congestion will remain an issue in the future.

For geographers there are a whole range of issues arising out of the growth of demand and the paralysis of congestion. Here, they are grouped into two categories. First are a series of questions surrounding how to provide solutions, second are the effects on future spatial patterns. In the past the solution to congestion was to provide more capacity by building more infrastructure. Such a response depended heavily on engineering solutions. As has been learned over the last few decades, the model of "predict and accommodate" has not worked well. It is now recognized that a multidisciplinary approach is required. It is recognized that there will still be a heavy reliance on engineering skills to design and construct infrastructure and systems, and to develop further technological innovations. However, transport policy and planning requires a broader perspective, one that considers different goals and alternatives, responds to different needs for mobility, and that seeks ways to manage demand.

Congestion is a phenomenon that is spatially bound. It takes place in specific locations with impacts at a multitude of scales, from a particular highway intersection that may delay traffic over a few hundred meters, to blockage in a port that may disrupt the flow of goods over half a continent. Each event produces a spatial response, from the car driver who searches out an alternative route to the shipper who selects a different mode or point of entry for succeeding shipments. Increased demand and the rising likelihoods of congestion will intensify new spatial responses and thus it appears very likely that new spatial flows and structures will come into being. They involve:

- **Demand management**. Concerns the conditions and in what types of locations travel demand can be modified. In a market context, when supply is fixed and demand increases, an upward price adjustment inevitably takes place. This is common in maritime and air transport with yield management strategies. However, many transport infrastructures, such as roads, are provided free of access, implying no cost changes as congestion levels increase. There is thus a growing need to provide incentives (or disincentives) and reassess the priority in the use of infrastructure, particularly in urban areas.
- **Concentration versus deconcentration**. Accessibility and infrastructure improvements usually lead to a concentration of activities, while congestion acts as a counteracting force to concentration since it creates various diseconomies. Already there is evidence of deconcentration in air transport by the growth in passengers and freight in some smaller airports. The density of economic and social activities and the related intensity of transport use imply a balance between the forces of concentration and deconcentration.
- **Economic and social impacts**. In a context where transport networks are increasingly synchronized congestion can create multiplying effects not just impacting costs, but also the reliability of transport systems. The economic and social impacts

of congestion remain a salient issue, particularly in developing countries where it can impede economic growth.

- **Passengers versus freight**. Congestion also raises the issue of the prioritization of passengers versus freight when they share transport infrastructure or when freight activities such as terminals or distribution centers are in proximity to locations where large numbers of passengers transit. This will require a careful assessment of the respective costs of congestion on specific passenger and freight transport systems and in which circumstances congestion exerts the most externalities. A salient issue concerns how freight distribution could be better integrated in the urban environment where passenger movements tend to dominate: the realm of city logistics.

Infrastructure

Regardless of the specific solutions to congestion that are considered, increasing demand is placing unprecedented requests for investment in transport infrastructures. A major question confronting all countries around the world is how to finance the construction and maintenance of transport infrastructures. As economies of scale are applied to transport systems, such as larger container ships or double-stacked rail corridors, capital requirements increase in proportion. Governments have traditionally been the primary source of funding in the transport sector, but the costs of keeping pace with the growth in demand are making it difficult for even the richest countries to countenance public funding on the scale required to meet expectations about the mobility of passengers and freight.

Capital requirements are particularly prevalent on both sides of the infrastructure life cycle spectrum. The highways in China and North America represent two salient cases. For China, the last decade has seen an impressive level of highway construction with the setting of a national highway network which totaled more than 85,000 km in 2012, the longest in the world. Comparatively, the American Interstate highway system of about 75,000 km is nearing a phase in its life cycle where a substantial amount of capital investment will be required to upgrade the system and maintain its operability, including thousands of aging highway bridges. While most of the Interstate is publicly funded, almost all Chinese highways were funded by private interests that are using tolls to recover their investments. Irrespective of the context, the issue of the role of private and public actors in transport infrastructure as well as pricing mechanisms will remain salient:

- **Public–private partnerships** and completely private enterprises are one set of solutions. For many developing countries this is the only solution, since public finances are inadequate to the task. Thus, in the future, a greater private involvement in the provision of transport infrastructure is to be expected. Several models are already well tested: BOT (Build–Operate–Transfer), where the private sector builds and operates a facility or system for a period of time, but then transfers it back to the government after an agreed period; BLT (Build–Lease–Transfer) where after building the facilities, it is leased for a fixed period for operation, and finally transferred back; ROT (Rehabilitate–Operate–Transfer) where the private party refurbishes an existing facility to be operated for a term prior to being turned back to the state.
- **Pricing**. Another approach that is gaining momentum is charging for use of transport infrastructure. Several segments of the transport system are privately owned and

operated, such as maritime shipping and air transportation, implying that pricing is generally set by market forces. Still, many transport infrastructures such as roads and airports are wholly or partially owned by the public sector. Pricing is becoming an important feature of transport planning in urban areas where common-use transport infrastructures are under stress. Whether it is cordon pricing, congestion pricing or tolling, drivers are being forced to pay for their use of roads and limited price elasticity has been observed so far. With growing concerns over the environment, charging for the externalities of transport modes is becoming a reality in many jurisdictions. It remains to be seen how effective these alternatives are and their effects on travel behavior.

The difficulties are not to be underestimated, however. Most transport infrastructure projects are long term, but are typified by the heaviest capital investment requirements being incurred over a short initial phase. Even if transport infrastructure can be built in phases most private enterprises cannot take a long-term perspective, because they need to cover their expenses and recover their capital investments over short periods of time. With the growing unwillingness or inability of the public sector to fund and provide transport infrastructure, new forms of infrastructure provision, maintenance and operation need to be achieved. This is where the financial sector, particularly long-term investment funds (such as pension funds) can be involved with a better synchronism between capital and time horizons of transport infrastructure projects.

Sustainability challenges

The issue of sustainability has become an increasing important consideration for the transport industry. It is now broadly recognized that there needs to be a balance between economic efficiency, social factors and the environment. Of these three, the issue of economic efficiency has always been to the forefront, and governments have been important in regulating social conditions (safety, security and working conditions). Despite the strong historic relationships between transport and the environment, the latter has tended to be overlooked by the industry. This is changing, and environmental issues are likely to play an ever more important role in the transport industry, particularly over four core dimensions:

- **Transport and climate change**. Transportation both influences and is impacted by climate change. Transport activities, particularly vehicles, account for 24 percent of carbon dioxide emissions worldwide. They are thus subject to regulatory pressures to improve their environmental performance in regard to the greenhouse gases they emit. Concomitantly, transportation activities can be negatively impacted by climate change. Severe weather occurrences have a disruptive effect on transport systems, particularly for air transportation which has become a crucial element of global and regional mobility. Potential impacts on infrastructure need to be assessed since infrastructures are built with an expected life cycle and climate change may reduce it or increase maintenance costs. The prospects of sea level rises are particularly problematic for coastal transport systems. The extent to which climate change is influenced by and will impact global transport systems, modes and terminals, needs to be seriously considered.
- **Transport and atmospheric pollution**. Air quality standards are being implemented with increasing rigor in more and more countries around the world. There are still

striking differences between regions and between the modes. For example, most of the countries of the developing world still have to go a long way to fixing and enforcing standards. However, the trend is towards greater control over emissions, which will have effects on modes and their respective competitiveness, particularly if a mode is subject to a greater degree of legislation than another.

- **Transport and water quality**. The contribution of transport to the pollution of rivers and oceans is considerable, and is only recently being addressed by international legislation. Considerable progress has been made in a number of areas such as ballast water, waste and oil spills. As legislation increases in its comprehensiveness, the more the transport industry is impacted. This is particularly evident in matters relating to dredging, where environmental constraints are placing a growing financial burden on ports that are seeking to deepen channels in order to keep pace with the growth of vessel size. Therefore it can be expected that these constraints may impact port competitiveness.
- **Transport and land take**. Increased demand for transport is already placing enormous pressures on the need for new infrastructures. Many of these transport facilities such as airports and ports require very large amounts of land for their own internal operations and for the external transport links that have to be provided. Rapid motorization in developing countries has resulted in the conversion of land to provide road infrastructure. This expanded scale of transport infrastructure questions the capacity of environmental systems to mitigate the disruptions and will likely have an impact on how transport infrastructure is designed.

Management of transport systems

The transportation industry is changing so significantly in form and function that the very important changes in the way it is organized and managed tend to be overlooked. Yet it is through different management practices that the spatial manifestations of the industry are expressed. It is perhaps easiest to see the changes in management through the lens of governance, where an industry that used to be largely managed and controlled by the public sector has become increasingly controlled by the private sector. The privatization of transport companies and infrastructures has been an important feature of the last decades, and is likely to continue further into the present century. However, there are still many issues about the role of the public sector in transportation, and deregulation which has prevailed could be reversed.

The growing role of the private sector over an industry that is becoming global and multi-functional has necessitated a shift in management and ownership relationships that are still evolving. They include:

- The emergence of **horizontally linked global corporations** that through a series of acquisitions and mergers have bought up similar operating companies in different markets. A good example is the global port terminal operators.
- The development of **vertically integrated corporations** that have grown by merger and acquisition to control several segments of the transport chain, namely modes and terminals.
- **Intermediaries** that provide transport services on a global scale, without direct ownership of infrastructure. Third-party logistics provider (3PL) companies operate in many markets and are major actors in the transport chain.

● **Alliances**, informal groupings of transport providers that pool resources and offer joint services between major global markets as the partners combine their respective regional networks.

At the same time transport is being increasingly integrated in global production systems. It is becoming an integral part of production and distribution chains. These management and business structures give rise to distinct patterns of spatial organization, with different operating practices. The operational interests of a vertically integrated enterprise is different than one horizontally linked. This highlights the need to understand the nature of the organization of the businesses involved in transport as a means of explaining existing patterns and predicting their future forms. The concentration of traffic (and resultant congestion) is as much explained by the organization of transport firms as it is by traditional explanations involving demand and capacity. In turn, the organization of the global firms themselves is shaped by conditions of local markets. A distinct geography of transport firms exists, a geography that is still ill understood.

Energy, safety and security

The macroeconomic and policy environment in which the transport sector evolves has substantially changed in recent years, bringing concerns over energy efficiency, safety and security, which before were rather secondary. With the significant increase in energy prices that took place in recent years, significant adjustments in transport modes may be expected. While technologies may make alternative fuel vehicles a commercial option to the internal combustion engine, the main question is the effect of higher prices on automobiles and trucks. As the costs are passed on to users, global production and distribution systems that depend upon cheap transport will be impacted. Energy, particularly the availability of oil, has been a salient factor in the development of transport systems. It is expected that an ongoing shift in the energy price structure as well as the commercial availability of alternate energy sources will cause a transition to more energy efficient modes such as rail or maritime shipping. It remains to be seen which forms of transport and mobility will take shape as the energy transition away from fossil fuels takes place.

Transportation safety issues are somewhat paradoxical. On one hand transportation modes and terminals are incrementally becoming safer as accident rates are declining. This is particularly the case for air transportation, in which safety performance has steadily improved in spite of a substantial growth of passengers being carried. Similar trends are observed for road transportation, particularly in developed countries, as fatality rates have declined. However, road transport safety remains a salient issue in developing countries where vehicle ridership is increasing and where enforcement of safety regulations is lacking. An enduring issue is therefore in light of growing mobility levels to ensure that transportation safety continues to improve through better modal and infrastructure design, operational practices and the enforcement of existing regulations.

Another prevalent matter concerns security practices that are now part of the business environment in which passenger and freight transport systems are evolving. Most of these measures are imposed by regulatory agencies and their consequences are often difficult to assess, but always involve additional costs and delays for transport operators. A balance between security measures and the efficient flow of passengers and freight will need to be achieved through a variety of regulatory, operational and technological innovations. The years since September 11, 2001, have clearly shown that security issues in transportation have played a more prevalent role.

Prospects for transport geography

Transport geography played a relatively small role in the field of transport studies, a field that has been dominated by engineers and economists. This was due in part to the needs of the industry being focused on providing infrastructures and technologies, at what costs and benefits and at what level of pricing. The contemporary industry is much more complex, with issues as varied as safety, aesthetics, working conditions, equity, deprivation, the environment and governance being necessary considerations. A much broader set of skills are required therefore, and transport studies have become a multidisciplinary field of application. Transport geography thus has opportunities to contribute to transport studies, transport planning and transport operations, in part because of the breadth of the approach and training. Still, transport geography, like the field of transportation in general, does not receive a level of attention in academia proportional to its economic and social importance.

It is also a fundamental fact that transport is a spatial activity. It has always been a space-adjusting service, but over the last few decades it has become increasingly global in scope. Contemporary transport operates at a wider range of scales than ever before. There are complex interactions between the local and the global. For example, the issues surrounding the expansion of an airport are usually decided at the local level, and the impacts are likely to be felt locally, namely its externalities such as noise and congestion. However, the effects on passenger and freight flows may have a global impact. The spatiality of transport and the many scale levels at which it operates are elements that are the particular concerns of transport geography. No other discipline has as its core interest the role of space in shaping human activities. The globalization of transport activities thus has represented unique opportunities in the development of transport geography.

One reason for the success of engineers and economists in transport studies and applications is that their training has been rigorous in the application of mathematics and multivariate statistics. They have demonstrated the ability to provide precise answers to the questions that decision makers have required – what to build, at what cost, with what cost effects. There has evolved a culture in the transport industry that unless it can be quantified it is of little value. Transport geography provides the quantitative skills in modeling, graph theory and multivariate statistics. However, there are newer techniques that provide geographers with opportunities to contribute to transport studies. GIS-T in particular should be an essential element in transport geography training. The multi-scalar, multivariate nature of the transport industry makes GIS-T an invaluable tool, and one that will raise the profile of transport geography in the transportation industry.

One of the key challenges in transport studies is data availability. Many times census and survey data are inadequate or unavailable in the form required. However, online availability of large datasets is increasingly offering a richer array of information to analyze transport issues. Knowledge of survey techniques and their limitations are also an important part of the transport geography toolkit. Many of the traditional tools and approaches are still relevant. They address problems that are frequently overlooked by other disciplines because of the lack of data or the inability to spatially represent these data. Questionnaires and interviews represent a vital source of information in many situations. Content analysis is extremely useful in providing quantified data from non-quantified sources. At the same time, fieldwork provides the opportunity to obtain detailed understanding of the particularities of the local conditions that cannot be obtained from reading texts and official documents.

A look back at the subject matter and topics covered in this book indicates an industry that is growing in significance and changing in the face of challenges and drivers of change. The issues that are achieving greater importance – sustainability, congestion, governance and management – are ones to which transport geography has the opportunity to contribute. As the transport industry becomes more complex, old approaches, focusing on a narrow range of factors, have to be replaced by more nuanced analysis and solutions. In the transport industry itself, in public planning, and in research institutions, the scope for transport geography is encouraging.

 **Glossary**

Many of the glossary terms are adapted from the Bureau of Transportation Statistics, the European Conference of Ministers of Transport, the Economic Commission for Europe, the Intermodal Association of North America and the Mineta Transportation Institute.

Access

The capacity to enter and exit a transport system. It is an absolute term implying that a location has access or does not.

Accessibility

The measure of the capacity of a location to be reached by, or to reach, different locations. The capacity and the structure of transport infrastructure are key elements in the determination of accessibility.

Aerodrome

A defined area on land or water (including any buildings, installations and equipment) intended to be used either wholly or in part for the arrival, departure and movement of aircraft. Aerodromes may include airports, heliports and other landing areas.

Aframax

A tanker of standard size between 75,000 and 115,000 dwt usually carrying half a million barrels of oil. The largest tanker size in the AFRA (Average Freight Rate Assessment) tanker rate system.

Agglomeration economies

See economies of agglomeration.

Air cargo

Total volume of freight, mail and express traffic transported by air. Includes the following: freight and express commodities of all kinds, including small package counter services, express services and priority reserved freight.

Air carrier

Commercial system of air transportation, consisting of domestic and international scheduled and charter service.

Air space

The segment of the atmosphere that is under the jurisdiction of a nation or under an international agreement for its use. They include two major components, one being land-based (takeoffs and landings) and the other air-based, mainly composed of air corridors. These corridors can cover altitudes up to 22,500 meters. Most commercial air transport services are limited to the use of predetermined corridors.

Air transportation

Includes companies that provide domestic and international passenger and freight services, and companies that operate airports and provide terminal facilities.

Airport

(1) An area of land or water that is used or intended to be used for the landing and takeoff of aircraft, and includes its buildings and facilities, if any. (2) A facility used primarily by conventional, fixed-wing aircraft. (3) A facility, either on land or water, where aircraft can take off and land. Usually consists of hard-surfaced landing strips, a control tower, hangars and accommodations for passengers and cargo. (4) A landing area regularly used by aircraft for receiving discharging passengers or cargo.

Alternative fuels

Low-polluting fuels which are used to propel a vehicle instead of high-sulfur diesel or gasoline. Examples include methanol, ethanol, propane or compressed natural gas, liquid natural gas, low-sulfur or "clean" diesel and electricity.

Amtrak

Operated by the National Railroad Passenger Corporation of Washington, DC. This rail system was created by President Nixon in 1970, and was given the responsibility for the operation of intercity, as distinct from suburban, passenger trains between points designated by the Secretary of Transportation.

Arterial street

A major thoroughfare, used primarily for through traffic rather than for access to adjacent land, that is characterized by high vehicular capacity and continuity of movement.

Association of Southeast Asian Nations (ASEAN)

Free trade area established on August 8, 1967, in Bangkok, Thailand, with the signing of the Bangkok Declaration. The members of ASEAN are Brunei Darussalam, Indonesia, Laos, Malaysia, Myanmar, Philippines, Singapore, Thailand and Vietnam. The Secretariat of the Association is located in Jakarta, Indonesia.

Average Vehicle Occupancy (AVO)

The number of people traveling by private passenger vehicles divided by the number of vehicles used.

Average Vehicle Rideship (AVR)

The ratio of all people traveling by any mode, including cars, buses, trains and bicycles (or telecommuting), in a given area during a given time period to the number of cars on the road. A key measure of the efficiency and effectiveness of a transportation network – the higher the AVR, the lower the level of energy consumption and air pollution.

Back haul

Traffic for the return movement of a car or container towards the point where the initial load originated or to handle a shipment in the direction of the light flow of traffic.

Balance of payments

A record of receipts from and payments to the rest of the world by a country's government and its residents. The balance of payments includes the international financial transactions of a country for commodities, services and capital transactions.

Balance of trade

The difference between a country's total imports and exports. If exports exceed imports, a positive balance of trade exists.

Baltic Dry Index (BDI)

Assessment of the average price to ship raw materials (such as coal, iron ore, cement and grains) on a number of shipping routes and by ship size. It is an indicator of the cost paid to ship raw materials on global markets and an important component of input costs. As such, the index is considered as a leading indicator (forward looking) of economic activity since it involves events taking place at the earlier stages of global commodity chains.

Barge

A non-motorized water vessel, usually flat-bottomed and towed or pushed by other craft, used for transporting freight. Dominantly used on river systems.

Barrel

A unit of volume equal to 42 US gallons (or 159 liters) at 60 degrees Fahrenheit, often used to measure volume in oil production, price, transportation and trade.

Base fare

The price charged to one adult for one transit ride; excludes transfer charges, zone charges, express service charges, peak-period surcharges and reduced fares.

Base period

The period between the morning and evening peak periods when transit service is generally scheduled on a constant interval. Also known as "off-peak period". The time of day during which vehicle requirements and schedules are not influenced by peak-period passenger volume demands (e.g. between morning and afternoon peak periods). At this time, transit riding is fairly constant and usually low to moderate in volume when compared with peak-period travel.

Berth

A specific segment of wharfage where a ship ties up alongside at a pier, quay, wharf or other structure that provides a breasting surface for the vessel. Typically, this structure is a stationary extension of an improved shore and intended to facilitate the transfer of cargo or passengers.

Bill of lading

A document that establishes the terms of a contract between a shipper and a transportation company. It serves as a document of title, a contract of carriage and a receipt for goods.

Block

A group of rail cars destined to the same location.

Break-bulk cargo

Refers to general cargo that has been packaged in some way with the use of bags, boxes or drums. This cargo tends to have numerous origins, destinations and clients. Before containerization, economies of scale were difficult to achieve with break-bulk cargo as the loading and unloading process was very labor-intensive and time-consuming.

British Thermal Unit (BTU)

The amount of energy required to raise the temperature of 1 pound of water 1 degree Fahrenheit (F) at or near 39.2 degrees F and 1 atmosphere of pressure.

Bulk cargo

Refers to freight, both dry or liquid, that is not packaged such as minerals (oil, coal, iron ore) and grains. It often requires the use of specialized ships such as oil tankers as well as specialized transshipment and storage facilities. Conventionally, this cargo has a single origin, destination and client. It is also prone to economies of scale.

Bulk carriers

All vessels designed to carry bulk cargo such as grain, fertilizers, ore and oil.

Bulk terminal

A purpose-designed berth or mooring for handling liquid or dry commodities, in unpackaged bulk form, such as oil, grain, ore and coal. Bulk terminals typically are installed with specialized cargo-handling equipment such as pipelines, conveyors, pneumatic evacuators, cranes with clamshell grabs, and rail lines to accommodate cargo-handling operations with ships or barges. Commodity-specific storage facilities such as grain silos, petroleum storage tanks and coal stock yards are also located at these terminals.

Bus (motorbus)

Any of several types of self-propelled vehicles, generally rubber-tired, intended for use on city streets, highways and busways, including but not limited to minibuses, forty- and thirty-foot buses, articulated buses, double-deck buses and electrically powered trolley buses, used by public entities to provide designated public transportation services and by private entities to provide transportation services including, but not limited to, specified public transportation services. Self-propelled, rubber-tired vehicles designed to look like antique or vintage trolleys are considered buses.

Bus, trolley

An electric, rubber-tired transit vehicle, manually steered, propelled by a motor drawing current through overhead wires from a central power source not on board the vehicle. Also known as "trolley coach" or "trackless trolley".

Bus lane

A street or highway lane intended primarily for buses, either all day or during specified periods, but sometimes also used by carpools meeting requirements set out in traffic laws.

Bus stop

A place where passengers can board or disembark from a bus, usually identified by a sign.

Cable car

An electric railway operating in mixed street traffic with unpowered, individually controlled transit vehicles propelled by moving cables located below the street surface and powered by engines or motors at a central location not on board the vehicle.

Cabotage

Transport between two terminals (a terminal of loading and a terminal of unloading) located in the same country irrespective of the country in which the mode providing the service is registered. Cabotage is often subject to restrictions and regulations. Under such circumstances, each nation reserves for its national carriers the right to move domestic freight or passenger traffic.

Canal

An artificial open waterway constructed to transport water, to irrigate or drain land, to connect two or more bodies of water, or to serve as a waterway for watercraft.

Capesize

Refers to a rather ill-defined standard which has the common characteristic of being incapable of using the Panama or Suez canals, not necessarily because of their tonnage, but because of their size. These ships serve deepwater terminals handling raw materials, such as iron ore and coal. As a result, "Capesize" vessels transit via Cape Horn (South America) or the Cape of Good Hope (South Africa). Their size ranges between 80,000 and 175,000 dwt.

Carbon dioxide (CO_2)

A colorless, odorless, non-poisonous gas that is a normal part of the ambient air. Carbon dioxide is a product of fossil fuel combustion.

Carbon monoxide (CO)

A colorless, odorless, highly toxic gas that is a normal by-product of incomplete fossil fuel combustion. Carbon monoxide, one of the major air pollutants, can be harmful in small amounts if breathed over a certain period of time.

Carpool

An arrangement where two or more people share the use and cost of privately owned automobiles in traveling to and from prearranged destinations together.

Carrier

The company moving the passengers or freight.

Catchment area

Area or region whose economic, political, cultural, social, etc. influence is felt over a larger area; it is the radius of action of a given point. In transportation, it consists in the area under influence of a focal point towards which centripetal fluxes converge; an interception zone of several carriers. Also labeled as area of influence or hinterland.

Centrality

Focus on the terminal as a point of origin and destination of traffic. Thus, centrality is linked with the generation and attraction of movements, which are related to the nature and the level of economic activities within the vicinity of the concerned terminal. The function of centrality also involves a significant amount of intermodal activities.

Charter

Originally meant a flight where a shipper contracted hire of an aircraft from an air carrier, but has usually come to mean any non-scheduled commercial service.

City logistics

The means over which freight distribution can take place in urban areas as well as the strategies that can improve its overall efficiency, such as mitigating congestion and environmental externalities.

Class I railroad

An American railroad with an annual gross operating revenue in excess of $250 million based on 1991 dollars.

Clean Air Act (CAA)

Federal legislation that sets national air quality standards.

Coach service

Transport service established for the carriage of passengers at special reduced passenger fares that are predicated on both the operation of specifically designed aircraft space and a reduction in the quality of service regularly and ordinarily provided.

Coal

A black or brownish-black solid, combustible substance formed by the partial decomposition of vegetable matter without access to air. The rank of coal, which includes anthracite, bituminous coal, subbituminous coal and lignite, is based on fixed carbon, volatile matter and heating value. Coal rank indicates the progressive alteration, or coalification, from lignite to anthracite. Lignite contains approximately 9 to 17 million British Thermal Units (BTU) per ton. The heat contents of subbituminous and bituminous coal range from 16 to 24 million BTU per ton, and from 19 to 30 million BTU per ton, respectively. Anthracite contains approximately 22 to 28 million BTU per ton.

Cold chain

A temperature-controlled supply chain linked to the material, equipment and procedures used to maintain specific shipments within the appropriate temperature range. Often relates to the distribution of food and pharmaceutical products.

Combi

A type of aircraft whose main deck is divided into two sections, one of which is fitted with seats and one which is used for cargo.

Commercial geography

Investigates the spatial characteristics of trade and transactions in terms of their cause, nature, origin and destination. It leans on the analysis of contracts and transactions.

Commodity

Resources that can be consumed and having no qualitative differentiation. They can be accumulated for a period of time (some are perishable while others can be virtually stored for centuries), exchanged as part of transactions or purchased on specific markets (such as futures market). Some commodities are fixed, implying that they cannot be transferred, except for the title. This includes land, mining, logging and fishing rights. In this context, the value of a fixed commodity is derived from the utility and the potential rate of extraction. Bulk commodities are commodities that can be transferred, which includes for instance grains, metals, livestock, oil, cotton, coffee, sugar and cocoa. Their value is derived from utility, supply and demand (market price).

Commodity chain (supply chain)

A functionally integrated network of production, trade and service activities that covers all the stages in a supply chain, from the transformation of raw materials, through intermediate manufacturing stages, to the delivery of a finished good to a market. The chain is conceptualized as a series of nodes, linked by various types of transactions, such as sales and intrafirm transfers. Each successive node within a commodity chain involves the acquisition or organization of inputs for the purpose of added value.

Common carrier

A transportation company engaged in the business of handling persons or freight for compensation and for all customers impartially.

Comparative advantages

The relative efficiencies with which countries (or any economic unit) can produce a product or service.

Compressed Natural Gas (CNG)

Natural gas which is comprised primarily of methane, compressed to a pressure at or above 2,400 pounds per square inch and stored in special high-pressure containers. It is used as a fuel for natural gas powered vehicles, mainly by buses.

Commuter

A person who travels regularly between home and work or school.

Commuter bus service

Fixed route bus service, characterized by service predominantly in one direction during peak periods, limited stops, use of multi-ride tickets and routes of extended length, usually between the central business district and outlying suburbs. Commuter bus service may also include other service, characterized by a limited route structure, limited stops and a coordinated relationship to another mode of transportation.

Commuter rail

Railroad local and regional passenger train operations between a central city, its suburbs and/or another central city. It may be either locomotive-hauled or self-propelled, and is characterized by multi-trip tickets, specific station-to-station fares, railroad employment practices, and usually only one or two stations in the central business district. Also known as "suburban rail".

Conference (liner)

An association of ship owners operating in the same trade route who operate under collective conditions such as tariff rates and shared capacity. They provide international liner services for the carriage of cargo on a particular route or routes within specified geographical limits and which has an agreement or arrangement within the framework of which they operate under uniform or common freight rates and any other agreed conditions with respect to the provision of liner services.

Congestion

Occurs when transport demand exceeds transport supply in a specific section of the transport system. Under such circumstances, each vehicle impairs the mobility of others. Urban congestion mainly concerns two domains of circulation, private and public, often sharing the same infrastructures.

Connecting carrier

A carrier that has a direct physical connection with another or forming a connecting link between two or more carriers.

Consignee

A person or company to whom commodities are shipped. Officially, the legal owner of the cargo.

Consolidated shipment

A method of shipping whereby an agent (freight forwarder or consolidator) combines individual consignments from various shippers into one shipment made to a destination agent, for the benefit of preferential rates. (Also called "groupage".) The consolidation is then de-consolidated by the destination agent into its original component consignments and made available to consignees. Consolidation provides shippers access to better rates than would be otherwise attainable.

Constant dollars

Figures where the effect of change in the purchasing power of the dollar has been removed. Usually the data are expressed in terms of dollars of a selected year or the average of a set of years.

Container

A large standard size metal box into which cargo is packed for shipment aboard specially configured transport modes such as oceangoing container ships. It is designed to be moved with common handling equipment enabling high speed intermodal transfers in economically large units between ships, rail cars, truck chassis and barges using a minimum of labor. The container, therefore, serves as the transfer unit rather than the cargo contained therein.

Container On Flatcar (COFC)

The movement of a container on a railroad flat car. This movement is made without the container being mounted on a chassis.

Container ship

A cargo vessel designed and constructed to transport, within specifically designed cells, portable tanks and freight containers which are lifted on and off with their contents intact. There are two types of container ships: full and partial. Full container ships are equipped with permanent container cells with little or no space for other types of cargo. Partial container ships are considered multi-purpose container vessels, where one or more but not all compartments are fitted with permanent container cells, and the remaining compartments are used for other types of cargo. This category also includes container/car carriers, container/rail car carriers and container/roll-on-roll-off vessels.

Containerization

Refers to the increasing and generalized use of the container as a means of freight transport. As a standard and versatile means, the container has greatly contributed to intermodal transportation of merchandise and its widespread use, therefore, is responsible for profound mutations in the transport sector. Through reduction of handling time, labor costs and packing costs, container transportation allows considerable increases in speed of rotation along a circuit and thus entails a better optimization of time and money.

Conventional car

A single platform flat car designed to carry a trailer or container. Containers can only be single stacked on a conventional car. Conventional cars are equipped with one or two stanchions, depending on length, for shipment of one or two trailers.

Corporate Average Fuel Economy (CAFE) Standards

CAFE standards were originally established by Congress for new automobiles, and later for light trucks, in Title V of the Motor Vehicle Information and Cost Savings Act (15 USC 1901, et seq.) with subsequent amendments. Under CAFE, automobile manufacturers are required by law to produce vehicle fleets with a composite sales-weighted fuel economy which cannot be lower than the CAFE standards in a given year, or for every vehicle that does not meet the standard, a fine of $5.00 is paid for every one-tenth of a mpg (mile per gallon) below the standard.

Corridor

A linear orientation of transport routes and flows connecting important locations that act as origins, destinations or points of transshipment. Corridors are multi-scalar entities depending on what types of flow are being investigated. Thus, they can be composed of streets, highways, transit routes, rail lines, maritime lines or air paths.

Costs (transport)

Monetary measure of what the transport provider must pay to produce transportation services; comes as fixed (infrastructure) and variable (operating). They depend on a variety of conditions related to geography, infrastructure, administrative barriers, energy, and on how passengers and freight are carried. Three major components, related to transactions, shipments and the friction of distance, impact on transport costs.

Cost–benefit analysis

A tool employed to evaluate a project by providing a set of values that are useful to determine its feasibility from an economic standpoint.

Costs–Insurance–Freight (CIF)

Price of a good is a uniform delivered price for all customers everywhere, with no spatially variable shipping price, which implies that the average shipping price is built into the price of a good. The CIF cost structure can be expanded to include several rate zones.

Cross-docking

A form of inventory management where goods are received at one door of the distribution center/sorting facility and shipped out through the other door on a very short amount of time without putting them in storage. It consequently contributes to the reduction of operating costs with an increase in the throughput and with a reduction of inventory levels.

Crude oil petroleum

A naturally occurring, oily, flammable liquid composed principally of hydrocarbons. Crude oil is occasionally found in springs or pools but usually is drilled from wells beneath the earth's surface.

Current dollars

The dollar value of a good or service in terms of prices current at the time the good or service is sold. This contrasts with the value of the good or service measured in constant dollars.

Deadhead

Miles and hours that a vehicle travels when out of revenue service. This includes leaving and returning to the garage, changing routes, etc., and when there is no reasonable expectation of carrying revenue passengers. However, it does not include charter service, school bus service, operator training, maintenance training, etc. For non-scheduled, non-fixed-route service (demand responsive), deadhead mileage also includes the travel between the dispatching point and passenger pick-up or drop-off.

Deadweight tons (dwt)

The lifting capacity of a ship expressed in long tons (2,240 lbs), including cargo, commodities and crew. Reflects the weight difference between a fully loaded and an unloaded ship.

Demand responsive

Non-fixed-route service utilizing vans or buses with passengers boarding and alighting at prearranged times at any location within the system's service area. Also called "Dial-a-Ride".

Demand (transport)

The expression of the transport needs, even if those needs are satisfied, fully, partially or not at all. Similar to transport supply, it is expressed in terms of number of people, volume or tons per unit of time and space.

Deregulation

Consists in a shift to a competitive economic climate by reorienting and/or suppressing regulatory mechanisms. Deregulation, however, does not necessarily refer to complete absence of free market regulation measures but rather to the promotion of competition-inducing measures (which can seek elimination of monopolies, for example). Particularly observed in the transport and telecommunications sectors.

Design capacity

A theoretical capacity of a transport infrastructure such as a road or terminal based on specific operating conditions.

Distribution center (freight)

Facility or a group of facilities that perform consolidation, warehousing, packaging, decomposition and other functions linked with handling freight. Their main purpose is to provide value-added services to freight and are a fundamental component of freight distribution. DCs are often in proximity to major transport routes or terminals. They can also perform light manufacturing activities such as assembly and labeling.

Dock

A feature built to handle ships. Can also refer to an enclosed port area used for maritime operations.

Double-stack

The movement of containers on articulated rail cars which enables one container to be stacked on another for better ride quality and car utilization.

Downtime

A period during which a vehicle or a whole system is inoperative because of repairs or maintenance.

Drayage

The movement of a container or trailer to or from the railroad intermodal terminal to or from the customer's facility for loading or unloading.

Dry bulk cargo

Cargo which may be loose, granular, free-flowing or solid, such as grain, coal and ore, and is shipped in bulk rather than in package form. Dry bulk cargo is usually handled by specialized mechanical handling equipment at specially designed dry bulk terminals.

Dunnage

Packaging materials used to keep cargo in place inside a container or transportation vehicle

Dwell time

The time a vehicle (bus, truck, train or ship) is allowed to load or unload passengers or freight at a terminal.

Dynamic routing

In demand–response transportation systems, the process of constantly modifying vehicle routes to accommodate service requests received after the vehicle began operations, as distinguished from predetermined routes assigned to a vehicle.

Economic evaluation (also called appraisal or analysis)

Refers to various methods for determining the value of a policy, project or program to help individuals, businesses and communities make decisions that involve tradeoffs. Economic evaluation is an important part of transportation decision making.

Economies of agglomeration

The benefits of having activities locate (cluster) next to another, such as the use of common infrastructures and services.

Economies of density

The benefits derived from the increasing density of features on the costs of accessing them. This could involve markets (e.g. consumption, labor) or resources (e.g. mining, agriculture).

Economies of scale

Cost reductions or productivity efficiencies achieved through size-increase. The outcome is a decrease in the unit cost of production associated with increasing output.

Economies of scope

Cost savings resulting from increasing the number of different goods or services produced.

Electronic Data Interchange (EDI)

Communication mode for inter- and intra-firm data exchange in the freight forwarding and logistics business.

Energy

The capacity for doing work as measured by the capability of doing work (potential energy) or the conversion of this capability to motion (kinetic energy). Energy has several forms, some of which are easily convertible and can be changed to another form useful for work. Electrical energy is usually measured in kilowatt hours, while heat energy is usually measured in British Thermal Units (BTU).

Energy intensity

In reference to transportation, the ratio of energy inputs to a process to the useful outputs from that process; for example, gallons of fuel per passenger-mile or BTU per ton-mile.

Environmental impact assessment

A process for carrying out an appraisal of the full potential effects of a development project on the physical environment.

Environmental management system

A set of procedures and techniques enabling an organization to reduce environmental impacts and increase its operating efficiency.

Ethanol

An alternative fuel; a liquid alcohol fuel with vapor heavier than air; produced from agricultural products such as corn, grain and sugar cane.

European Union (EU)

Became the EU after the signing of the Maastricht Treaty in November 1993; was formerly the European Community (EC). A regional trade block composed of 27 European states. Its core institutions are known as the "institutional triangle" composed of the European Parliament (Strasbourg), the Commission (Brussels) and the EU Council (Brussels). Also of great notoriety is the European Bank which manages the common currency.

Exclusive right of way

A highway or other facility that can only be used by buses or other transit vehicles.

Externality (external cost)

Economic cost not normally taken into account in markets or in decisions by market players.

Fare

The price paid by the user of a transport service at the moment of use.

Fare elasticity

The extent to which ridership responds to fare increases or decreases.

Fare structure

The system set up to determine how much is to be paid by various passengers using a transit system at any given time.

Feeder

Short sea shipping service which connects at least two ports in order for the freight (generally containers) to be consolidated or redistributed to or from a deep sea service in one of these ports. By extension, this concept may be used for inland transport services and air transportation.

Ferryboat

A boat providing fixed-route service across a body of water, which can be short or long distance.

Fixed cost

Costs that do not vary with the quantity shipped in the short run, i.e. costs that must be paid upfront to begin producing transportation services.

Fixed route

Service provided on a repetitive, fixed-schedule basis along a specific route with vehicles stopping to pick up and deliver passengers or freight to specific locations; each fixed-route trip serves the same origins and destinations, unlike demand responsive. The terms apply to many modes of transportation, including public transit, air services and maritime services.

Flag state

Country of registry of a seagoing vessel. A seagoing vessel is subject to the maritime regulations in respect of manning scales, safety standards and consular representation abroad of its country of registration.

Flat car

A freight car having a floor without any housing or body above. Frequently used to carry containers and/or trailers or oversized/odd-shaped commodities. The three types of flat cars used in intermodal are conventional, spine and stack cars.

Fleet

The vehicles in a transport system. Usually, "fleet" refers to highway vehicles, rail vehicles as well as ships.

Foreland

A maritime space with which a port performs commercial relationships. It includes overseas customers with which the port undertakes commercial exchanges.

Forwarding agent/freight forwarder

Intermediary who arranges for the carriage of goods and/or associated services on behalf of a shipper.

Fourth-Party Logistics Provider (4PL)

Integrates the resources of producers, retailers and third-party logistics providers to build a system-wide improvement in supply chain management. They are non-asset based meaning that they mainly provide organizational expertise.

Free trade zone

A port or an area designated by the government of a country for duty-free entry of any non-prohibited goods. Merchandise may be stored, displayed, used for manufacturing, etc., within the zone and re-exported without duties.

Freight On Board (FOB; or Free On Board)

The price of a good is the combination of the factory costs and the shipping costs from the factory to the consumer. The consumer pays for the freight transport costs. Consequently, the price of a commodity will vary according to transportation costs.

Freight consignee and handlers

Freight consignees are independent of shippers or producers. They are commissioned by the latter to accomplish all transport operations including storage, transport, management, sometimes re-expedition, etc. from origin to final destination. The notion of freight handler is broader. It comprises any actor involved in transport of freight from origin to destination including transport terminals and subcontractual services, for instance.

Freight distribution center

See distribution center.

Freight forwarder

An individual or company that accepts Less than Truckload (LTL) or Less than Carload (LCL) shipments from shippers and combines them into carload or truckload lots. Carriers collect small shipments to be cumulatively consolidated and transported, relying upon a single or several modes of transportation to a given destination. Functions performed by a freight forwarder may include: receiving small shipments (e.g. less than container load) from consignors, consolidating them into larger lots, contracting with carriers for transport between ports of embarkation and debarkation, conducting documentation transactions and arranging delivery of shipments to the consignees.

Freight village

A concentration (or a cluster) of freight-related activities within a specific area, commonly built for such a purpose, master-planned and managed. These activities include distribution centers, warehouses and storage areas, transport terminals, offices and other facilities supporting those activities, such as public utilities, parking space and even hotels and restaurants. Although a freight village can be serviced by a single mode, intermodal facilities can offer direct access to global and regional markets.

Fringe parking

An area for parking usually located outside the Central Business District (CBD) and most often used by suburban residents who work or shop downtown. Commonly corresponds to an access point of a transit system, such as a rail or subway station.

Fuel cell

A device that produces electrical energy directly from the controlled electrochemical oxidation of the fuel, commonly hydrogen. It does not contain an intermediate heat cycle, as do most other electrical generation techniques.

Gasohol

A blend of motor gasoline (leaded or unleaded) and alcohol (generally ethanol but sometimes methanol) limited to 10 percent by volume of alcohol. Gasohol is included in finished leaded and unleaded motor gasoline.

Gasoline

A complex mixture of relatively volatile hydrocarbons, with or without small quantities of additives, obtained by blending appropriate refinery streams to form a fuel suitable for use in spark ignition engines. Motor gasoline includes both leaded or unleaded grades of finished motor gasoline, blending components and gasohol.

Gateway

A location offering accessibility to a large system of circulation of freight, passengers and/or information. Gateways reap advantage of a favorable physical location such as highway junctions, confluence of rivers and seaboards, and have been the object of a significant accumulation of transport infrastructures such as terminals and their links. A gateway generally commands the entrance to and the exit from its catchment area. In other words, it is a pivotal point for the entrance and the exit of merchandise in a region, a country or a continent. Gateways tend to be locations where intermodal transfers are performed.

General cargo

General cargo consists of those products or commodities such as timber, structural steel, rolled newsprint, concrete forms and agricultural equipment that are not conducive to packaging or unitization. Break-bulk cargo (e.g. packaged products such as lubricants and cereal) are often regarded as a subdivision of general cargo.

Geographic Information System (GIS)

A special-purpose system composed of hardware and software in which a common spatial coordinate system is the primary means of reference. GIS contain subsystems for: data input; data storage, retrieval and representation; data management, transformation and analysis; and data reporting and product generation.

GIS-T

Acronym for transportation-oriented Geographic Information Systems.

Graph theory

A branch of mathematics concerned about how networks can be encoded and their properties measured.

Great circle distance

The shortest path between two points on a sphere. The circumference inferred out of these two points divides the earth in two equal parts, thus the great circle. The great circle distance is useful to establish the shortest path to use when traveling at the intercontinental air and maritime level. The great circle route follows the sphericity of the globe, any shortest route is the one following the curve of the planet, along the parallels.

Green logistics

Supply chain management practices and strategies that reduce the environmental and energy footprint of freight distribution. They focus on material handling, waste management, packaging and transport.

Gross Domestic Product (GDP)

A measure of the total value of goods and services produced by a domestic economy during a given period, usually one year. Obtained by adding the value contributed by each sector of the economy in the form of profits, compensation to employees and depreciation (consumption of capital). Only domestic production is included, not income arising from investments and possessions owned abroad, hence the use of the word domestic.

Gross National Product (GNP)

The total market value of goods and services produced during a given period by labor and capital supplied by residents of a country, regardless of where the labor and capital are located. GNP differs from GDP primarily by including the capital income that residents earn from investments abroad and excluding the capital income that nonresidents earn from domestic investment.

Gross register tonnage

The total cargo space available for a ship to carry commercial cargo. It excludes non-cargo revenue space, such as the engine room and stores.

Handy and handymax

Traditionally the workhorses of the dry bulk market, the Handy and more recent Handymax types remain popular ships with less than 50,000 dwt. This category is also used to define small-sized oil tankers.

Haulage, carrier/merchant

Carrier haulage is an inland container movement (to or from a port terminal) done by the ocean shipping company, often through a parent company. The carrier is liable if the merchandise is lost or damaged during transport, or if there is a delay. Merchant haulage is when the importer or the exporter assumes the transport of the container to or from a port terminal. The merchant is liable if the cargo is lost or damaged. One of the main advantages of merchant haulage is that it gives importers and exporters more flexibility in the timing of inland distribution. However, the merchant must pick up and bring back the container at a predesignated location and time.

Headway

Time interval between vehicles moving in the same direction on a particular route.

Heavy rail

An electric railway with the capacity for a "heavy volume" of traffic and characterized by exclusive rights of way, multi-car trains, high speed and rapid acceleration, sophisticated signaling and high platform loading.

High-Occupancy-Vehicle Lane (HOV)

A highway or road lane reserved to vehicles that have a specific level of occupancy, with at least one passenger. Often used to alleviate congestion and favor carpooling.

Hinterland

Land space over which a transport terminal, such as a port, sells its services and interacts with its clients. It accounts for the regional market share that a terminal has relative to a set of other terminals servicing this region. It regroups all the customers directly bounded to the terminal. The terminal, depending on its nature, serves as a place of convergence for the traffic coming by roads, railways or by sea/fluvial feeders.

Hub

Central point for the collection, sorting, transshipment and distribution of goods and passengers for a particular area. This concept comes from a term used in air transport for passengers as well as freight. It describes collection and distribution through a single point such as the "Hub-and-Spoke" concept. Hubs tend to be transmodal (transfers within the same mode) locations.

Infrastructure

Capital goods that are not directly consumed and serve as support to the functions of a society (individuals and corporations). (1) In transport systems, all the fixed components, such as rights of way, tracks, signal equipment, terminals, parking lots, bus stops, maintenance facilities, etc. (2) In transportation planning, all the relevant elements of the environment in which a transportation system operates.

Integrated carriers

Carriers that have both air and ground fleets; or other combinations, such as sea, rail and truck. Since they usually handle thousands of small parcels an hour, they are less expensive and offer more diverse services than regular carriers.

Intermediacy

Focus on the terminal as an intermediate point in the flows of passengers or freight. This term is applied to the frequent occurrence of places gaining advantage because they are between other places. The ability to exploit transshipment has been an important feature of many terminals.

Intermodal terminal

A terminal that can accommodate several modes of transportation. They increasingly tend to be specializing at handling specific types of passengers or freight traffic, while they may share the same infrastructures.

Intermodal transport

The movement of goods in one and the same loading unit or road vehicle, which uses successively two or more modes of transport without handling the goods themselves in changing modes. Enables cargo to be consolidated into economically large units (containers, bulk grain rail cars, etc.), optimizing the use of specialized intermodal handling equipment to effect high-speed cargo transfer between ships, barges, rail cars and truck chassis using a minimum of labor to increase logistic flexibility, reduce consignment delivery times and minimize operating costs.

Intermodalism

A system of transport whereby two or more modes of transport are used to transport the same loading unit or truck in an integrated manner, without loading or unloading, in a transport chain. Typically used in three contexts: (1) most narrowly, it refers to containerization, piggyback service or other technologies that provide the seamless movement of goods and people by more than one mode of transport; (2) more broadly, intermodalism refers to the provision of connections between different modes, such as adequate highways to ports or bus feeder services to rail transit; (3) in its broadest interpretation, intermodalism refers to a holistic view of transportation in which individual modes work together or within their own niches to provide the user with the best choices of service, and in which the consequences on all modes of policies for a single mode are considered. This view has been called balanced, integrated or comprehensive transportation in the past.

International Air Transportation Association (IATA)

Established in 1945, a trade association serving airlines, passengers, shippers, travel agents and governments. The association promotes safety, standardization in forms (baggage checks, tickets, weight bills), and aids in establishing international airfares.

International Civil Aviation Organization (ICAO)

A specialized agency of the United Nations whose objective is to develop the principles and techniques of international air navigation and to foster planning and development of international civil air transport.

International Commercial Terms (Incoterms)

Pre-defined commercial contract terms which stipulate exactly which party owns cargo over the course of a shipment, as well as who bears responsibility for transporting the cargo.

International Maritime Organization (IMO)

Established as a specialized agency of the United Nations in 1948. The IMO facilitates cooperation on technical matters affecting merchant shipping and traffic, including improved maritime safety and prevention of marine pollution. Headquarters are in London, England.

International Organization for Standardization (ISO)

Worldwide federation of national standards bodies from some 100 countries, one from each country. ISO is a non-governmental organization established in 1947. The mission of ISO is to promote the development of standardization and related activities in the world with a view to facilitating the international exchange of goods and services, and to developing cooperation in the spheres of intellectual, scientific, technological and economic activity. ISO's work results in international agreements which are published as International Standards.

Jet stream

A migrating stream of high speed winds present at high altitudes.

Jitney

Privately owned, small or medium-sized vehicle usually operated on a fixed route but not on a fixed schedule.

Just-in-time

The principle of production and inventory management in which goods arrive when needed for production or consumption. Warehousing tends to be minimal or nonexistent, but in all cases much more efficient and more limited in duration.

Knot, nautical

The unit of speed equivalent to one nautical mile: 6,080.20 feet per hour or 1.85 kilometers per hour.

Lading

Refers to the freight shipped; the contents of a shipment.

Landbridge

An intermodal connection between two ocean carriers separated by a land mass, linked together in a seamless transaction by a land carrier.

Landed cost

The dollar per barrel price of crude oil at the port of discharge. Included are the charges associated with the purchase, transporting and insuring of a cargo from the purchase point to the port of discharge. Not included are charges incurred at the discharge port (e.g. import tariffs or fees, wharfage charges and demurrage charges).

Layover time

Time built into a schedule between arrival at the end of a route and the departure for the return trip, used for the recovery of delays and preparation for the return trip.

Less than Truckload (LTL)

A shipment that would not by itself fill the truck to capacity by weight or volume.

Level of service

(1) A set of characteristics that indicates the quality and quantity of transportation service provided, including characteristics that are quantifiable and those that are difficult to quantify. (2) For highway systems, a qualitative rating of the effectiveness of a highway or highway facility in serving traffic, in terms of operating conditions. A rating of traffic flow ranging from A (excellent) through F (heavily congested), and compares actual or projected traffic volume with the maximum capacity of the intersection or road in question. (3) For paratransit, a variety of measures meant to denote the quality of service provided, generally in terms of total travel time or a specific component of total travel time. (4) For pedestrians, sets of area occupancy classifications to connect the design of pedestrian facilities with levels of service.

Light-Rail Transit (LRT)

Fixed guideway transportation mode that typically operates on city streets and draws its electric power from overhead wires; includes streetcars, trolley cars and tramways. Differs from heavy rail – which has a separated right of way, and includes commuter and intercity rail – in that it has lighter passenger capacity per hour and more closely spaced stops.

Lighter-Aboard-Ship (LASH)

A type of barge-carrying vessel equipped with an overhead crane capable of lifting barges of a common size and stowing them into cellular slots in athwartship position. LASH is an all-water technology analogous to containerization.

Line haul costs

Costs that vary with distance shipped, i.e. costs of moving goods and people once they are loaded on the vehicles.

Liner

Derived from the term "line traffic", which denotes operation along definite routes on the basis of definite, fixed schedules. A liner thus is a vessel that engages in this kind of transportation, which usually involves the haulage of general cargo as distinct from bulk cargo.

Liquefied Natural Gas (LNG)

An alternative fuel; a natural gas cooled to below its boiling point of −260 degrees Fahrenheit so that it becomes a liquid; stored in a vacuum-type container at very low temperatures and under moderate pressure. LNG vapor is lighter than air.

Load factor

The ratio of passengers or freight actually carried versus the total passenger or freight capacity of a vehicle or a route.

Logistic zone

Grouping of activities dealing with freight transportation (freight forwarders, shippers, transport operators, customs) and related services (storage, maintenance and repair) within a defined area.

Logistics

The process of designing and managing the supply chain in the wider sense. The chain can extend from the delivery of supplies for manufacturing, through the management of materials at the plant, delivery to warehouses and distribution centers, sorting, handling, packaging and final distribution to point of consumption. A more fitted meaning consists in the set of all operations required for goods (material or nonmaterial) to be made available on markets or to specific destinations.

Long ton

2,240 pounds.

Maglev – magnetic levitation

Technology enabling trains to move at high speed above a guideway on a cushion generated by magnetic force.

Manifest

A list of the goods being transported by a carrier.

Maritime routes

Corridors of a few kilometers in width trying to avoid the discontinuities of land transport by linking ports, the main elements of the maritime/land interface. Maritime routes are a function of obligatory points of passage, which are strategic places, of physical constraints (coasts, winds, marine currents, depth, reefs, ice) and of political borders. As a result, maritime routes draw arcs on the water surface of the earth as intercontinental maritime transportation tries to follow the great circle distance.

Maritime terminal

A designated area of a port, which includes but is not limited to wharves, warehouses, covered and/or open storage spaces, cold storage plants, grain elevators and/or bulk cargo loading and/or unloading structures, landings and receiving stations, used for the transmission, care and convenience of cargo and/or passengers in the interchange of same between land and water carriers or between two water carriers.

Market area

The surface over which a demand offered at a specific location is expressed. Commonly, a customer is assumed to go to a location where a product or service can be acquired or a part or a finished good has to be shipped from the place of production to the place of consumption.

Materials management

Considers all the activities related in the manufacturing of commodities in all their stages of production along a supply chain. It includes production and marketing activities such as production planning, demand forecasting, purchasing and inventory management. It must ensure that the requirements of supply chains are met by dealing with a wide array of parts for assembly and raw materials, including packaging (for transport and retailing) and, ultimately, recycling discarded commodities. All these activities are assumed to be inducing physical distribution demands.

MERCOSUR

A trade alliance between Argentina, Brazil, Paraguay and Uruguay, with Chile and Bolivia as associate members.

Methanol

An alternative fuel; a liquid alcohol fuel with vapor heavier than air; primarily produced from natural gas.

Microbridge

A cargo movement in which the water carrier provides a through service between an inland point and the port of load/discharge.

Minibridge

A joint water, rail or truck container move on a single Bill of Lading for a through route from a foreign port to a US port destination through an intermediate US port or the reverse.

Mobility

Refers to a movement of people or freight. It can have different levels linked to the speed, capacity and efficiency of movements.

Modal share

The percentage of total passengers or freight moved by a particular type of transportation.

Modal split (share)

(1) The proportion of total person trips that uses each of various specified modes of transportation. (2) The process of separating total person trips into the modes of travel used. (3) A term that describes how many people use alternative forms of transportation. It is frequently used to describe the percentage of people who use private automobiles, as opposed to the percentage who use public transportation.

Mode, transport

The physical way a movement is performed.

Model

An analytical tool (often mathematical) used by transportation planners to assist in making forecasts of land use, economic activity, travel activity and their effects on the quality of resources such as land, air and water.

Monorail

An electric railway in which a rail car or train of cars is suspended from or straddles a guideway formed by a single beam or rail. Most monorails are either heavy rail or automated guideway systems.

Motorway/highway

Road, specially designed and built for motor traffic, which does not serve properties bordering on it, and which: (a) is provided, except at special points or temporarily, with separate carriageways for the two directions of traffic, separated from each other, either by a dividing strip not intended for traffic or exceptionally by other means; (b) does not cross at level with any road, railway or tramway track or footpath; (c) is specially signposted as a motorway and is reserved for specific categories of road motor vehicles. Entry and exit lanes of motorways are included irrespectively of the location of the signposts. Urban motorways are also included.

Multimodal platform

A physical converging point where freight and/or passenger transshipment takes place between different modes of transportation, usually a transport terminal.

National transportation system

An intermodal system consisting of all forms of transportation in a unified, interconnected manner to reduce energy consumption and air pollution while promoting economic development and supporting the nation's position in international commerce. The NTS includes the National Highway System (NHS), public transportation and access to ports and airports.

North American Free Trade Agreement (NAFTA)

Came into force on January 1, 1994. NAFTA binds Canada, the United States and Mexico over respect of a series of common economics rules. Beside the liberalization of exchange of goods and services, the NAFTA regulates investments, intellectual property, public markets and the non-tariff barrier. The NAFTA is a result of a tradition of trade negotiations between Canada and the US that became explicit with the 1989 Free Trade Agreement (FTA) and the 1991 Canada–US Trade Agreement (CUSTA).

Net tonnage

The net or register tonnage of a vessel is the remainder after deducting from the gross tonnage of the vessel the tonnage of crew spaces, master's accommodation, navigation spaces, allowance for propelling power, etc. It is expressed in tons of 100 cubic feet.

Network

Framework of routes within a system of locations, identified as nodes. A route is a single link between two nodes that are part of a larger network that can refer to tangible routes such as roads and rails, or less tangible routes such as air and sea corridors.

Network analysis

The pattern of transportation systems, the location of routes or rails, the location of intersections, nodes and terminals that can be considered as a network. However, on the analytic side, more attention is paid to the whole system rather than to single routes or terminals. Network analysis aims at identifying flows, shortest distances between two given points or the less expensive road to take for transporting goods between those points. To facilitate the task, networks have been approximated by the use of the graph theory relying on topology.

Nitrogen Oxides

A product of combustion of fossil fuels whose production increases with the temperature of the process. It can become an air pollutant if concentrations are excessive.

Ocean bill of lading

A receipt for the cargo and a contract for transportation between a shipper and the ocean carrier. It may also be used as an instrument of ownership which can be bought, sold or traded while the goods are in transit.

Oceanic airspace

Airspace over the oceans of the world, considered international airspace, where oceanic separation and procedures per the International Civil Aviation Organization are applied. Responsibility for the provision of air traffic control service in this airspace is delegated to various countries, based generally upon geographic proximity and the availability of the required resources.

Offpeak period

Non-rush periods of the day when travel activity is generally lower and less transit service is scheduled. Also called "base period".

Offshoring

The transfer of an organizational production function to another country, whether the work is outsourced or stays within the same corporation.

Offshore hub

A port terminal that dominantly serves transmodal operations, implying limited connections in relation to its total traffic with its hinterland. They are mainly used for feedering, relay and interlining between maritime shipping routes. The term offshore can be misleading as many ports performing this function are located at standard port locations.

Operating cost

Costs that vary with the quantity shipped in the short run. (1) Fixed operating cost: refers to expenditures that are independent of the amount of use. For a car, it would involve costs such as insurance costs, fees for license and registration, depreciation and finance charges. (2) Variable operating cost: expenditures that are dependent on the amount of use. For a car, it would involve costs such as the cost of gasoline, oil, tires and other maintenance.

Organisation for Economic Cooperation and Development (OECD)

In 1961 this replaced the Organisation for European Economic Cooperation (OEEC), created in 1948 to facilitate postwar reconstruction of Europe via American aid. It acts as a policy leveling forum where government representatives of member states seek to harmonize economic policies touching such sectors as commerce, industry, cooperation, foreign aid and agriculture.

Outsourcing

The practice of having some activities that used to be performed within a corporation done by another corporation. It often enables costs to be reduced and the business to focus on core competencies by outsourcing low productivity tasks to a subcontractor.

Pallet

A raised platform, normally made of wood, facilitating the handling of goods. Pallets are of standard dimensions.

Panamax

A maritime standard corresponding to about 65,000 deadweight tons or 4,200 TEU. It refers to a ship with dimensions that allow it to pass through the Panama canal: maximum length 295 m, maximum beam overall 32.25 m, maximum draught 13.50 m.

Pandemic

An epidemic of infectious disease that spreads through human populations across a large area, even worldwide.

Park and ride

An access mode to transit in which patrons drive private automobiles or ride bicycles to a transit station, stop or carpool/vanpool waiting area and park the vehicle in the area provided for the purpose. They then ride the transit system or take a car- or vanpool to their destinations.

Particulates

Carbon particles formed by partial oxidation and reduction of the hydrocarbon fuel. Also included are trace quantities of metal oxides and nitrides, originating from engine wear, component degradation and inorganic fuel additives. In the transportation sector, particulates are emitted mainly from diesel engines.

Passenger-km (or passenger-mile)

The total number of kilometers (miles) traveled by passengers on vehicles; determined by multiplying the number of unlinked passenger trips times the average length of their trips.

Payload

Weight of commodity being hauled. Includes packaging, pallets, banding, etc., but does not include the truck, truck body, etc.

Peak oil

A theory concerning oil production initially published by the geophysicist King Hubbert in 1956, that assumes due to the finite nature of oil reserves that production will at some point reach maximum output. Once peak production has been reached, production declines and prices go up until oil resources are depleted or too costly to have a widespread use.

Peak period (hour)

Represents a time period of high usage of a transport system. For transit, it refers to morning and afternoon time periods when ridership is at its highest.

Peak/base ratio

The number of vehicles operated in passenger or freight service during the peak period divided by the number operated during the base period.

Pendulum service

Involves a set of sequential port calls along a maritime range, commonly including a transoceanic service from ports in another range and structured as a continuous loop. They are almost exclusively used for container transportation with the purpose of servicing a market by balancing the number of port calls and the frequency of services.

Physical distribution

The collective term for the range of activities involved in the movement of goods from points of production to final points of sale and consumption. It must ensure that the mobility requirements of supply chains are entirely met. Physical distribution comprises all the functions of movement and handling of goods, particularly transportation services (trucking, freight rail, air freight, inland waterways, marine shipping and pipelines), transshipment and warehousing services (e.g. consignment, storage, inventory management), trade, wholesale and, in principle, retail. Conventionally, all these activities are assumed to be derived from materials management demands.

Piggyback trailers

Trailers designed for quick loading on rail cars.

Pipeline

A continuous pipe conduit, complete with such equipment as valves, compressor stations, communications systems and meters for transporting natural and/or supplemental gas from one point to another, usually from a point in or beyond the producing field or processing plant to another pipeline or to points of utilization. Also refers to a company operating such facilities.

Planning

Refers to a process that allows people's needs, preferences and values to be reflected in decisions. Planning occurs at many different levels, from day-to-day decisions made by individuals and families, to major decisions made by governments and businesses that have comprehensive, long-term impacts on society. Management can be considered a short-term form of planning, while planning can be considered a longer-term form of management.

Platform/modular manufacturing

Strategy in which a multinational corporation retains its core competencies, namely its research and development, retailing, marketing and distribution, while subcontracting much of the manufacturing to the lowest bidders.

Policy (transport)

The development of a set of constructs and propositions that are established to achieve particular objectives relating to social, economic and environmental development and the functioning and performance of the transport system.

Port

A harbor area in which are located marine terminal facilities for transferring cargo between ships and land transportation.

Port authority

An entity of state or local government that owns, operates or otherwise provides wharf, dock and other marine terminal investments at ports.

Port holding

An entity, commonly private, that owns or leases port terminals in a variety of locations. It is also known as a port terminal operator.

Port of entry

A port at which foreign goods are admitted into the receiving country. Also refers to an air terminal or land access point (customs) where foreign passengers and freight can enter a country.

Primary transportation

Conveyance of large shipments of petroleum, raw materials and refined products usually by pipeline, barge or oceangoing vessel. All crude oil transportation is primary, including the small amounts moved by truck. All refined product transportation by pipeline, barge or oceangoing vessel is primary transportation.

Product life cycle

Defined as the period that starts with the initial product design (research and development) and ends with the withdrawal of the product from the marketplace. A product life cycle is characterized by specific stages, including research, development, introduction, maturity, decline and obsolescence.

Propane

An alternative fuel; a liquid petroleum gas (LPG) which is stored under moderate pressure and with vapor heavier than air; produced as a by-product of natural gas and oil production.

Public transportation

Passenger transportation services, usually local in scope, that is available to any person who pays a prescribed fare. It operates on established schedules along designated routes or lines with specific stops and is designed to move relatively large numbers of people at one time.

Radio Frequency Identification Device (RFID)

Technology that uses small devices attached to objects that transmit data to a receiver. An alternative to bar coding used for identification and tracking purposes, notably for items shipped in units (boxes, containers, etc.), but can also be attached to an individual item. Main technical advantages include data storage capacity, read/write capability and no line-of-sight requirements during scanning.

Rail, commuter

Railroad local and regional passenger train operations between a central city, its suburbs and/or another central city. It may be either locomotive-hauled or self-propelled, and is characterized by multi-trip tickets, specific station-to-station fares, railroad employment practices and usually only one or two stations in the central business district. Also known as "suburban rail".

Rail, heavy

An electric railway with the capacity for a "heavy volume" of traffic and characterized by exclusive rights of way, multi-car trains, high speed and rapid acceleration, sophisticated signaling and high platform loading. Also known as "rapid rail", "subway", "elevated (railway)" or "metropolitan railway (metro)".

Rail, high speed

A rail transportation system with exclusive right of way which serves densely traveled corridors at speeds of 124 miles per hour (200 km/hr) and greater.

Rail, light

An electric railway with a "light volume" traffic capacity compared to heavy rail. Light rail may use shared or exclusive rights of way, high or low platform loading and multi-car trains or single cars. Also known as "streetcar", "trolley car" and "tramway".

Railroad

All forms of non-highway ground transportation that run on rails or electro-magnetic guideways, including: (1) Commuter or other short-haul rail passenger service in a metropolitan or suburban area, and (2) high speed ground transportation systems that connect metropolitan areas, without regard to whether they use new technologies not associated with traditional railroads. Such a term does not include rapid transit operations within an urban area that are not connected to the general railroad system of transportation.

Rapid transit

Rail or motorbus transit service operating completely separately from all modes of transportation on an exclusive right of way.

Rate

The price of transportation services paid by the consumer. It is the negotiated monetary cost of moving a passenger or a unit of freight between a specific origin and destination. Rates are often visible to the consumer since transport providers must provide this information to secure transactions.

Reefer ship

General cargo ship with 80 percent or more insulated cargo space.

Ridesharing

A form of transportation, other than public transit, in which more than one person shares the use of the vehicle, such as a van or car, to make a trip. Also known as "carpooling" or "vanpooling".

Ridership

The number of rides taken by people using a public transportation system in a given time period.

Road train

A tractor unit pulling two or more trailers linked together.

Roll-on-roll-off (RORO) Vessel

Ships that are especially designed to carry wheeled containers, trailers or other wheeled cargo, and use the roll-on-roll-off method for loading and unloading. Main method of transporting automobiles on international markets.

Rolling stock

The vehicles used in a transit system, including buses and rail cars.

Rubber wheel/tire interchange

Containers or trailers that are interchanged between two railroads by means of drayage.

Semi-trailer

A non-powered vehicle for the carriage of goods, intended to be coupled to a motor vehicle in such a way that a substantial part of its weight and of its load is borne by the motor vehicle.

Shipper

The company sending goods.

Short sea shipping

Commercial waterborne transportation that does not transit an ocean. It is an alternative form of commercial transportation that utilizes inland and coastal waterways to move commercial freight from major domestic ports to its destination.

Shunting

Operation related to moving a rail vehicle or set of rail vehicles within a railway instal-lations (station, depot, workshop, marshaling yard, etc.). It mainly concerns the assembly and disassembly of unit trains.

Shuttle

A public or private vehicle that travels back and forth over a particular route, especially a short route or one that provides connections between transportation systems, employment centers, etc.

Silk Road

Historical trade route linking the Eastern Mediterranean basin to Central and East Asia. Named because of the prized commodity, namely silk, as well as tea and jade, that was carried from China. Was operational between the first century BC and the sixteenth century.

Single-Occupant Vehicle (SOV)

A vehicle with one occupant, the driver, who is sometimes referred to as a "drive alone".

Site

The geographical characteristics of a specific location.

Situation

The relationships a location has in regard to other locations.

Source loading

Refers to the loading of a shipment, commonly in a container, at the location where the goods it carries are produced. The shipment remains untouched until it reaches its desti-nation, thus conferring a level of integrity in the supply chain.

Spatial interaction

A realized movement of people, freight or information between an origin and a destination. It is a transport demand/supply relationship expressed over a geographical space. Spatial interactions cover a wide variety of movements such as journeys to work, migrations, tourism, the usage of public facilities, the transmission of information or capital, the market areas of retailing activities, international trade and freight distribution.

Spatial structure

The manner in which space is organized by the cumulative locations of infrastructure, economic activities and their relations.

Steel wheel interchange

Containers or trailers that are interchanged between two railroads while on the railroad flatcar.

Suezmax

Standard that represents the limitations of the Suez Canal. Before 1967, the Suez Canal could only accommodate tanker ships with a maximum of 80,000 dwt. The canal was closed between 1967 and 1975 because of the Israel–Arab conflict. Once it reopened in 1975, the Suezmax capacity went to 150,000 dwt. An enlargement to enable the canal to accommodate 200,000 dwt tankers is being considered.

Supply chain

See commodity chain.

Supply Chain Management (SCM)

The management of the whole commodity/supply chain, from suppliers, manufacturers, retailers and the final customers. To achieve higher productivity and better returns, SCM mainly tries to reduce inventory, increase transaction speeds and satisfy the needs of the customers in terms of cost, quantity, quality and delivery as much as possible.

Supply (transport)

The capacity of transportation infrastructures and modes, generally over a geographically defined transport system and for a specific period of time. Therefore, supply is expressed in terms of infrastructures (capacity), services (frequency) and networks. The number of passengers, volume (for liquids or containerized traffic) or mass (for freight) that can be transported per unit of time and space is commonly used to quantify transport supply.

Sustainable development

Development that meets the needs of the present without compromising the ability of future generations to meet their own needs.

Tanker

An oceangoing ship specially designed to haul liquid bulk cargo in world trade, particularly oil.

Tare weight

(a) The weight of a container and the material used for packing. (b) As applied to a car/trailer, the weight of the car/trailer exclusive of its contents.

Tariff

A general term for any listing of rates or charges. The tariffs most frequently encountered in foreign trade are: tariffs of international transportation companies operating on sea, land and in the air; tariffs of international cable, radio and telephone companies; and the customs tariffs of the various countries that list goods that are duty-free and those subject to import duty, giving the rate of duty in each case.

Telecommuting

Using information and telecommunication technologies to perform work at a location away from the traditional office location and environment.

Terminal

Any location where freight and passengers either originate, terminate or are handled in the transportation process. Terminals are central and intermediate locations in the movements of passengers and freight. They often require specific facilities to accommodate the traffic they handle.

Terminal costs

Costs of loading and unloading. They do not vary with distance shipped.

Third-Party Logistics Provider (3PL)

An asset-based company that offers logistics and supply chain management services to its customers (manufacturers and retailers). It commonly owns distribution centers and transport modes.

Threshold

The minimum and vital market size required to support a given type of economic activity. A mean number of passengers per trip can be identified to sustain profitability of a coach line, for example. A threshold thus rests on a level of demand and can play a determining role in organizing both freight and passenger transport structures on the basis of demographic dynamics, geographic relations to markets and intensity of economic activities.

Ton

A unit of measurement of weight, frequently used in freight transport statistics. A metric ton is equivalent to 1,000 kilograms or 2,205 pounds. A short ton is equivalent to 2,000 pounds or 0.908 metric tons (in the United States the term ton is commonly used but implies short ton). A long ton, a term not as frequently used, is equivalent to 2,240 pounds or 1.06 metric tons.

Ton-km (or ton-mile)

Measure expressing the realized freight transport demand. Although both the passenger-km and ton-km are most commonly used to measure realized demand, the measure can equally apply for transport supply.

Track gauge

The distance between the internal sides of rails on a railway line. The standard gauge is generally 1.435 m. Other gauges are used, for instance, in Spain and Portugal (1.676 m) or in the Russian Federation (1.524 m).

Trailer on Flat Car (TOFC)

A rail trailer or container mounted on a chassis that is transported on a rail car. Also known as piggyback.

Tramp

An oceangoing vessel that does not operate along a definite route or on a fixed schedule, but rather calls at any port where cargo is available.

Transactions

In the business domain, a transaction is synonymous with exchange and refers to a commercial operation. Generally, before a transaction, there are some negotiations. Transactions generate varying costs, depending on the stakes, the competition, the context of the economic market, etc.

Transaction costs

Costs required for gathering information, negotiating and enforcing contracts, letters of credit and transactions. Often referred to as the cost of doing business.

Transit system

An organization (public or private) providing local or regional multi-occupancy-vehicle passenger service. Organizations that provide service under contract to another agency are generally not counted as separate systems.

Transloading

The transshipment of loads from truck to rail and vice versa. It is done to exploit the respective advantages of trucking and rail, namely to avoid long distance trucking. Also refers to the moving of the contents of a container, say a 40-foot maritime container, into another container, such as a 53-foot domestic container, or a regular truckload.

Transmodal transportation

The movements of passengers or freight within the same mode of transport. Although "pure" transmodal transportation rarely exists and an intermodal operation is often required (e.g. ship to dockside to ship), the purpose is to ensure continuity within the network.

Transport geography

Subdiscipline of geography concerned with movements of freight, people and information. It seeks to link spatial constraints and attributes with the origin, the destination, the extent, the nature and the purpose of movements.

Transportability

The ease of movement of passengers, freight or information. It is related to transport costs as well as to the attributes of what is being transported (fragility, perishability, price). Political factors can also influence transportability such as laws, regulations, borders and tariffs. When transportability is high, activities are less constrained by distance.

Transshipment

The transfer of goods from one carrier to another and/or from one mode to the other.

Trip assignment

In planning, a process by which trips, described by mode, purpose, origin, destination and time of day, are allocated among the paths or routes in a network by one of a number of models.

Trip generation

In planning, the determination or prediction of the number of trips produced by and attracted to each zone.

Twenty-foot Equivalent Unit (TEU)

A standard unit based on an ISO container of 20-foot long (6.10 m), used as a statistical measure of traffic flows or capacities. One standard 40-foot ISO Series 1 container equals 2 TEUs.

Ultra Large Crude Carriers (ULCC)

Tanker ships from 300,000 to 550,000 dwt in size. Used for carrying crude oil on long-haul routes from the Persian Gulf to Europe, America and East Asia, via the Cape of Good Hope or the Strait of Malacca. The enormous size of these vessels requires custom-built terminals.

Unit load

Packages loaded on a pallet, in a crate or any other way that enables them to be handled as a unit.

Unlinked passenger trips

The number of passengers who board public transportation vehicles. A passenger is counted each time he/she boards a vehicle even though he/she may be on the same journey from origin to destination.

Upstream/downstream

Refers to the relative location of a given activity along a supply chain.

Urban form

The spatial imprint of an urban transport system as well as the adjacent physical infrastructures and socioeconomic activities. Jointly, they confer a level of spatial arrangement to cities.

Variable cost

A cost that varies in relation to the level of operational activity.

Very Large Crude Carrier (VLCC)

Crude oil carrying ships of between 150,000 and 320,000 deadweight tons. They offer good flexibility since many terminals can accommodate their draft. They are used in ports that have depth limitations, mainly around the Mediterranean, West Africa and the North Sea. They can be ballasted through the Suez Canal.

Vessel

Every description of watercraft, used or capable of being used as a means of transportation on the water.

Vessel sharing agreement

Agreement between two or more ocean carriers in which a number of container slots are reserved on particular vessels for each of the participants (right to book slots and obligation of the other carrier to carry the containers). Used to create operational efficiencies across carriers with more port calls and higher frequency of service.

Warehouse

A place for the reception, delivery, consolidation, distribution and storage of freight.

Waterway

River, canal, lake or other stretch of water that by natural or man-made features is suitable for navigation.

Waybill

A document covering a shipment and showing the forwarding and receiving station, the names of consignor and consignee, the car initials and number, the routing, the description and weight of the commodity, instructions for special services, the rate, total charges, advances and waybill reference for previous services and the amount prepaid.

Weight

Gross: the weight of the goods including packing, wrappers or containers, both internal and external. The total weight as shipped. Net: the weight of the goods themselves without the inclusion of any wrapper. Tare: the weight of the packaging or container. Weight/Measurement Ton: in many cases, a rate is shown per weight/measurement ton, carrier's option. This means that the rate will be assessed on either a weight ton or measurement ton basis, whichever will yield the carrier the greater revenue. Weight Ton: metric measure equals 1,000 kilograms; in English measure a short ton is 2,000 pounds, a long ton is 2,240 pounds.

Wharf

A landing place where vessels may tie up for loading and unloading of cargo.

World Bank

A financial body, part of the United Nations system. It was created in 1944 at the Bretton Woods financial and monetary conference. First loans helped finance reconstruction of Western Europe and Japan following the Second World War, but today the World Bank has considerably broadened its presence throughout the globe, lending to countries of Africa, Asia, Central Europe, Latin America, the Middle East and the former Soviet Union. Its priority is to lend capital to governments of developing countries to promote economic growth through financing of large infrastructure projects, economic reform packages and technical assistance. It thus has vested interests in a number of developing countries worldwide. Loans are also aimed at encouraging private sector development. Presently, the World Bank is composed of four main branches: the International Bank for Reconstruction and Development (IBRD), the Multilateral Investment Guarantee Agency (MIGA), the International Development Agency (IDA) and the International Financial Society (IFS).

World Trade Organization (WTO)

Was established on January 1, 1995, as a result of the Uruguay Round negotiations (1986–94) under the General Agreement on Tariffs and Trade (GATT). The seat of the WTO is located in Geneva, Switzerland. It performs various functions including administering WTO trade agreements, organizing forums for trade negotiations, handling trade disputes, monitoring national trade policies, providing technical assistance and training for developing countries, and cooperating with other international organizations.

Yard

A system of auxiliary tracks used exclusively for the classification of passenger or freight cars according to commodity or destination; assembling of cars for train movement; storage of cars; or repair of equipment.

Yield management (transportation)

The process of managing the usage price of a transport asset, such as the fare paid by users, in view of changes in the demand. The goal of such an approach is to maximize profit in the context where the transport supply is fixed. Commonly used in air transportation.

Index

Note: Page numbers in **bold** type refer to **figures**; page numbers followed by 'P' refer to photographs